Computational Aspects of Complex Analysis

NATO ADVANCED STUDY INSTITUTES SERIES

Proceedings of the Advanced Study Institute Programme, which aims at the dissemination of advanced knowledge and the formation of contacts among scientists from different countries

The series is published by an international board of publishers in conjunction with NATO Scientific Affairs Division

A	Life Sciences	Plenum Publishing Corporation
B	Physics	London and New York
C	Mathematical and Physical Sciences	D. Reidel Publishing Company Dordrecht, Boston and London
D	Behavioural and Social Sciences	Martinus Nijhoff Publishers
E	Engineering and Materials Sciences	The Hague, London and Boston
F	Computer and Systems Sciences	Springer Verlag Heidelberg
G	Ecological Sciences	

Series C – Mathematical and Physical Sciences

Volume 102 - Computational Aspects of Complex Analysis

Computational Aspects of Complex Analysis

Proceedings of the NATO Advanced Study Institute held at Braunlage, Harz, Germany, July 26 - August 6, 1982

edited by

H. WERNER
University of Bonn, West Germany

L. WUYTACK
Universitaire Instelling Antwerpen, Belgium

E. NG
Jet Propulsion Laboratories, Pasadena, Calif., U.S.A.

and

H. J. BÜNGER
University of Bonn, West Germany

D. Reidel Publishing Company

Dordrecht / Boston / Lancaster

Published in cooperation with NATO Scientific Affairs Division

Library of Congress Cataloging in Publication Data

NATO Advanced Study Institute (1982 : Braunlage, Germany)
Computational aspects of complex analysis.

(NATO advanced study institutes series. Series C, Mathematical and physical sciences ; v. 102)
Includes index.
1. Mathematical analysis–Data processing–Congresses.
2. Functions of complex variables–Data processing–Congresses.
I. Werner, H. (Helmut), 1931– . II. Title. III. Series.
QA300.N29 1982 515'.028'54 83-3216

ISBN-13: 978-94-009-7123-3 e-ISBN-13: 978-94-009-7121-9
DOI: 10.1007/978-94-009-7121-9

Published by D. Reidel Publishing Company
P.O. Box 17, 3300 AA Dordrecht, Holland

Sold and distributed in the U.S.A. and Canada
by Kluwer Boston Inc.,
190 Old Derby Street, Hingham, MA 02043, U.S.A.

In all other countries, sold and distributed
by Kluwer Academic Publishers Group,
P.O. Box 322, 3300 AH Dordrecht, Holland

D. Reidel Publishing Company is a member of the Kluwer Academic Publishers Group

Softcover reprint of the hardcover 1st edition 1983

TABLE OF CONTENTS

PREFACE

The NATO Advanced Study Institute (ASI) on "Computational Aspects of Complex Analysis" was held at Braunlage/Harz (Germany) from July 26 to August 6, 1982. These proceedings contain the invited lectures presented at this institute, the aim of which was to bring together scientists from pure and applied mathematics as well as computer scientists. The main topics were problems dealing with approximation and interpolation by polynomial and rational functions (in particular Padé approximation), numerical methods for the solution of algebraic equations and differential equations, the large field of conformal mapping, aspects of computer implementation of complex arithmetic and calculations based on complex variable techniques.

The sessions on short communications not only provided a platform for the presentation of contributions by the participants of the ASI but also the opportunity to discuss the material more thoroughly, to bring up open problems and to point out the interrelationship of the above mentioned topics. Quite naturally the short communications grouped around the topics of the main lectures. The stimulating atmosphere caused many discussions to continue privately for hours. Even out of the social program there emanated two short communications by L. Wuytack and L. Trefethen, which are included at the end of these proceedings.

We gratefully appreciate the support of the International Advisory Committee that was formed by L. Collatz, Germany, C. Brezinski, France, G. Golub, U.S.A., P. Henrici, Switzerland, J. van Hulzen, the Netherlands, O. Skovgaard, Denmark, I. Sneddon, United Kingdom, and J. Todd, U.S.A.

The ASI and the publication of these proceedings were only made possible by the active participation and cooperation of all lecturers. The ASI was supported financially by the NATO Scientific Affairs Division. The sympathetic understanding of its directors Dr. M. di Lullo and Dr. C. Sinclair for our scientific intentions and its realisation was of great help. To all of them we are most thankful.

Especially we thank Mrs. Haverkamp for the expert typing of most of the manuscript and Mrs. E. Becker for all the secretarial help. The book also pays its tribute to the advent of computerized

H. Werner et al. (eds.), Computational Aspects of Complex Analysis, vii–viii.

mathematical typesetting, as can be seen from the text of the last lecture of these proceedings.

The MARITIM HOTEL at Braunlage provided a very stimulating and pleasant atmosphere in which scientific communication could grow.

Last but not least we wish to express our sincere gratitude to all participants for their contributions in helping to make the ASI a successful and pleasant event.

Bonn, Antwerp, Pasadena
December 1982

H. Werner, L. Wuytack, E. Ng, H. J. Bünger

OUTLINES OF PADE APPROXIMATION

Claude Brezinski

University of Lille I - France

Abstract : In the first section Padé approximants are defined and some properties are given. The second section deals with their algebraic properties and, in particular, with their connection to orthogonal polynomials and quadrature methods. Recursive schemes for computing sequences of Padé approximants are derived. The third section is devoted to some results concerning the convergence problem. Some applications and extensions are given in the last section.

1. DEFINITION AND PROPERTIES

Let f be a formal power series

$$f(t) = \sum_{i=o}^{\infty} c_i \, t^i$$

We shall look for a rational fraction whose numerator has the *exact* degree p and whose denominator has the *exact* degree q such that its power series expansion agrees with f as far as possible, that is

$$f(t) - \sum_{i=o}^{p} a_i t^i \Big/ \sum_{i=o}^{q} b_i t^i = O(t^{p+q+1}).$$

Multiplying by the denominator we get

H. Werner et al. (eds.), Computational Aspects of Complex Analysis, 1–50.

$$\left(\sum_{i=o}^{\infty} c_i t^i\right)\left(\sum_{i=o}^{q} b_i t^i\right) - \left(\sum_{i=o}^{p} a_i t^i\right) = O(t^{p+q+1})$$

Identifying the coefficients of the terms in $t^o, t^1, \ldots, t^{p+q}$ we obtain

degree 0 $\quad a_o = c_o b_o$

degree 1 $\quad a_1 = c_1 b_o + c_o b_1$

....

degree p $\quad a_p = c_p b_o + c_{p-1} b_1 + \ldots + c_{p-q} b_q$

degree p+1 $\quad 0 = c_{p+1} b_o + c_p b_1 + \ldots + c_{p-q+1} b_q$

....

degree p+q $\quad 0 = c_{p+q} b_o + c_{p+q-1} b_1 + \ldots + c_p b_q$

with the convention that $c_i = 0$ if $i < 0$.

Since the numerator and the denominator of a rational fraction are defined apart from a multiplying factor we can set $b_o=1$. Thus $b_1, \ldots, b_q$ are given as the solution of the system of linear equations obtained from the preceding relations (degrees p+1 to p+q) :

$$\begin{pmatrix} c_{p-q+1} & c_{p-q+2} & \cdots & c_p \\ c_{p-q+2} & c_{p-q+3} & \cdots & c_{p+1} \\ \cdots & \cdots & \cdots & \cdots \\ c_p & c_{p+1} & \cdots & c_{p+q-1} \end{pmatrix} \begin{pmatrix} b_q \\ b_{q-1} \\ \vdots \\ b_1 \end{pmatrix} = - \begin{pmatrix} c_{p+1} \\ c_{p+2} \\ \vdots \\ c_{p+q} \end{pmatrix}$$

Having obtained the b_i's the first p+1 equations give $a_o, \ldots, a_p$. When such a rational fraction exists and is *irreductible* it is called a *Padé approximant* of the series f. It will be denoted by

$$[p/q]_f(t).$$

Such rational fractions were named after the French mathematician Henri Padé (1863-1953) who studied them in details in his thesis in 1892. However Padé is not their discoverer.

In 1731, rational fractions which are Padé approximants can be found in a letter of an unknown English mathematician

Georges Anderson to William Jones (1675-1749). Such rational fractions are also given by Leonhard Euler (1707-1783) in a letter dated 1751 to the astronomer Tobias Mayer (1723-1762). But neither Anderson nor Euler can be credited with the discovery of Padé approximants since they were not aware of their fundamental property of matching the series up to the term whose degree is equal to the sum of the degrees of the numerator and of the denominator.

The first mathematician to be conscious of this property was Joseph Louis Lagrange (1736-1813) in a paper dated 1776 and dealing with the solution of differential equations by means of continued fractions. Transforming the convergents of these continued fractions into rational fractions by using their recurrence relationship he claims that they match the series "up to the power of x inclusively which is the sum of the highest powers of x in the numerator and in the denominator". This is the birth certificate of Padé approximants !

Many very well known mathematicians have contributed to the theory of Padé approximants ; among them : Carl Gustav Jacob Jacobi (1804-1851), Leopold Kronecker (1823-1891), Bernhard Riemann (1826-1866) and Georg Frobenius (1849-1917).

In his thesis, passed in 1892 under Charles Hermite (1822-1901), Padé studied systematically such rational fractions. He arranged them in a double entry table now known as the Padé table.

$$\begin{matrix} [0/0] & [0/1] & [0/2] & \ldots \\ [1/0] & [1/1] & [1/2] & \ldots \\ [2/0] & [2/1] & [2/2] & \ldots \\ \vdots & \vdots & \vdots & \end{matrix}$$

and he studied the so-called block structure of this table which occurs when the system of linear equations defining the b_i's is singular for some p and q.

Since 1965 a growing interest for Padé approximants appeared in theoretical physics, chemistry, electronics, numerical analysis, ... Several international conferences were organized (17, 47, 54) and several books were written (4, 5, 14, 25).

Padé approximants are connected with continued fractions (29, 32, 42, 50), orthogonal polynomials (9, 21, 22, 44, 56) and convergence acceleration methods (8, 10, 53) so that chapters on them can usually be found in books on these topics. (See (15) for a more complete history).

Let us now study the algebraic properties of Padé approximants. First let us set :

$$H_k(u_n) = \begin{vmatrix} u_n & u_{n+1} & \cdots\cdots & u_{n+k-1} \\ u_{n+1} & u_{n+2} & \cdots\cdots & u_{n+k} \\ \cdots & \cdots & \cdots & \cdots \\ u_{n+k-1} & u_{n+k} & \cdots\cdots & u_{n+2k-2} \end{vmatrix}$$

Such determinants are called Hankel determinants. Thus from the definition of Padé approximants we immediatly have

Theorem 1 : A necessary and sufficient condition that [p/q] exists is that $H_q(c_{p-q+1}) \neq 0$. If [p/q] exists then it is unique.

If $H_q(c_{p-q+1}) = 0$ then the system of linear equations defining [p/q] can have infinitely many solutions or no solution. Thus, in the first case, a rational fraction [p/q] with a numerator of degree less than p or a denominator of degree less than q exists such that

$$f(t) - [p/q]_f(t) = O(t^{p+q+1}).$$

Such a rational fraction is called a Padé form by some authors (25). In the second case neither a Padé approximant nor a Padé form exists. The solution of systems of linear equations with an Hankel determinant has been studied in great details by A. Draux (21, 22).

If, ∀ p, q ≥ 0, $H_q(c_{p-q+1}) \neq 0$ then the Padé table is said to be normal ; otherwise it is non-normal. The structure of the Padé table in the non-normal case was studied by Padé who proved the

Theorem 2 : Let R(t) = Q(t)/P(t) be an irreductible rational fraction with a numerator of degree m and a denominator of degree n such that

$$f(t)\,P(t) - Q(t) = O(t^{m+n+k+1}) \qquad k \geq 0$$

Then

$$[m+i/n+j]_f(t) \equiv R(t) \qquad i, j = 0, \ldots, k$$

and no other Padé approximant is identical with R if k is maximum

Thus identical Padé approximants can only appear in square blocks of the Padé table. They cannot appear in blocks having a different shape ; this structure is known as the block structure of the Padé table.

We shall now assume, for the sake of simplicity, that the Padé tables we consider are normal. The reader interested by the non-normal case is referred to the literature on the subject.

Let us now give a determinantal expression for [p/q] which is due to Jacobi (1846).

Property 1 :

$$[p/q]_f(t) = \frac{\begin{vmatrix} t^q S_{p-q}(t) & t^{q-1} S_{p-q+1}(t) & \cdots & S_p(t) \\ c_{p-q+1} & c_{p-q+2} & \cdots & c_{p+1} \\ \cdots & \cdots & \cdots & \cdots \\ c_p & c_{p+1} & \cdots & c_{p+q} \end{vmatrix}}{\begin{vmatrix} t^q & t^{q-1} & \cdots & 1 \\ c_{p-q+1} & c_{p-q+2} & \cdots & c_{p+1} \\ \cdots & \cdots & \cdots & \cdots \\ c_p & c_{p+1} & \cdots & c_{p+q} \end{vmatrix}}$$

with $S_k(t) = \sum_{i=o}^{k} c_i t^i$ for $k \geq 0$ and $S_k(t) = 0$ for $k < 0$.

Proof : the b_i's are given by the system

$$\begin{pmatrix} 0 & \cdots & 0 & 1 \\ c_{p-q+1} & \cdots & c_p & c_{p+1} \\ \cdots & \cdots & \cdots & \cdots \\ c_p & \cdots & c_{p+q-1} & c_{p+q} \end{pmatrix} \begin{pmatrix} b_q \\ b_{q-1} \\ \vdots \\ b_o \end{pmatrix} = \begin{pmatrix} 1 \\ 0 \\ \vdots \\ 0 \end{pmatrix}$$

and thus

$$b_q t^q + \ldots + b_1 t + b_o = (-1)^q \begin{vmatrix} t^q & \cdots & 1 \\ c_{p-q+1} & \cdots & c_{p+1} \\ \cdots & \cdots & \cdots \\ c_p & \cdots & c_{p+q} \end{vmatrix} / H_q(c_{p-q+1}).$$

$$\sum_{i=o}^{p} a_i t^i = f(t) \sum_{i=o}^{q} b_i t^i + 0(t^{p+q+1})$$

$$= (-1)^q \begin{vmatrix} t^q f(t) & \cdots\cdots\cdots & f(t) \\ c_{p-q+1} & \cdots\cdots\cdots & c_{p+1} \\ \cdots\cdots & \cdots\cdots\cdots & \cdots\cdots \\ c_p & \cdots\cdots\cdots & c_{p+q} \end{vmatrix} / H_q(c_{p-q+1}) + 0(t^{p+q+1})$$

Since the numerator of [p/q] has degree p the result follows. ▮

Let us now consider the series g formally defined by

$$f(t)\ g(t) = 1.$$

If we set $g(t) = \sum_{i=o}^{\infty} d_i t^i$ then the d_i's are given by

$$c_o d_o = 1$$

$$c_o d_i + c_1 d_{i-1} + \ldots + c_i d_o = 0 \qquad i \geq 1$$

Thus g exists if and only if $c_o = f(0) \neq 0$. g is called the reciprocal series of f.

<u>Property 2</u> : If $f(0) \neq 0$ then $\forall$ p, q $\geq$ 0, $[p/q]_f(t)\ [q/p]_g(t)=1$.

Proof : We set $[p/q]_f(t) = P(t)/Q(t)$. Then

$$f(t)\ Q(t) - P(t) = 0(t^{p+q+1})$$

Multiplying by g we get

$$Q(t) - g(t)\ P(t) = 0(t^{p+q+1})$$

and thus, due to the uniqueness of Padé approximants (theorem 1)

$$Q(t)\ /\ P(t) = [q/p]_g(t)$$ ▮

<u>Property 3</u> : Let $h(t) = t^k f(t)$ then $[p+k/q]_h(t) = t^k [p/q]_f(t)$.

Proof : we set $[p/q]_f\ (t) = P(t)/Q(t)$.

$$P(t) - f(t)\ Q(t) = 0(t^{p+q+1})$$

and thus

$$t^k\ P(t) - h(t)\ Q(t) = 0(t^{p+k+q+1})$$

Consequently $t^k P(t)/Q(t)$ is the Padé approximant $[p+k/q]$ of the series h and the result is proved. ∎

Property 4 : If $f(0) = \ldots = f^{(k-1)}(0) = 0$ we set $h(t) = t^{-k} f(t)$. Then

$$[p/q]_h(t) = t^{-k}[p+k/q]_f(t)$$

Proof : Since $f(t) = t^k h(t)$ then property 3 applies. ∎

Property 5 : Let R be a polynomial of degree k. If $p \geq q+k$ then

$$[p/q]_{f+R}(t) = [p/q]_f(t) + R(t).$$

Proof : We set $[p/q]_f(t) = P(t)/Q(t)$. Then

$$\frac{P(t)}{Q(t)} + R(t) = \frac{P(t)+Q(t)R(t)}{Q(t)} = f(t) + R(t) + 0(t^{p+q+1}).$$

If $p \geq q+k$ then the degree of the numerator is less or equal to p and thus, by the uniqueness property, $P(t)/Q(t)+R(t)=[p/q]_{f+R}(t)$. ∎

Property 6 : Let $g(t) = \dfrac{A+Bf(t)}{C+Df(t)}$ where A, B, C and D are constants such that $C+Df(0) \neq 0$.

$$[p/p]_g(t) = \frac{A+B[p/p]_f(t)}{C+D[p/p]_f(t)}$$

Proof : we set $[p/p]_f(t) = P(t)/Q(t)$. Then

$$\frac{A+B[p/p]_f(t)}{C+D[p/p]_f(t)} = \frac{AQ(t)+BP(t)}{CQ(t)+DP(t)} = \frac{AQ(t)+BQ(t)[f(t)+0(t^{2p+1})]}{CQ(t)+DQ(t)[f(t)+0(t^{2p+1})]}$$

$$= \frac{A+Bf(t)+0(t^{2p+1})}{C+Df(t)+0(t^{2p+1})} = \frac{A+Bf(t)}{C+Df(t)} + 0(t^{2p+1})$$

if the constant term of the denominator $C+Df(0)$ is different from zero. ∎

Another obvious property is

Property 7 : Let $g(t) = af(t)$, $a \neq 0$. Then $[p/q]_g(t)=a[p/q]_f(t)$.

All the preceding properties deal with transformations of the series. They are called value transformations.

We shall now apply some transformations to the variable t. Such transformations are called argument transformations. The results have been gathered into the following property :

Property 8 :

a) Let $g(t) = f(at)$, $a \neq 0$. Then $[p/q]_g(t) = [p/q]_f(at)$.

b) Let $g(t) = f(t^k)$, $k > 0$. Then, $\forall$ i, j such that $i+j \leq k-1$

$$[pk+i/qk+j]_g(t) = [p/q]_f(t^k).$$

c) Let $T(t) = At^k/R(t)$ where $A \neq 0$ and where R is a polynomial of degree $k > 0$ such that $R(0) \neq 0$. Let $g(t) = f(T(t))$. Then $\forall$ i, j such that $i+j \leq k-1$

$$[pk+i/pk+j]_g(t) = [p/p]_f(T(t)).$$

Proof :

a) is trivial

b) We set $[p/q]_f(t) = P(t)/Q(t)$. Then

$$f(t^k)Q(t^k) - P(t^k) = g(t)Q(t^k)-P(t^k) = O((t^k)^{p+q+1})=O(t^{pk+qk+k}).$$

But $Q(t^k)$ has degree qk with respect to t and $P(t^k)$ has degree pk and the result follows by theorem 2.

c) This property is due to Lubinsky (38) and the proof will not be given since it is a little bit more difficult. ∎

Let us end this section by the consistency property :

Theorem 3 : Let f be the power series expansion of a rational fraction whose numerator has degree p and whose denominator has degree q. Then $\forall$ i, j ≥ 0

$$[p+i/q+j]_f(t) \equiv f(t).$$

Proof : Let $f(t) = P(t)/Q(t)$. Then $\forall$ i, j ≥ 0

$$P(t)/Q(t) = f(t) + O(t^{p+i+q+j+1})$$

which shows that $P(t)/Q(t) = [p+i/q+j]_f(t)$ by the uniqueness argument. ∎

A more complete study of the properties of Padé approximants can be found in the books listed in the references.

2. ALGEBRAIC THEORY

If only one Padé approximant has to be computed then the simplest method is to solve the system of linear equations given in section one. However in practice this is an unusual situation and in most of the cases one has to compute a *sequence* of Padé approximants. The computation of *any* sequence of Padé approximants can be made by using a recursive method. All these recursive methods come out from the theory of formal orthogonal polynomials even in the non-normal case.

Thus in this section we shall first study some properties of formal orthogonal polynomials. Then we shall indicate the connection with Padé approximants and Gaussian quadrature methods. Finally recursive methods for computing Padé approximants will be derived from the theory.

As we shall see along this section the theory of formal orthogonal polynomials provides a natural and unified basis for the complete algebraic theory of Padé approximants.

2.1. Formal orthogonal polynomials

Let (c_n) be a sequence of real numbers. We shall define a linear functional c acting on the vector space of real polynomials by

$$c(x^i) = c_i \qquad i = 0, 1, \ldots$$

Definition 1 : The family of polynomials $\{P_k\}$ is said to be a family of formal orthogonal polynomials with respect to the functional c (or with respect to the sequence (c_n)) if $\forall\ k \geq 0$, P_k has the exact degree k and if

$$c(x^i P_k(x)) = 0 \qquad \text{for } i = 0, \ldots, k-1$$

The preceding relations, which are called the orthogonality relations, completely define the polynomials P_k. If we set

$$P_k(x) = a_o + a_1 x + \ldots + a_k x^k$$

they give

$$c(P_k(x)) = a_o c_o + a_1 c_1 + \ldots + a_k c_k = 0$$

$$c(xP_k(x)) = a_o c_1 + a_1 c_2 + \ldots + a_k c_{k+1} = 0$$

$$\ldots\ldots\ldots\ldots\ldots\ldots\ldots\ldots\ldots\ldots\ldots\ldots$$

$$c(x^{k-1} P_k(x)) = a_o c_{k-1} + a_1 c_k + \ldots + a_k c_{2k-1} = 0$$

The orthogonality relations are still satisfied if P_k is multiplied by a constant different from zero. Thus P_k is defined apart from a multiplying factor. We can choose it such that $\forall$ k, $a_k = 1$: P_k is said to be monic. Thus P_k, if it exists, will have the exact degree k and its coefficients will be given as the solution of the system

$$\begin{pmatrix} c_0 & c_1 & \cdots\cdots & c_{k-1} \\ c_1 & c_2 & \cdots\cdots & c_k \\ \cdots & \cdots & \cdots & \cdots \\ c_{k-1} & c_k & \cdots\cdots & c_{2k-2} \end{pmatrix} \begin{pmatrix} a_0 \\ a_1 \\ \vdots \\ a_{k-1} \end{pmatrix} = - \begin{pmatrix} c_k \\ c_{k+1} \\ \vdots \\ c_{2k-1} \end{pmatrix}$$

As we can see this system is very much similar to the system defining the denominators of Padé approximants and, as above, we have the

Theorem 4 : A necessary and sufficient condition that P_k exists is that $H_k(c_0) \neq 0$. If P_k exists then it is unique (apart from a multiplying factor).

We shall now assume that $\forall$ k, $H_k(c_0) \neq 0$. In that case the functional c is said to be definite. However it must be noticed that a theory of formal orthogonal polynomials can be constructed even if this condition is not satisfied and that all the properties mentionned below can be extended to this case (22).

Let us now examine some properties of formal orthogonal polynomials. We must first note that the orthogonality relations imply that $c(p(x)P_k(x)) = 0$ for every polynomial p of degree less than k, that $c(p(x)P_k(x)) \neq 0$ for every polynomial of exact degree k and that $\forall$ n $\neq$ k, $c(P_n(x)P_k(x)) = 0$.

We have the

Property 9 :

$$P_0(x) = 1$$

$$P_k(x) = \begin{vmatrix} c_0 & c_1 & \cdots\cdots & c_k \\ c_1 & c_2 & \cdots\cdots & c_{k+1} \\ \cdots & \cdots & \cdots & \cdots \\ c_{k-1} & c_k & \cdots\cdots & c_{2k-1} \\ 1 & x & \cdots\cdots & x^k \end{vmatrix} \Big/ H_k(c_0)$$

Proof

$$c(x^i P_k(x)) = \begin{vmatrix} c_o & c_1 & \cdots\cdots & c_k \\ \cdots & \cdots & \cdots & \cdots \\ c_{k-1} & c_k & \cdots\cdots & c_{2k-1} \\ c_i & c_{i+1} & \cdots & c_{i+k} \end{vmatrix} / H_k(c_o) = 0 \text{ for } i=0,\ldots,k-1$$

since two rows of the numerator are identical. The property follows from theorem 4. The multiplying factor is equal to $1/H_k(c_o)$ so that the coefficient of x^k be unity. ∎

The formal orthogonal polynomials we defined possess most of the properties of the classical orthogonal polynomials (that is when $c_i = \int_a^b x^i d\alpha(x)$ with α bounded and non decreassing in [a,b]).

First they satisfy a three terms recurrence relationship

Theorem 5 : the polynomials $\{P_k\}$ satisfy

$$P_{k+1}(x) = (A_{k+1}x + B_{k+1})P_k(x) - C_{k+1}P_{k-1}(x) \qquad k = 0, 1, \ldots$$

with $P_{-1}(x) = 0$ and $P_o(x)$ = an arbitrary non zero constant . The coefficients are given by

$$A_{k+1} = h_{k+1}/\beta_k \quad B_{k+1} = -h_{k+1}\alpha_k/(h_k\beta_k) \quad C_{k+1} = \beta_{k-1}h_{k+1}/(\beta_k h_{k-1})$$

with

$$\alpha_k = c(xP_k^2(x)) \quad \beta_k = c(xP_k(x)P_{k+1}(x)) \quad h_k = c(P_k^2(x)).$$

The proof is the same as in the classical case. If the polynomials P_k are monic then

$$A_{k+1} = 1 \qquad B_{k+1} = -\alpha_k/h_k \qquad C_{k+1} = h_k/h_{k-1}$$

Since the polynomials $\{P_k\}$ satisfy a three terms recurrence relationship they also satisfy the so-called Christoffel - Darboux identify which is one of its leading consequences in the classical case :

Theorem 6 : For $k \geq 0$

$$\frac{t_k}{h_k t_{k+1}} [P_{k+1}(x)P_k(t) - P_{k+1}(t)P_k(x)] = (x-t)\sum_{i=o}^{k} h_i^{-1} P_i(x)P_i(t)$$

where t_k is the coefficient of x^k in P_k.

Let us now define Q_k as

$$Q_k(t) = c\left(\frac{P_k(x)-P_k(t)}{x-t}\right)$$

where t is a parameter and where c acts on the variable x. It is easy to see that Q_k is a polynomial of degree k-1 in t. The polynomials $\{Q_k\}$ are said to be associated with the polynomials $\{P_k\}$. From the definition we have the :

<u>Property 10</u> :

$$Q_0(t) = 0$$

$$Q_k(t) = \begin{vmatrix} c_o & c_1 & c_2 & \cdots & c_k \\ \cdots & & & & \\ c_{k-1} & c_k & c_{k+1} & \cdots & c_{2k-1} \\ 0 & c_o & c_o t+c_1 & \cdots & (c_o t^{k-1}+c_1 t^{k-2}+\ldots+c_{k-1}) \end{vmatrix} / H_k(c_o)$$

Let us now study some properties of the associated polynomials.

<u>Theorem 7</u> : The polynomials $\{Q_k\}$ satisfy the same recurrence relationship as the polynomials $\{P_k\}$ with the initial conditions $Q_{-1}(x) = -1$, $Q_0(x) = 0$ and $C_1 = A_1\, c\, (P_0(x))$.

Proof : Let us write the recurrence relationship for the polynomials P_k with x and t as variables. Let us subtract the second relation from the first one and divide by x-t. Let us use the identity

$$\frac{xP_k(x)-tP_k(t)}{x-t} = t\,\frac{P_k(x)-P_k(t)}{x-t} + P_k(x)$$

Applying c and using $c(P_k(x)) = 0$ for $k > 0$ we obtain

$$Q_{k+1}(t) = (A_{k+1}t + B_{k+1})Q_k(t) - C_{k+1}Q_{k-1}(t) \qquad k = 1, 2, \ldots$$

The definition of Q_k gives $Q_0(t) = 0$ and $Q_1(t) = A_1\, c\, (P_0(x))$ and, from the recurrence relationship with $k = 0$ we get $Q_1(t)=-C_1Q_{-1}(t)$. Thus, since C_1 can be arbitrarily chosen, we can set $Q_{-1}(t)=-1$ and $C_1 = A_1\, c\, (P_0(x))$. ∎

Since the polynomial $\{Q_k\}$ satisfy the same recurrence relation as the polynomials $\{P_k\}$ they also satisfy the Christoffel-Darboux identity.

Another important property is

<u>Property 11</u> : For $k \geq 0$

$$P_k(x)Q_{k+1}(x) - Q_k(x)P_{k+1}(x) = A_{k+1}\, h_k$$

Proof : Let us write the recurrence relation for P_{k+1} and multiply it by Q_k. Let us write the recurrence relation for Q_{k+1} and multiply it by P_k. Let us subtract the first relation from the second one. We get

$$P_k(x)Q_{k+1}(x) - Q_k(x)P_{k+1}(x) = C_{k+1}[P_{k-1}(x)Q_k(x)-Q_{k-1}(x)P_k(x)].$$

Thus for $k > 0$

$$P_k(x)Q_{k+1}(x)-Q_k(x)P_{k+1}(x)=C_{k+1}C_k\ldots C_2[P_0(x)Q_1(x)-Q_0(x)P_1(x)]$$

$$= \frac{A_{k+1}h_k}{A_1\, h_0} P_0(x)Q_1(x).$$

Using the definitions of A_1, h_0, P_0 and Q_1 we get the result. The result also obviously holds for $k \geqq 0$.∎

For more properties of formal orthogonal polynomials in the definite case and for their matrix interpretation see (14). The non-definite case has been treated by Draux (21, 22).

The last result I would like to mention is the

<u>Theorem 8</u> : If c is definite then $\forall$ k, P_k and P_{k+1} have no common zero, neither Q_k and Q_{k+1} nor P_k and Q_k.

Proof : It is a direct consequence of property 11. If x is a zero of P_k and P_{k+1} then the left hand side of the relation of property 11 is equal to zero while the right hand side is different from zero. The proof is the same for the two other pairs of polynomials.∎

2.2. Gaussian quadrature methods and Padé approximants

We shall now study the connection between formal orthogonal polynomials and Padé approximants. The link between both subjects is realized via Gaussian quadrature formulas.

We set

$$\tilde{P}_k(t) = t^k P_k(t^{-1}) \qquad \tilde{Q}_k(t) = t^{k-1} Q_k(t^{-1})$$

The passage from P_k to $\tilde{P}_k$ (and from Q_k to $\tilde{Q}_k$) is obtained just by reversing the numbering of the coefficients.

From properties 1, 9 and 10 we immediately get the

Theorem 9 :

$$[k-1/k]_f(t) = \tilde{Q}_k(t)/\tilde{P}_k(t)$$

We shall see later how to connect the other Padé approximants of the table to formal orthogonal polynomials. But, for the moment, let us have a look at formal Gaussian quadrature methods. The whole process is based on a very simple but fundamental result :

Theorem 10 :

$$f(t) = c\left(\frac{1}{1-xt}\right)$$

Proof : We formally have

$$(1-xt)^{-1} = 1 + xt + x^2t^2 + \ldots$$

Thus $c((1-xt)^{-1}) = c(1) + c(x)t + c(x^2)t^2 + \ldots = c_0 + c_1t + c_2t^2 + \ldots$ ∎

Thus the computation of f(t), for a fixed value of t, can be viewed as a formal integration process since, in the classical case when $c_i = \int_a^b x^i d\alpha(x)$ we have

$$f(t) = \int_a^b \frac{d\alpha(x)}{1-xt}$$

Such an idea goes back to Wall (50, chap. 11). See also Wynn (56).

The approximate computation of such an integral is usually realized in numerical analysis by a so-called interpolatory quadrature method where the function to be integrated ($(1-xt)^{-1}$ in our case) is replaced by an interpolating polynomial. If the interpolation nodes are distinct the Lagrange interpolation polynomial is used. If some interpolation points coincide the Hermite interpolation polynomial has to be considered.

For k arbitrary interpolation points the quadrature method is exact on $\mathcal{P}_{k-1}$ the vector space of polynomials of degree less than or equal to k-1. If the interpolation nodes are the zeros of the polynomial of degree k belonging to the family of orthogonal polynomials with respect to the functional c, then the quadrature method is exact on $\mathcal{P}_{2k-1}$. In that case we have a Gaussian quadrature method.

Let us first express the Hermite interpolation polynomial of $(1-xt)^{-1}$.

<u>Theorem 11</u> : Let $P_k(x) = (x-x_1)^{k_1} \dots (x-x_n)^{k_n}$ where $x_1,\dots,x_n$ are distinct and $k = k_1 +\dots+ k_n$. The polynomial

$$P(x) = \frac{1}{1-xt}\left(1 - t^k \frac{P_k(x)}{\tilde{P}_k(t)}\right)$$

is the Hermite interpolation polynomial of $(1-xt)^{-1}$ such that

$$P^{(j)}(x_i) = \frac{d^j}{dx^j}\left(\frac{1}{1-xt}\right)_{x=x_i} \qquad i=1,\dots,n \text{ and } j=0,\dots,k_i-1.$$

Proof : We first have to prove that P has the exact degree k-1. We set

$$P_k(x) = a_o + \dots + a_k x^k \qquad a_k \neq 0$$

Then

$$\tilde{P}_k(t)-t^k P_k(x) = a_1 t^{k-1}(1-xt)+a_2 t^{k-2}(1-x^2t^2)+\dots+a_k(1-x^k t^k)$$

and thus

$$\tilde{P}_k(t)P(x) = a_1 t^{k-1}+a_2 t^{k-2}(1+xt)+\dots+a_k(1+xt+\dots+x^{k-1}t^{k-1}).$$

Let us now prove that P satisfies the interpolation conditions. We have

$$P^{(j)}(x) = \frac{d^j}{dx^j}\left(\frac{1}{1-xt}\right) - \frac{t^k}{\tilde{P}_k(t)} \frac{d^j}{dx^j}\left(\frac{P_k(x)}{1-xt}\right)$$

Since $P_k^{(j)}(x_i) = 0$ for $i = 1, \dots, n$ and $j = 0, \dots, k_i-1$ then

$$\frac{d^j}{dx^j}\left(\frac{P_k(x)}{1-xt}\right)_{x=x_i} = 0$$

and thus

$$P^{(j)}(x_i) = \frac{d^j}{dx^j}\left(\frac{1}{1-xt}\right)_{x=x_i}$$

Since the Hermite interpolation polynomial is unique the theorem is proved.∎

This theorem has two important consequences. The first one is :

Theorem 12 :

$$c(P(x)) = [k-1/k]_f(t)$$

Proof :

$$c(P(x)) = \frac{1}{\tilde{P}_k(t)}\, c\left(\frac{\tilde{P}_k(t)-t^k P_k(x)}{1-xt}\right) = \frac{t^{k-1}}{\tilde{P}_k(t)}\, c\left(\frac{P_k(t^{-1})-P_k(x)}{t^{-1}-x}\right)$$

$$= \frac{t^{k-1}Q_k(t^{-1})}{\tilde{P}_k(t)} = \tilde{Q}_k(t)/\tilde{P}_k(t)$$

and the result follows by theorem 9.∎

The second consequence of theorem 11 is an expression for the error :

Theorem 13 :

$$f(t) - [k-1/k]_f(t) = \frac{t^{2k}}{\tilde{P}_k(t)}\, c\left(\frac{x^k P_k(x)}{1-xt}\right) = \frac{t^{2k}}{\tilde{P}_k^2(t)}\, c\left(\frac{P_k^2(x)}{1-xt}\right)$$

Proof : Applying the functional c to the expression of P we get

$$c(P(x)) = c\left(\frac{1}{1-xt}\right) - \frac{t^k}{\tilde{P}_k(t)}\, c\left(\frac{P_k(x)}{1-xt}\right)$$

or

$$f(t) - [k-1/k]_f(t) = \frac{t^k}{\tilde{P}_k(t)}\, c\left(\frac{P_k(x)}{1-xt}\right)$$

But

$$c\left(\frac{P_k(x)}{1-xt}\right) = c\left(P_k(x)\left(1 + xt + \ldots + x^{k-1}t^{k-1} + \frac{x^k t^k}{1-xt}\right)\right)$$

and the first expression follows since $c(x^i P_k(x))=0$ for $i=0,..,k-1$.

In the proof of theorem 11 we saw that $(P_k(x)-P_k(t^{-1}))/(1-xt)$ is a polynomial of degree k-1 in x. Thus

$$c\left(\frac{P_k(x)-P_k(t^{-1})}{1-xt} \quad P_k(x)\right) = 0 = c\left(\frac{P_k^2(x)}{1-xt}\right) - P_k(t^{-1}) \; c\left(\frac{P_k(x)}{1-xt}\right)$$

or

$$c\left(\frac{P_k(x)}{1-xt}\right) = \frac{t^k}{\tilde{P}_k(t)} \; c\left(\frac{P_k^2(x)}{1-xt}\right)$$ and the second expression is proved. ∎

As a consequence of the first expression of the error let us mention

$$f(t) - [k-1/k]_f(t) = \frac{H_{k+1}(c_o)}{H_k(c_o)} \; t^{2k} + 0(t^{2k+1})$$

Moreover we have

$$f(t) - [k-1/k]_f(t) = \frac{t^{2k}}{\tilde{P}_k(t)} \sum_{i=o}^{\infty} d_{k+i} t^i$$

with $d_j = c(x^j P_k(x))$.

2.3. Adjacent systems of orthogonal polynomials

Let us come back to the recurrence relationship of monic orthogonal polynomials. We set

$$B_{k+1} = -q_{k+1} - e_k \qquad\qquad C_{k+1} = q_k \, e_k$$

These relations, together with the initial condition $e_o = 0$, completely define the sequences (q_k) and (e_k) if $\forall\, k \geq 1$, $q_k \neq 0$. For consistency we must add the convention that $q_o \, e_o = c_o$.

Let us set

$$P_{k+1}(x) = x \, \bar{P}_k(x) - q_{k+1} \, P_k(x) \qquad\qquad k = 0, 1, \ldots$$

Then the recurrence relation gives

$$\bar{P}_k(x) = P_k(x) - e_k \, \bar{P}_{k-1}(x) \qquad\qquad k = 1, 2, \ldots$$

But from the first identity we have $\bar{P}_o(x) = P_o(x) = 1$ and thus the second relation holds for k = 0 with $\bar{P}_{-1}(x) = 0$.

From the two preceding relations it is easy to see that $\forall k$, $\bar{P}_k$ is a polynomial of exact degree k and that the two families $\{P_k\}$ and $\{\bar{P}_k\}$ can be constructed simultaneously by an alternate use of these two relations. Moreover we have

$$c(x\, x^i\, \bar{P}_k(x)) = c(x^i P_{k+1}(x)) + q_{k+1}\, c(x^i P_k(x)) = 0 \text{ for } i=0,\ldots,k-1$$

Thus the polynomials $\{\bar{P}_k\}$ form a family of orthogonal polynomials with respect to the linear functional $\bar{c}$ defined by

$$\bar{c}(x^i) = c(x^{i+1}) = c_{i+1} \qquad i = 0, 1, \ldots$$

In other words the polynomials $\{\bar{P}_k\}$ are orthogonal with respect to the sequence $c_1, c_2, \ldots$ The families $\{P_k\}$ and $\{\bar{P}_k\}$ are called adjacent (or contiguous (44)) systems of orthogonal polynomials.

It follows that the polynomials $\{\bar{P}_k\}$ satisfy a three terms recurrence relationship which can be written as

$$\bar{P}_{k+1}(x) = (x-\bar{q}_{k+1}-\bar{e}_k)\, \bar{P}_k(x) - \bar{q}_k\, \bar{e}_k\, \bar{P}_{k-1}(x) \qquad k = 0, 1, \ldots$$

with $\bar{P}_o(x) = 1$, $\bar{P}_{-1}(x) = 0$, $\bar{e}_o = 0$ and the convention that $\bar{q}_o\bar{e}_o = c_1$.

We have

$$\bar{P}_{k+1}(x) = P_{k+1}(x) - e_{k+1}\, \bar{P}_k(x) = x\, \bar{P}_k(x) - q_{k+1}P_k(x) - e_{k+1}\bar{P}_k(x)$$

$$= x\, \bar{P}_k(x) - q_{k+1}(\bar{P}_k(x) + e_k\bar{P}_{k-1}(x)) - e_{k+1}\, \bar{P}_k(x)$$

and finally we obtain

$$\bar{P}_{k+1}(x) = (x-q_{k+1}-e_{k+1})\, \bar{P}_k(x) - e_k q_{k+1}\, \bar{P}_{k-1}(x).$$

By identification we get

$$\bar{q}_{k+1} + \bar{e}_k = q_{k+1} + e_{k+1}$$

$$\bar{q}_k\, \bar{e}_k = q_{k+1}\, e_k$$

and thus the sequences $(\bar{q}_k)$ and $(\bar{e}_k)$ can be computed from (q_k) and (e_k).

Obviously the same process can be indefinitely repeated and families of orthogonal polynomials with respect to the sequences $c_n, c_{n+1}, \ldots$ can be constructed. But, before entering into this general process, it would be better to change our notations. We shall set $e_k^{(0)} = e_k$, $q_k^{(0)} = q_k$, $P_k^{(0)} = P_k$, $c^{(0)} = c$, $e_k^{(1)} = \bar{e}_k$, $q_k^{(1)} = \bar{q}_k$, $P_k^{(1)} = \bar{P}_k$ and $c^{(1)} = \bar{c}$.

In general we shall denote by $\{P_k^{(n)}\}$ the family of orthogonal polynomials with respect to the linear functional $c^{(n)}$ defined by

$$c^{(n)}(x^i) = c_{n+i} \qquad i = 0, 1, \ldots$$

We shall ensure the existence of all these families by the assumption that $\forall\, n, k \geq 0$, $H_k(c_n) \neq 0$. (For the general case see Draux (22)).

The recurrence relationship of the family $\{P_k^{(n)}\}$ will be written as

$$P_{k+1}^{(n)}(x) = (x-q_{k+1}^{(n)}-e_k^{(n)})P_k^{(n)}(x)-q_k^{(n)}e_k^{(n)}P_{k-1}^{(n)}(x) \qquad k=0,1,.. \quad (1)$$

with $P_{-1}^{(n)}(x) = 0$, $P_0^{(n)}(x) = 1$, $e_0^{(n)} = 0$ and the convention that $q_0^{(n)}e_0^{(n)} = c_n$

We have

$$P_{k+1}^{(n)}(x) = x\, P_k^{(n+1)}(x) - q_{k+1}^{(n)}\, P_k^{(n)}(x) \qquad k = 0, 1, \ldots \quad (2)$$

$$P_k^{(n+1)}(x) = P_k^{(n)}(x) - e_k^{(n)}\, P_{k-1}^{(n+1)}(x) \qquad k = 0, 1, \ldots \quad (3)$$

Moreover the sequences $(q_k^{(n)})$ and $(e_k^{(n)})$ satisfy

$$q_{k+1}^{(n)} + e_{k+1}^{(n)} = q_{k+1}^{(n+1)} + e_k^{(n+1)}$$

$$q_{k+1}^{(n)}\, e_k^{(n)} = q_k^{(n+1)}\, e_k^{(n+1)}$$

This algorithm is called the q-d algorithm. It has been studied in details by Rutishauser (45). The previous derivation of this algorithm is due to Stiefel (48). Many results on the q-d algorithm have been obtained by Henrici (27) (see also (29, Vols 1 and 2)).

Usually the $q_k^{(n)}$ and the $e_k^{(n)}$ are displayed in a two dimensional array as follows

$$\begin{array}{cccc}
e_0^{(o)} & & & \\
 & q_1^{(o)} & & \\
e_0^{(1)} & & e_1^{(o)} & \\
 & q_1^{(1)} & & q_2^{(o)} \\
e_0^{(2)} & & e_1^{(1)} & \\
 & q_1^{(2)} & & q_2^{(1)} \\
e_0^{(3)} & & e_1^{(2)} & \\
 & q_1^{(3)} & & q_2^{(2)} \\
\vdots & & e_1^{(3)} & \\
 & \vdots & & q_2^{(3)} \\
 & & \vdots & \\
 & & & \vdots \quad \ddots
\end{array}$$

All the entries of this array can be computed if the first diagonal is known. They can also be obtained from the first two columns $e_0^{(n)} = 0$ and $q_1^{(n)} = c_{n+1}/c_n$.

Let us give determinantal expressions for $q_k^{(n)}$ and $e_k^{(n)}$. Setting x to zero we have

$$e_k^{(n)} P_{k-1}^{(n+1)}(0) = P_k^{(n)}(0) - P_k^{(n+1)}(0)$$

$$P_k^{(n)}(0) = -q_k^{(n)} P_{k-1}^{(n)}(0)$$

Using the determinantal expression of $P_k^{(n)}$ (property 9) we find

$$e_k^{(n)} = \frac{H_{k-1}^{(n+1)}}{H_{k-1}^{(n+2)}} \frac{H_k^{(n)} H_k^{(n+2)} - [H_k^{(n+1)}]^2}{H_k^{(n)} H_k^{(n+1)}}$$

$$q_k^{(n)} = \frac{H_k^{(n+1)} H_{k-1}^{(n)}}{H_k^{(n)} H_{k-1}^{(n+1)}}$$

with $H_k^{(n)} = H_k(c_n)$.

But $q_k^{(n)} e_k^{(n)}$ is equal to the coefficient C_{k+1} of the recurrence relationship of $\{P_k^{(n)}\}$. Using the results of theorem 5 we obtain

$$e_k^{(n)} = \frac{H_{k-1}^{(n+1)} H_{k+1}^{(n)}}{H_k^{(n)} H_k^{(n+1)}}$$

Thus our assumption on Hankel determinants ensures that all the entries in the q-d table exist and are different from zero.

If we compare the two expressions for $e_k^{(n)}$ we get

$$H_k^{(n)} H_k^{(n+2)} - [H_k^{(n+1)}]^2 = H_{k+1}^{(n)} H_{k-1}^{(n+2)} \qquad n, k = 0, 1, \ldots$$

with $H_o^{(n)} = 1$.

This is the well known recurrence relation for Hankel determinants which is usually obtained via Sylvester indentity.

In section 1 the convention that $c_i = 0$ for $i < 0$ was made. The same convention permits to define $H_k^{(n)}$ for $n \geq -k+1$ and thus $q_k^{(n)}$ and $e_k^{(n)}$ for $n \geq -k+1$. The same is also true for the families $\{P_k^{(n)}\}$ if $H_k^{(n)} \neq 0$ for $k = 0, 1, \ldots$ and $n = -k+1, -k+2, \ldots$

We shall now assume that this condition is satisfied. In that case c is said to be completely definite.

We have the :

<u>Theorem 14</u> : If c is completely definite then $\forall\, k \geq 0$ and $\forall\, n \geq -k+1$, $P_k^{(n)}$ has no common zero with $P_k^{(n+1)}$, $P_{k-1}^{(n+1)}$, $P_k^{(n-1)}$, $P_{k+1}^{(n-1)}$, $P_{k-1}^{(n)}$ and $P_{k+1}^{(n)}$. Moreover $P_k^{(n)}(0) \neq 0$.

Proof : Let us assume that x is such that $P_k^{(n)}(x) = P_k^{(n+1)}(x) = 0$. Then

$$P_{k+1}^{(n)}(x) = x\, P_k^{(n+1)}(x) - q_{k+1}^{(n)}\, P_k^{(n)}(x) = 0.$$

Thus x will also be a zero of $P_{k+1}^{(n)}$ which is impossible by theorem 8. The other proofs are similar. From the determinantal formula for $P_k^{(n)}$ we have $H_k^{(n)} P_k^{(n)}(0) = (-1)^k H_k^{(n+1)}$ which shows that $P_k^{(n)}(0) \neq 0$. ∎

We can obviously define the polynomials $\{Q_k^{(n)}\}$ associated with the polynomials $\{P_k^{(n)}\}$ by

$$Q_k^{(n)}(t) = c^{(n)}\left(\frac{P_k^{(n)}(x) - P_k^{(n)}(t)}{x-t}\right)$$

These polynomials satisfy relations similar to those satisfied by the polynomials $\{P_k^{(n)}\}$ but we shall not give them to shorten the exposition.

From the three basic preceding relations, from the determinantal expressions of $e_k^{(n)}$ and $q_k^{(n)}$ and from the recurrence relation of Hankel determinants we can obtain five other relations linking together three adjacent polynomials :

$$H_k^{(n+2)} H_k^{(n)} P_k^{(n)}(x) = [H_k^{(n+1)}]^2\, P_k^{(n+1)}(x) + H_{k+1}^{(n)} H_{k-1}^{(n+2)} P_{k-1}^{(n+2)}(x) \qquad (4)$$

$$x H_k^{(n)} H_{k+1}^{(n-1)} H_k^{(n+1)} P_k^{(n+1)}(x) = [x H_k^{(n+1)} H_{k+1}^{(n-1)} + H_{k+1}^{(n)} H_k^{(n)}] H_k^{(n)} P_k^{(n)}(x) - H_{k+1}^{(n)} H_k^{(n+1)} H_k^{(n-1)} P_k^{(n-1)}(x) \qquad (5)$$

$$H_k^{(n)} H_k^{(n+1)} H_{k+1}^{(n-1)} P_{k+1}^{(n-1)}(x) = [x H_{k+1}^{(n-1)} H_k^{(n+1)} - H_k^{(n)} H_{k+1}^{(n)}] H_k^{(n)} P_k^{(n)}(x) - x H_{k+1}^{(n-1)} H_{k+1}^{(n)} H_{k-1}^{(n+1)} P_{k-1}^{(n+1)}(x) \qquad (6)$$

$$H_k^{(n+1)} H_{k+1}^{(n-1)} P_{k+1}^{(n-1)}(x) = x H_{k+1}^{(n-1)} H_k^{(n+1)} P_k^{(n+1)}(x) - H_{k+1}^{(n)} H_k^{(n)} P_k^{(n)}(x) \qquad (7)$$

$$[H_k^{(n+1)}]^2 H_{k+1}^{(n-2)} P_{k+1}^{(n-2)}(x) = [xH_{k-1}^{(n+2)} H_{k+2}^{(n-2)} - xH_{k+1}^{(n-2)} H_k^{(n+2)} + H_k^{(n+1)} H_{k+1}^{(n-1)}] H_k^{(n)} P_k^{(n)}(x) - x^2 [H_{k+1}^{(n-1)}]^2 H_{k-1}^{(n+2)} P_{k-1}^{(n+2)}(x) \quad (8)$$

Of course it is possible to obtain many other relations by combining these expressions but these eight relations will be sufficient to compute recursively any sequence of Padé approximants as we shall see in the next section. Similar relations hold for the associated polynomials $\{Q_k^{(n)}\}$.

2.4. Recursive computation of Padé approximants

Let us first connect Padé approximants with adjacent systems of orthogonal polynomials. We set

$$f_n(t) = \sum_{i=n}^{\infty} c_i t^{i-n}.$$

From theorem 9 we have

$$\tilde{Q}_k^{(n+1)}(t)/\tilde{P}_k^{(n+1)}(t) = [k-1/k]_{f_{n+1}}(t) = f_{n+1}(t) + 0(t^{2k})$$

Thus

$$\sum_{i=o}^{n} c_i t^i + t^{n+1} \tilde{Q}_k^{(n+1)}/\tilde{P}_k^{(n+1)}(t) = \sum_{i=o}^{n} c_i t^i + t^{n+1} f_{n+1}(t) + 0(t^{n+2k+1})$$

$$= f(t) + 0(t^{n+2k+1})$$

Moreover the left hand side is a rational fraction with a numerator of degree n+k and a denominator of degree k. Thus, due to the unicity of Padé approximants, we have the :

<u>Theorem 15</u> : $\forall\, k \geq -n$

$$[n+k/k]_f(t) = \sum_{i=o}^{n} c_i t^i + t^{n+1} \tilde{Q}_k^{(n+1)}(t)/\tilde{P}_k^{(n+1)}(t).$$

For simplicity we set

$$N = [n+k-1/k]$$

$$W = [n+k/k-1] \qquad C = [n+k/k] \qquad E = [n+k/k+1]$$

$$S = [n+k+1/k]$$

From an immediate generalisation of property 11 we have

$$P_k^{(n)}(t)\, Q_{k+1}^{(n)}(t) - Q_k^{(n)}(t)\, P_{k+1}^{(n)}(t) = H_{k+1}^{(n)}/H_k^{(n)}$$

or

$$\tilde{P}_k^{(n)}(t)\, \tilde{Q}_{k+1}^{(n)}(t) - \tilde{Q}_k^{(n)}(t)\, \tilde{P}_{k+1}^{(n)}(t) = t^{2k}\, H_{k+1}^{(n)}/H_k^{(n)}$$

Thus, dividing by $\tilde{P}_k^{(n)}(t)\ \tilde{P}_{k+1}^{(n)}(t)$, we get :

$$N-E = -\frac{H_{k+1}^{(n)}}{H_k^{(n)}\tilde{P}_k^{(n)}(t)\tilde{P}_{k+1}^{(n)}(t)}\, t^{n+2k} \tag{9}$$

From (1) we have

$$(E-N)\ \tilde{P}_{k+1}^{(n)}(t) = (C-N)\ \tilde{P}_k^{(n+1)}(t)$$

and we obtain

$$N-C = -\frac{H_{k+1}^{(n)}}{H_k^{(n)}\tilde{P}_k^{(n)}(t)\tilde{P}_k^{(n+1)}(t)}\, t^{n+2k} \tag{10}$$

Moreover

$$(N-C) - (N-E) = E-C$$

and, again from (1), we get

$$E-C = \frac{H_{k+1}^{(n+1)}}{H_k^{(n+1)}\tilde{P}_{k+1}^{(n)}(t)\tilde{P}_k^{(n+1)}(t)}\, t^{n+2k+1} \tag{11}$$

Finally we have, from the relations given in section 2.3 :

$$E-S = \frac{[H_{k+1}^{(n+1)}]^2}{H_{k+1}^{(n)}H_k^{(n+2)}\tilde{P}_{k+1}^{(n)}(t)\tilde{P}_k^{(n+2)}(t)}\, t^{n+2k+2} \tag{12}$$

These four relations will be useful later. For the moment if we combine them and use (2) we obtain :

$$(N-C)^{-1} + (S-C)^{-1} = (E-C)^{-1} + (W-C)^{-1} \tag{13}$$

This is the so-called cross rule which has been obtained by Wynn (55) from the ε-algorithm and its connection with Shanks' transformation of sequences and the Padé table.

Together with the initial conditions :

$$[-1/q]_f(t) = 0 \qquad [p/-1]_f(t) = \infty$$

$$[0\ /q]_f(t) = 1/[q/0]_g(t) \qquad [p\ /0]_f(t) = \sum_{i=o}^{p} c_i t^i$$

This relation can be used to compute recursively the whole Padé table in the normal case. Wynn's cross rule has been extended to non-normal Padé tables by Cordellier (19). His extended cross rule is as follows

$$(N_i - C)^{-1} + (S_i - C)^{-1} = (E_i - C)^{-1} + (W_i - C)^{-1} \qquad i = 1,\dots,r$$

where

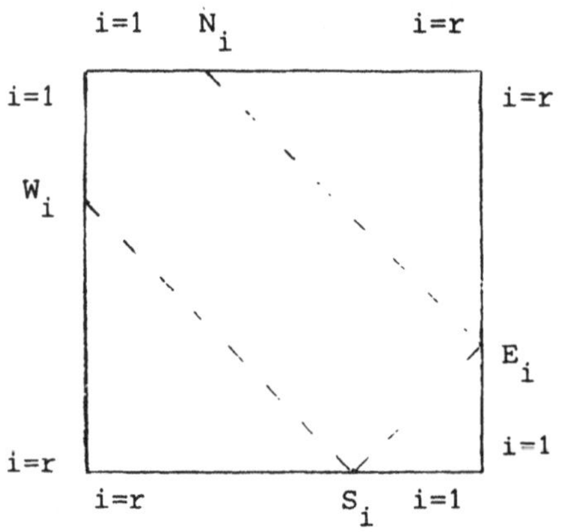

The disadvantage of Wynn's cross rule is that it can only be used for the computation of the whole Padé table. It is impossible to use it only to obtain some sequence of Padé approximants. For this purpose we shall use the relations (1) to (8) given in section 2.3. We set

$$[n+k/k]_f(t) = \tilde{N}_k^{(n+1)}(t)/\tilde{P}_k^{(n+1)}(t) = \sum_{i=o}^{n+k} a_i^{(k,n+1)} t^i / \sum_{i=o}^{k} b_i^{(k,n+1)} t^i \qquad (14)$$

Since the polynomials $P_k^{(n)}$ are monic we have

$$b_o^{(k,n)} = 1 \qquad a_o^{(k,n)} = c_o$$

$$b_k^{(k,n)} = (-1)^k H_k^{(n+1)}/H_k^{(n)} \qquad a_{n+k-1}^{(k,n)} = (-1)^k H_{k+1}^{(n-1)}/H_k^{(n)} .$$

Thus relation (7) becomes

$$\frac{\tilde{N}_{k+1}^{(n)}(t)}{\tilde{P}_{k+1}^{(n)}(t)} = \frac{a_{n+k}^{(k,n+1)}\,\tilde{N}_k^{(n+2)}(t) - t\,a_{n+k+1}^{(k,n+2)}\,\tilde{N}_k^{(n+1)}(t)}{a_{n+k}^{(k,n+1)}\,\tilde{P}_k^{(n+2)}(t) - t\,a_{n+k+1}^{(k,n+2)}\,\tilde{P}_k^{(n+1)}(t)}$$

If the approximants C and S are known, then this relation gives the approximant E.

On the same way relation (4) becomes

$$\frac{\tilde{N}_k^{(n)}(t)}{\tilde{P}_k^{(n)}(t)} = \frac{a_{n+k}^{(k-1,n+2)}\tilde{N}_k^{(n+1)}(t) - a_{n+k}^{(k,n+1)}\tilde{N}_{k-1}^{(n+2)}(t)}{a_{n+k}^{(k-1,n+2)}\tilde{P}_k^{(n+1)}(t) - a_{n+k}^{(k,n+1)}\tilde{P}_{k-1}^{(n+2)}(t)}$$

Thus, if the approximants C and W are known, the approximant N can be computed.

From the relations given in the preceding section and also from some other relations that were not given here, it is quite easy to obtain a complete collection of identities relating three adjacent Padé approximants. In all of these identities the knowledge of two Padé approximants allows to compute the third one. Thus any sequence of Padé approximants can be recursively calculated.

A Fortran conversational program using these identities is given in (14). All these identities have been extended to the non-normal case by Draux (22).

We saw in this section that the theory of formal orthogonal polynomials unifies the algebraic theory of Padé approximants. Old and new identities between adjacent Padé approximants derive from this unified framework thus setting in order a quite obscure situation. Recursive methods such as those of Baker (3), Longman (36), Pindor (43), Watson (51) and Brezinski (7) appear as by-products of the theory. Moreover, interesting results such as the error term of theorem 13 and the link with Hermite interpolation have been obtained.

3. CONVERGENCE RESULTS

We first have to define what is meant by convergence of Padé approximants. Let $([p/q]_f(t))$ be a sequence of Padé approximants where p or/and q tends to infinity. The convergence will consist in the study of the convergence of this sequence to the function whose series f is the Taylor expansion when t belongs to some domain of the complex plane.

The convergence behaviour of Padé approximants is a difficult problem. It has only been studied for some particular classes of functions and for some particular sequences of approximants. Before entering into more details let us give some examples showing the difficulties encountered. Many of the examples are borrowed from the book of Bender and Orszag (6).

3.1. Access to the convergence theory

Let us first consider the function $f(t) = (10+t)/(1-t^2)$ the coefficients of whose power series expansion are $c_{2i} = 10$ and $c_{2i+1}=1$. The series converges for $|t| < 1$. We have

$$[k/1]_f(t) = \sum_{i=0}^{k-1} c_i t^i + \frac{c_k t^k}{1-c_{k+1} t/c_k}$$

When k is odd, $[k/1]$ has a simple pole at $t = 1/10$ and thus the sequence $([k/1])$ does not converge to f in $|t| < 1$ when k goes to infinity.

This example shows that the convergence of a sequence of Padé approximants can be affected by the poles that is by the zeros of the denominators. To prove the convergence of a sequence of Padé approximants in a domain D of the complex plane it will be necessary to prove that the extraneous poles (that is the poles which do not approximante poles of the function f) move out of D.

A more supprising situation is that the zeros of the numerators can prevent convergence. To illustrate this fact let us consider the reciprocal series of the previous example $g(t) = (1-t^2)/(10+t)$. We have

$$[1/k]_g(t) = 1/[k/1]_f(t)$$

Since $[1/2k+1]_g\ (0.1) = 0$ and $g(0.1) \neq 0$ then the sequence $([1/k]_g)$ cannot converge in the domain $|t| < 10$ where the series does.

Thus a sequence of Padé approximants can diverge in the convergence domain of the series. But the contrary can also be true : let us consider the series $f(t) = t - t^2/2 + t^3/3 - \ldots$ which converges to Log (1+t) for $|t| \leq 1$, $t \neq -1$. For $t = 2$ we have

$$\text{Log } 3 = 1.09861228866811 0\ldots$$

k	$\sum_{i=0}^{2k} c_i t^i$	$[k/k]_f(2)$
1	$0.260\ 10^1$	1.14
2	$0.506\ 10^1$	1.101
3	$0.126\ 10^2$	1.0988
4	$0.375\ 10^2$	1.098625
5	$0.121\ 10^3$	1.0986132
6	$0.410\ 10^3$	1.09861235
7	$0.142\ 10^4$	1.098612293
8	$0.504\ 10^4$	1.0986122890
9	$0.181\ 10^5$	1.098612288692
10	$0.655\ 10^5$	1.0986122886698
11	$0.239\ 10^6$	1.09861228866823

The convergence of diagonal sequences of Padé approximants $([n+k/k])_{k\in\mathbb{N}}$ can be investigated using the connection with orthogonal polynomials since the denominators of three successive approximants satisfy a recurrence relationship. The same is true for any other sequence whose members are adjacent Padé approximants. In some cases the asymptotic behaviour of the successive numerators and denominators can be deduced from the recurrence relations (1) to (8) given in the preceding section.

Let us give an example to illustrate our purpose. We consider the series

$$f(t) = \frac{1}{4} - \frac{t}{16} + \frac{t^2}{32} - \frac{5t^3}{256} + \ldots = \frac{1}{4}\sum_{i=0}^{\infty} \frac{(-1)^i}{(i+1)!}\cdot\frac{\Gamma(i+1/2)}{\Gamma(1/2)}\, t^i$$

For $|t| < 1$ this series converges to the function

$$f(t) = [\sqrt{1+t} - 1]/2t$$

Let us consider the sequence of Padé approximants $C_o(t)=[0/0]_f(t)$, $C_1(t)=[0/1]_f(t)$, $C_2(t)=[1/1]_f(t)$, $C_3(t)=[1/2]_f(t)$,... We set $C_k(t) = A_k(t)/B_k(t)$. Then the relations (1) and (2) of section 2.3 show that

$$B_o(t) = 1 \qquad\qquad B_1(t) = 1 + t/4$$

$$B_{k+1}(t) = B_k(t) + \frac{t}{4} B_{k-1}(t) \qquad\qquad k = 1, 2, \ldots$$

and that the A_k's satisfy the same relation with the initializations

$$A_o(t) = A_1(t) = 1/4$$

Thus the A_k's and the B_k's satisfy a second order difference equation with constant coefficients whose solutions are

$$A_k(t) = \frac{1}{4\sqrt{1+t}} \left[\left(\frac{1+\sqrt{1+t}}{2}\right)^{k+1} - \left(\frac{1-\sqrt{1+t}}{2}\right)^{k+1} \right]$$

$$B_k(t) = \frac{1}{\sqrt{1+t}} \left[\left(\frac{1+\sqrt{1+t}}{2}\right)^{k+2} - \left(\frac{1-\sqrt{1+t}}{2}\right)^{k+2} \right]$$

Thus

$$C_k(t) = \frac{A_k(t)}{B_k(t)} = \frac{1}{2(1+\sqrt{1+t})} \quad \left(1+0\left(\frac{1-\sqrt{1+t}}{1+\sqrt{1+t}}\right)^{k+1}\right).$$

When $|\arg(1+t)| < \pi$ then $|1-\sqrt{1+t}| < |1+\sqrt{1+t}|$ and it follows that

$$\lim_{k\to\infty} C_k(t) = \frac{1}{2(1+\sqrt{1+t})} = \frac{\sqrt{1+t} - 1}{2t} = f(t)$$

Thus the sequence (C_k) of Padé approximants converges to f in the complex plane except on the cut from -1 to $-\infty$. When $t < -1$, $\arg(1+t) = \pi$ and $|1 - \sqrt{1+t}| = |1 + \sqrt{1+t}|$: the sequence (C_k) oscillates and does not converge.

Although very simple (since the coefficients of the recurrence relationship do not depend on k) this example has the merit to show how a convergence study can be conducted in some cases. Some more complicated examples are treated in the book of Bender and Orszag already mentionned (6).

To end this section let us come back to $f(t) = \mathrm{Log}(1+t)$. For $t = 1$ the series converges and we get the following results

$$\text{Log } 2 = 0.6931471805599453 \ldots$$

k	$S_{2k}(1) = \sum_{i=0}^{2k} c_i t^i$	$[k/k]_f(1)$
1	0.830	0.7
2	0.783	0.6933
3	0.759	0.693152
4	0.745	0.69314733
5	0.736	0.6931471849
6	0.730	0.69314718068
7	0.725	0.693147180563
8	0.721	0.69314718056000
9	0.718	0.6931471805599485
10	0.716	0.6931471805599454

The error of the sequence $(S_{2k}(1))$ behaves as $1/2k$ and it can be proved that the error of the sequence $([k/k]_f(1))$ varies as 0.029^k. Thus the achievement of a precision of 10^{-16} with the series needs 10^{16} terms while the same precision can be obtained with the [10/10] Padé approximant using only 21 terms. The convergence of the sequence of the partial sums of the series has been accelerated.

3.2. General results

Some very general convergence theorems for Padé approximants can be indirectely deduced from the uniform boundeness theorem. This theorem is the foundation to prove the convergence of Gaussian quadrature methods and to study summation methods. As Padé approximants are related to both subjects then their convergence can be also studied on the same basis.

Let $S_n(t) = \sum_{i=0}^{n} c_i t^i$ be the partial sums of the series f.

Then, using the same notations as in (14), we have from property 1 :

$$\tilde{N}_k^{(n+1)}(t) = \sum_{i=0}^{k} b_i^{(k,n+1)} t^i S_{n+k-i}(t).$$

Thus

$$[n+k/k]_f(t) = \sum_{i=0}^{k} B_i^{(k,n)} S_{n+i}(t)$$

with

$$B_i^{(k,n)} = b_{k-i}^{(k,n+1)} \, t^{k-i} / \tilde{P}_k^{(n+1)}(t) \quad \text{and} \quad \sum_{i=0}^{k} B_i^{(k,n)} = 1$$

Written in this form the sequence transformation $(S_n(t)) \to ([n+k/k]_f(t))$ when n or k goes to infinity appears as a linear summation process. Then the convergence of the sequence $([n+k/k])$ is regulated by Toeplitz theorem which is a consequence of the uniform boundeness theorem :

Theorem 16 : Let D be the domain of convergence of the series f. Let us assume that $t \in D$.

$\forall\, k \geq 0$ fixed, the sequence $([n+k/k]_f(t))_{n\in\mathbb{N}}$ converges to f(t) if a constant M_k exists such that

$$\forall\, n \geq 0 \qquad \sum_{i=0}^{k} |B_i^{(k,n)}| \leq M_k$$

$\forall\, n \geq -1$ fixed, the sequence $([n+k/k]_f(t))_{k\in\mathbb{N}}$ converges to f(t) if a constant L_n exists such that

$$\forall\, k \geq 0 \qquad \sum_{i=0}^{k} |B_i^{(k,n)}| \leq L_n$$

and if $\forall\, i \geq 0 \qquad \lim_{k\to\infty} B_i^{(k,n)} = 0$

An immediate consequence of this theorem is the following result (since in that case the B_i's are positive) :

Theorem 17 : Let $x_i^{(k,n+1)}$ be the roots of $P_k^{(n+1)}$. Let us assume that $t\in D$ and $t\geq 0$. If $\forall$ i and $\forall$ n, $x_i^{(k,n+1)} \leq 0$ then $\forall\, k \geq 0$ fixed the sequence $([n+k/k]_f(t))_{n\in\mathbb{N}}$ converges to f(t).

A convergence result for the diagonals can be deduced from a theorem due to Wimp (52) which follows from Toeplitz theorem :

Theorem 18 : Let us assume that $t \in D$. If $\exists\, a > 0$ such that $\forall$ k and $\forall$ i, $t\, x_i^{(k,n+1)} \in [-a, 0]$ then $\forall\, n \geq -1$ fixed the sequence $([n+k/k]_f(t))_{k\in\mathbb{N}}$ converges to f(t).

Let us now come to a convergence result which is a consequence of the convergence of Gaussian quadrature methods studied by Stieltjes in 1884. It is a classical result in numerical analysis that Gaussian quadrature methods converge to $\int_a^b g(x)\,d\alpha(x)$ when g is continuous on [a, b] and α is bounded and nondecreasing in [a, b]. Thus we immediately have the :

Theorem 19 : Let us assume that $f(t) = \int_a^b \frac{d\alpha(x)}{1-xt} = c_0+c_1t+c_2t^2+\ldots$ where α is bounded and nondecreasing in $[a, b]$ (a or/and b can be infinite). Then $\forall$ $n \geq -1$ fixed and $\forall$ $t \notin I$ and real

$$\lim_{k\to\infty} [n+k/k]_f(t) = f(t)$$

where $I = [b^{-1}, a^{-1}]$ if $0 \notin]a,b[$ and $I = (-\infty,a^{-1}] \cup [b^{-1},+\infty)$ otherwise.

Let us remark that theorem 18 also applies. For example when a = 0 we know that $\forall$ k and $\forall$ i, $x_i^{(k,n+1)} \in [0, b]$. Thus $\forall$ $n \geq -1$ fixed, the sequence $([n+k/k]_f(t))_{k\in\mathbb{N}}$ converges to f(t), $\forall$ $t \leq 0$. It must be noticed that theorem 19 insures the convergence $\forall$ $t < b^{-1}$. Such a case will be studied in details in the next section.

Before ending this section let us give two important results. The first one is due to Montessus de Ballore [40] :

Theorem 20 : Let f be a meromorphic function in the disc $|t| < R$ and having in that disc n simple poles and no other singularity. Then $\forall$ $|t| < R$ and different from the poles the sequence $([k/n]_f(t))_{k\in\mathbb{N}}$ converges to f uniformly on every compact subset.

The second result is a consequence of the Stieltjes-Vitali theorem (32, p. 194) :

Theorem 21 : Let (m_i) and (n_i) be sequences of non negative integers such that

$$\lim_{i\to\infty} \max(m_i, n_i) = \infty$$

Let $R_i(t) = [m_i/n_i]_f(t)$ and let D be a domain of the complex plane containing the origin. Then

- (R_i) converges uniformly on every compact subset of D iff $\{R_i^1(t)\}$ is uniformly bounded on every compact subset of D.

- If (R_i) converges uniformly on every compact subset of D, then $f(t) = \lim_{i\to\infty} R_i(t)$ is holomorphic in D and the series is the Taylor expansion of the function f about the origin.

The problem of convergence for Padé approximants is a difficult problem. General theorems as those given in this section are hard to use in practical cases. When uniform convergence is impossible to prove weaker types of convergence are studied : ponctual convergence, convergence almost everywhere, convergence in measure or in capacity. See Lubinsky (38) for a detailed account.

Up to now the convergence has been almost completely solved only in the case of Stieltjes series. Progresses towards the convergence behaviour of sequences of Padé approximants for Pólya frequency series (also called series of class S or totally positive series) have been recently made by Léopold (34). See also (26).

3.3. Stieltjes series

The series $f(t) = \sum_{i=0}^{\infty} c_i t^i$ is said to be a Stieltjes series if

$$c_i = \int_0^{\infty} x^i d\alpha(x) \qquad i = 0, 1, \ldots$$

where α is bounded and non decreasing in $[0, \infty)$. Such a series is the formal expansion of the function

$$f(t) = \int_0^{\infty} \frac{d\alpha(x)}{1-xt}$$

If R is the radius of convergence of the series (which can be equal to zero) then the preceding integral runs from 0 to R^{-1} and the function f is defined in the whole complex plane cut along $[R, \infty)$. We shall assume that α takes infinitely many different values so that f is not a rational function.

The sequence (c_n) is called a sequence of moments of Stieltjes and we shall write $(c_n) \in MS$. The reverse problem is called the Stieltjes moment problem. It is a fundamental problem for the convergence of Padé approximants ; it can be formulated as follows : let (c_n) be a given sequence of real numbers ; what are the conditions that $(c_n) \in MS$? The answer to this question was given by Stieltjes in 1894 (49) :

Theorem 22 : A necessary and sufficient condition for the Stieltjes moment problem to have a solution with α bounded, non decreasing in $[0, \infty)$ and taking infinitely many different values is that $\forall\, k \geq 1,\ H_k(c_o) > 0$ and $H_k(c_1) > 0$.

The proof that we shall not give, uses results on quadratic forms with infinitely many variables.

A very crucial point is that the preceding condition does not insure the uniqueness of the solution α. This fact will cause some trouble for the convergence of Padé approximants since a Stieltjes series could be the asymptotic expansion of several functions. If the solution to the moment problem is unique it is said to be determinate. No necessary and sufficient condition for the Stieltjes moment problem to be determinate is known but several sufficient conditions are. The most popular, due to Carleman (18), is that $\sum_{i=1}^{\infty} c_i^{-1/2i} = \infty$.

Before going to properties of Padé approximants for Stieltjes series let us mention a fundamental result :

Theorem 23 : If f is a non rational Stieltjes series then ∀k and ∀ n, $H_k^{(n)} > 0$.

Proof :

$$c_{n+i} = \int_0^{\infty} x^{n+i} d\alpha(x) = \int_0^{\infty} x^i d\alpha_n(n)$$

with $d\alpha_n(x) = x^n d\alpha(x)$. Since α is bounded and non decreasing in $[0,\infty)$ and takes infinitely many different values, the same is true for α_n and the result follows from theorem 22.∎

Other properties of MS sequences are given in (14). An immediate consequence of this result is the :

Theorem 24 : The Padé table of a non rational Stieltjes series is normal.

We shall now prove some inequalities between Padé approximants of a Stieltjes series when t ∈ (-∞, R[. Since

$$c_i = \int_0^{1/R} x^i d\alpha(x)$$

where α is bounded and non decreasing in $[0, R^{-1}]$ we known from the theory of classical orthogonal polynomials that the roots of $P_k^{(n)}$ are simple, real and in $[0, R^{-1}]$. Thus the roots of $\tilde{P}_k^{(n)}$ belong to $[R, \infty)$ and ∀ k and n

$$\forall\, t \in (-\infty, R[\qquad \tilde{P}_k^{(n)}(t) > 0, \quad \tilde{Q}_k^{(n)}(t) > 0$$

From (9) we immediately obtain

$$[n+k-1/k]_f(t) - [n+k/k+1]_f(t) \leq 0 \qquad \forall\, t \in\,]0, R[$$

$$(-1)^n([n+k-1/k]_f(t) - [n+k/k+1]_f(t)) \leq 0\ \forall\, t\ (-\infty, 0[$$

From an immediate generalisation of the error term (theorem 13) we have

$$f(t) - [n+k/k+1]_f(t) \geq 0 \qquad \forall\, t \in\,]0, R[$$

$$(-1)^n (f(t) - [n+k/k+1]_f(t)) \geq 0\ \ \forall\, t \in (-\infty, 0[$$

These two results can be, of course, combined together to give inequalities on the errors. Moreover similar inequalities can be obtained from relations (10), (11) and (12).

Analogous inequalities hold for the upper half of the Padé table since the series g defined by $f(t)(1+tg(t)) = c_o$ is also a Stieltjes series (for a proof see (4)) and since, from properties 2 and 3

$$[p/q]_f(t)(1+t[q-1/p]_g(t)) = c_o$$

All the inequalities that can be obtained in that way are gathered in the following theorem :

Theorem 25 : Let f be a non rational Stieltjes series. Then

$\forall k \geq 0$, $\forall n \geq -1$ and $\forall t \in]0, R[$ we have :

$$0 \leq f(t) - [n+k+1/k+1]_f(t) \leq f(t) - [n+k/k]_f(t)$$

$$0 \leq f(t) - [n+k+1/k+1]_f(t) \leq f(t) - [n+k+2/k]_f(t)$$

$$0 \leq f(t) - [n+k+1/k]_f(t) \leq f(t) - [n+k/k]_f(t)$$

$$0 \leq f(t) - [n+k+1/k+1]_f(t) \leq f(t) - [n+k+1/k]_f(t)$$

$$0 \leq f(t) - [k+1/n+k+2]_f(t) \leq f(t) - [k/n+k+1]_f(t)$$

$$0 \leq f(t) - [k+1/n+k+1]_f(t) \leq f(t) - [k/n+k+2]_f(t)$$

$$0 \leq f(t) - [k/n+k+2]_f(t) \leq f(t) - [k/n+k+1]_f(t)$$

$$0 \leq f(t) - [k+1/n+k+1]_f(t) \leq f(t) - [k/n+k+2]_f(t)$$

$\forall k \geq 0$, $\forall n \geq 0$ and $\forall t \in (-\infty, 0[$ we have :

$$0 \leq (-1)^n([n+k+1/k+1]_f(t) - f(t)) \leq (-1)^n([n+k/k]_f(t) - f(t))$$

$$0 \leq (-1)^n([n+k+1/k+1]_f(t) - f(t)) \leq (-1)^n([n+k+2/k]_f(t)-f(t))$$

$\forall k \geq 0$, $\forall n \geq -1$ and $\forall t \in (-\infty, 0[$ we have

$$0 \leq (-1)^n(f(t) - [k+1/n+k+2]_f(t)) \leq (-1)^n(f(t)-[k/n+k+1]_f(t))$$

$$0 \leq (-1)^n(f(t) - [k+1/n+k+2]_f(t)) \leq (-1)^n(f(t)-[k/n+k+3]_f(t))$$

$\forall k \geq 0$ and $\forall t \in (-\infty, 0[$ we have

$$[k/k]_f(t) \geq f(t) \geq [k-1/k]_f(t)$$

$$[k/k+1]_f(t) \geq [k+1/k]_f(t)$$

These inequalities can be visualized on the following diagram where the arrows indicate the sense of the inequalities that is to say that they go from the term with the greatest error to the term with the smallest error. The diagram is for $t \in]0, R[$

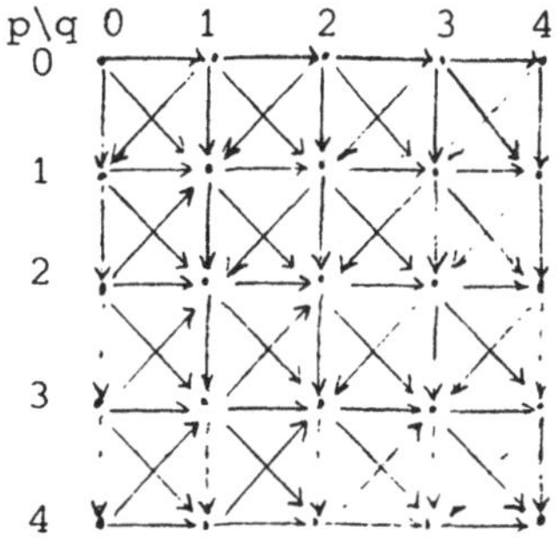

In practice when only k+1 coefficients of the series are known then it is only possible to compute the triangular part of the Padé table formed with the approximants $[m/n]$ such that $m+n \leq k$. The preceding inequalities show that the best Padé approximant (that is the one with the smallest error $\forall\ t \in]0, R[$) is the approximant $[p/p]$ when $k = 2p$ and is the approximant $[p+1/p]$ when $k = 2p+1$. Thus, when k increases, the sequence of best Padé approximant on $]0, R[$ is the sequence $[0/0]$, $[1/0]$, $[1/1]$, $[2/1]$, $[2/2]$, $[3/2]$, ...

The preceding inequalities also show that $\forall\ t \in (-\infty, 0[$ the sequence $([k/k])$ is monotonically decreasing while the sequence $([k/k+1])$ is monotonically increasing. More precisely we have

$$[0/1]_f(t) \leq [1/2]_f(t) \leq \ldots \leq [k/k+1]_f(t) \leq \ldots \leq f(t) \leq \ldots \leq [k/k]_f(t)$$

$$\leq \ldots \leq [1/1]_f(t) \leq [0/0]_f(t)$$

Thus these two sequences have a limit and

$$\lim_{k\to\infty} [k/k+1]_f(t) \leq f(t) \leq \lim_{k\to\infty} [k/k]_f(t)$$

If both limits are equal then the Stieltjes moment problem is determinate and there is only one single function having the series as its Taylor expansion. Reciprocally if both limits are different then the Stieltjes moment problem is indeterminate and several functions correspond to the same expansion.

When $n \geq -1$ these results still hold. When $n < 1$ the monotonicity property of the sequence $([n+k/k])_{k\in\mathbb{N}}$ only holds for k sufficiently large.

If we now assume that the radius of convergence R of the series f is strictly positive then, $\forall$ n

$$c_n = \int_0^{1/R} x^n d\alpha(x) \leq c_0 \, R^{-n}$$

and $\liminf_{n\to\infty} c_n^{-1/2n} = R^{1/2} > 0$ which shows that Carleman's condition is satisfied. Thus we have the

Theorem 26 : Let f be a non rational Stieltjes series with a stricty positive radius of convergence. Then $\forall \, t \in (-\infty, R[$

$$\forall \, n \geq 0 \text{ fixed} \quad \lim_{k\to\infty} [n+k/k]_f(t) = \lim_{k\to\infty} [k/n+k]_f(t) = f(t)$$

$$\forall \, k \geq 0 \text{ fixed} \quad \lim_{n\to\infty} [n+k/k]_f(t) = \lim_{n\to\infty} [k/n+k]_f(t) = f(t)$$

When $t \in [-R, R[$ the partial sums $f_n(t) = \sum_{i=0}^{n} c_i t^i$ converge to f and it can be proved that the convergence is accelerated (11). More precisely we have the :

Theorem 27 : Let f be a non rational Stieltjes series with a strictly positive radius of convergence R. Then, $\forall \, t \in [-R, R[$

$$\forall \, k \geq 0 \text{ fixed } [n+k/k]_f(t)-f(t) = o(f_{n+2k}(t)-f(t)) \text{ when } n\to\infty$$

$$\forall \, n \geq 0 \text{ fixed } [n+k/k]_f(t)-f(t) = o(f_{n+2k}(t)-f(t)) \text{ when } k\to\infty$$

As an example of such a situation we saw, in the previous section, the case of the series Log(1+t).

To end this section let us mention that it has been proved that, $\forall \, n \geq -1$, the sequence $([n+k/k])_{k\in\mathbb{N}}$ converge uniformly to an analytic function in every subset of the complex plane cut from R to $+\infty$.

3.4. Pólya frequency series

We shall say that a series f is a Pólya frequency series if it is the Taylor expansion of the function

$$f(t) = A \, e^{\gamma t} \prod_{i=1}^{\infty} \frac{1+\alpha_i t}{1-\beta_i t}$$

where A, γ, α_i and β_i are positive constants and where the series $\sum_{i=1}^{\infty} (\alpha_i+\beta_i)$ converges.

The following result is due to Arms and Edrei (2) :

Theorem 28 : A necessary and sufficient condition that $f(t) = \sum_{i=0}^{\infty} c_i t^i$ is a Pólya frequency series is that $\forall\ k \geq 1$ and $\forall\ n \geq 0$

$$(-1)^{k(k-1)/2}\ H_k(c_{n-k+1}) \geq 0$$

In that case the sequence (c_n) is said to be totally positive. Properties of such sequences have been studied in details by Karlin (33).

The convergence of Padé approximants is given by the theorem (2) :

Theorem 29 : Let f be a non rational Pólya frequency series. Its Padé table is normal. Moreover if $([m_k/n_k]_f(t) = P_k(t)/Q_k(k))$ is a sequence of Padé approximants of f such that

$$\lim_{k\to\infty} m_k/n_k = \lambda > 0$$

then

$$\lim_{k\to\infty} P_k(t) = A\ e^{\frac{\lambda}{\lambda+1}\gamma t} \prod_{i=1}^{\infty} (1+\alpha_i t)$$

$$\lim_{k\to\infty} Q_k(t) = e^{-\frac{1}{\lambda+1}\gamma t} \prod_{i=1}^{\infty} (1-\beta_i t)$$

and the convergence is uniform on every compact subset of the complex plane. If $\gamma = 0$ the condition on m_k/n_k can be removed. Thus the sequence $([m_k/n_k])$ converges uniformly to f on every compact subset of $\mathbb{C}\backslash\{\beta_i^{-1}\}$.

Very fine results on the localisation of the zeros and poles of Padé approximants of a Pólya frequency series have been recently obtained by Léopold (34). (See also (26)). These results extend those of Saff and Varga (46) for the exponential function which is a particular case of such series (let us mention that the first results for the exponential function were given by Padé in his thesis (41)). Léopold has also obtained inequalities for the numerators and the denominators of Padé approximants of Pólya frequency series and inequalities for the approximants themselves.

4. EXTENSIONS AND APPLICATIONS

In this section we shall introduce the notion of Padé-type approximation which generalizes that of Padé approximation. We shall also give an outline of some applications to numerical analysis.

4.1. Padé-type approximants

Let v be an arbitrary polynomial of degree k. We define w as

$$w(t) = c\left(\frac{v(x) - v(t)}{x - t}\right)$$

where t is a parameter and where c acts on the variable x. w is a polynomial of degree k-1 in t.

Let $x_1,\ldots,x_n$ be the roots of v with multiplicities $k_1,\ldots,k_n$ such that $k = k_1+\ldots+k_n$. The Hermite interpolation polynomial P of $(1-xt)^{-1}$ is still given by theorem 11 where P_k is replaced by v. Thus, by theorem 12

$$c(P) = \tilde{w}(t)/\tilde{v}(t)$$

where $\tilde{v}(t) = t^k v(t^{-1})$ anc $\tilde{w}(t) = t^{k-1} w(t^{-1})$. Let us make use of the notation

$$\tilde{w}(t)/\tilde{v}(t) = (k-1/k)_f(t)$$

From the proof of theorem 13 we immediately get the :

<u>Theorem 30</u> :

$$f(t) - (k-1/k)_f(t) = \frac{t^k}{\tilde{v}(t)} c\left(\frac{v(x)}{1-xt}\right)$$

Thus (k-1/k) is the ratio of a polynomial of degree k-1 by a polynomial of degree k whose expansion in ascending powers of t coincides with f up to the term of degree k-1 inclusively. Such rational approximants exhibit a property quite similar to the fundamental property of Padé approximants. It is the reason why they have been called Padé-type approximants. v is called the generating polynomial of the approximant.

It is of course possible to construct Padé-type approximants with various degrees in the numerator and in the denominator. Let $\tilde{v}$ be an arbitrary polynomial of degree q at most :

$$\tilde{v}(t) = b_o + b_1 t + \ldots + b_q t^q$$

Let us define $\tilde{w}$ as

$$\tilde{w}(t) = b_o\, S_p(t) + b_1 t\, S_{p-1}(t) + \ldots + b_q t^q\, S_{p-q}(t)$$

with $S_k(t) = \sum_{i=o}^{k} c_i t^i$ for $k \geq 0$ and $S_k(t) = 0$ for $k < 0$. $\tilde{w}$ is a polynomial of degree less or equal to p. We have

$$\tilde{w}(t)-f(t)\tilde{v}(t)=b_o[f(t)+0(t^{p+1})]+\ldots+b_q t^q[f(t)+0(t^{p-q+1})]$$
$$-b_o f(t)-\ldots-b_q t^q f(t)$$
$$=0(t^{p+1})$$

Thus we are able to construct rational approximants, denoted by $(p/q)_f(t)$, with a numerator of degree at most p and a denominator of degree at most q such that

$$f(t) - (p/q)_f(t) = 0(t^{p+1})$$

We can see from above that the construction of (p/q) makes use of $c_o,\ldots,c_p$ while that of [p/q] needs $c_o,\ldots,c_{p+q}$. Thus we might compare [p/q] with (p+q/q). Both approximants will use $c_o,\ldots,c_{p+q}$ and will match the series up to the term of degree p+q included. Thus from the algebraic point of view both kind of approximants are comparable since they need the same coefficients of the series to attain the same order of approximation.

The great advantage of Padé-type approximants over Padé approximants lies in the free choice of the denominator that is in the free choice of the poles which may lead to a better approximation of the function. To illustrate this fact let us compare the Padé approximant [1/2] for $f(t) = t^{-1}\text{Log}(1+t)$ with the Padé-type approximant (3/4) constructed from a generating polynomial v with roots $x_i = -1/i$ for $i = 1,\ldots,4$. The relative errors are (13) :

t	$[1/2]_f(t)$	$(3/4)_f(t)$
-0.8	$-0.27\ 10^{-1}$	$0.29\ 10^{-1}$
-0.5	$-0.12\ 10^{-2}$	$0.73\ 10^{-3}$
0.1	$-0.46\ 10^{-6}$	$0.92\ 10^{-7}$
0.5	$-0.15\ 10^{-3}$	$0.56\ 10^{-5}$
1.0	$-0.12\ 10^{-2}$	$-0.13\ 10^{-3}$
1.5	$-0.35\ 10^{-2}$	$-0.77\ 10^{-3}$
2.0	$-0.70\ 10^{-2}$	$-0.21\ 10^{-2}$
4.0	$-0.27\ 10^{-1}$	$-0.14\ 10^{-1}$

We shall not dwell on the algebraic properties of Padé-type approximants since they are quite similar to those of Padé approximants apart from some restrictions on their generating polynomials. These properties are stated in (13) and (14).

The main open question on Padé-type approximants is the "best" choice of the poles that is the choice of the roots x_i of the generating polynomials. Some attempts to solve this difficult problem for some particular cases have been made (39). Another problem connected with the choice of the x_i's is the problem of convergence of Padé-type approximants.

An answer to this problem can be again obtained from theorem 16. Theorems 17 and 18 still hold where now $x_i^{(k,n+1)}$ designates a zero of the generating polynomial of the approximants $(n+k/k)$.

Let us give two more results. D is the domain of convergence of the series.

Theorem 31 : Let (x_i) be a sequence of negative numbers converging to zero and let $v_k(x) = (x-x_1)\ldots(x-x_k)$ be the generating polynomial of $(n+k/k)_f$. Then $\forall$ n ≥ -1 fixed, $\forall$ t $\in$ D and t ≥ 0 the sequence $((n+k/k)_f(t))_{k\in\mathbb{N}}$ converges to f(t).

Proof : As we previously saw, $(n+k/k)$ can be written as

$$(n+k/k)_f(t) = \sum_{i=o}^{k} B_i^{(k,n)} S_{n+i}(t).$$

We have to prove that, under the assumptions, the two last conditions of theorem 16 are satisfied. Since $v_{k+1}(x) = (x-x_{k+1})v_k(x)$ it is easy to see that (the upper index n have been droped for simplicity since n is fixed) :

$$B_o^{(k+1)} = -x_{k+1} t\, B_o^{(k)}/(1-x_{k+1}t)$$

$$B_i^{(k+1)} = (B_{i-1}^{(k)} - x_{k+1} t\, B_i^{(k)})/(1-x_{k+1}t) \qquad i = 1, \ldots, k$$

$$B_{k+1}^{(k+1)} = B_k^{(k)}/(1-x_{k+1}t)$$

with $B_o^{(o)} = 1$

If $t \geq 0$, $x_i \leq 0$ and $\lim_{i\to\infty} x_i = 0$ then $B_i^{(k)} \geq 0$ and $\sum_{i=o}^{k} |B_i^{(k)}| = 1$.
Moreover $\lim_{k\to\infty} B_o^{(k)} = 0$. Let us assume that $\lim_{k\to\infty} B_{i-1}^{(k)} = 0$. Then

$$\lim_{k\to\infty} x_{k+1} t(B_i^{(k)} - B_i^{(k+1)}) + B_i^{(k+1)} = 0$$

If $\exists$ M such that $\forall$ k, $B_i^{(k)} \leq M$ then $\lim_{k\to\infty} B_i^{(k)} = 0$. We have to prove the existence of such an M. If M does not exist then, since $B_i^{(k)} \geq 0$, it is impossible that $\sum_{i=0}^{k} B_i^{(k)} = 1$. Thus the result follows from theorem 16. ∎

The following theorem is an immediate consequence of results due to Wimp (52) by using Newton's relations between the coefficients and the zeros of a polynomial :

Theorem 32 : Let (x_i) be a sequence of numbers and let $v_k(x) = (x-x_1)\ldots(x-x_k)$ be the generating polynomial of $(n+k/k)_f$. If $\forall\, t \in D' \subseteq D$ and $\forall\, i$, $t\,x_i \geq 0$ and $tx_i \neq 1$ and if the series $\sum_{i=1}^{\infty} tx_i/(1-tx_i)$ converges then $\forall\, n \geq 0$ fixed and $\forall\, t \in D'$

$$\lim_{k\to\infty} (n+k/k)_f(t) = f(t).$$

It must be noticed that the preceding conditions imply that the sequence (x_i) converges to zero.

To end this section I would like to mention that the notion of Padé-type approximants have been extended to double power series (12, 14), series of functions (14, 21, 22) and power series with matrix coefficients (23).

4.2. Laplace transform inversion

Let f be the Laplace transform of g defined by

$$f(t) = \int_0^{\infty} g(x)\, e^{-xt} dx.$$

The problem of the Laplace transform inversion consists in finding g when f is known. From the numerical point out of view f can be known either by its Taylor expansion or by its values at some points.

When f is known by its Taylor expansion (or at least by its first terms) then it can be approximated by a Padé approximant where the degree of the denominator is greater than the degree of the numerator since f tends to zero when t tends to infinity.

After perfoming the partial fraction decomposition of the Padé approximant (which can be avoided by using a method due to Longman and Sharir (37)) it is inverted.

Padé-type approximants can be used instead of Padé approximants.

The advantage is that the poles are known and thus the partial fraction decomposition is trivial.

If the zeros x_i of the generating polynomial v are simple then

$$(k-1/k)_f(t) = \sum_{i=1}^{k} \frac{A_i^{(k)}}{1-x_i t} = - \sum_{i=1}^{k} \frac{A_i^{(k)} x_i^{-1}}{t-x_i^{-1}}$$

where $A_i^{(k)} = w(x_i)/v'(x_i)$. The inverse $\tilde{g}$ of $(k-1/k)$ is given by

$$\tilde{g}(x) = - \sum_{i=1}^{k} A_i^{(k)} x_i^{-1} e^{x/x_i}$$

For example if

$$f(t) = t^{-1} \exp(-t/(1+at)^{1/2})$$

with a = 0.1, then with k = 2, $x_1 = -1/4$, $x_2 = -1/3$ we get

$$\tilde{g}(x) = 1 + 10.4\ e^{-4x} - 10.8\ e^{-3x}$$

If we compare the results obtained from (1/2) and from [2/3] we get :

x	Padé-type approximant	Padé approximant	exact value
0	0.6	-0.8182	0
0.5	-0.0023	0.2675	0.0274
1.0	0.6528	0.7049	0.5475
1.5	0.9058	0.8811	0.9290
2.0	0.9767	0.9521	0.9944
2.5	0.9945	0.9807	0.9997

Thus the result given by the inversion of (1/2) are better than those obtained from [2/3] which makes use of $c_0, \ldots, c_5$ instead of c_0 and c_1.

For this application of Padé-type approximants the "best" choice of the x_i's is an open problem. However some theoretical results on this method have been recently obtained (30).

4.3. A-acceptable approximations to the exponential

Let us consider the Cauchy problem

$$y'(x) = f(x,y(x)) \qquad x \in [a,b]$$

$$y(a) = y_0$$

The integration of this differential equation by a numerical method will produce approximations y_n of the exact solution $y(x_n)$ at points $x_n = a+nh$ where h is the step size.

Let us consider the differential equation

$$y'(x) = -\lambda y(x)$$
$$y(0) = 1$$

where λ is a complex number with $Re(\lambda) > 0$. The solution is

$$y(x) = e^{-\lambda x}$$

and the condition on λ involves that

$$\lim_{x \to \infty} y(x) = 0$$

A numerical method for integrating differential equations will be said A-stable if, when it is applied to the preceding differential equation, $\forall\ h\lambda$ such that $Re(h\lambda) > 0$

$$\lim_{n \to \infty} y_n = 0$$

It is well known (see, for example, (28)) that one step linear methods produce approximate values satisfying

$$y_{n+1} = R(h\lambda)\ y_n$$

where R is a polynomial if the method is explicit and a rational fraction if the method is implicit. Since

$$y_n = [R(h\lambda)]^n$$

the method will be A-stable if and only if, $\forall$ t such that $Re(t)>0$

$$|R(t)| < 1$$

Obviously this condition cannot be satisfied by explicit one step methods. Moreover

$$y(x_{n+1}) = e^{-h\lambda} y(x_n)$$

If the one step method has order p then

$$y_n - y(x_n) = 0(h^p)$$

and we shall have

$$R(h\lambda) - e^{-h\lambda} = O(h^{p+1})$$

Thus Padé and Padé-type approximants are candidates for such rational approximations. The question arises to known whether the condition for A-stability is satisfied or not. By using the maximum modulus theorem it can be shown that this condition is equivalent to (1) :

- $\forall\ t \in \mathbb{R} \qquad |R(it)| \le 1$

- $\lim_{|t|\to\infty} |R(t)| \le 1$

- R analytic in $Re(t) > 0$.

Rational approximations to the exponential function which satisfy these conditions are said to be A - acceptable.

The Padé approximants $[k/k]$, $[k-1/k]$ and $[k-2/k]$ are A-acceptable. An enormous literature on this subject appeared these last few years. An account of the recent advances in the theory has been recently given by Iserles (31).

Let us now investigate the case of Padé-type approximants. When the degree of the denominator is greater than the degree of the numerator then the second condition for A-acceptability is satisfied. If the zeros of the generating polynomial have negative real parts the third condition is also fulfilled. The first condition is more difficult to handle. However the following result holds :

<u>Theorem 33</u> : Let $(k/n+k)$ be a Padé-type approximant to $f(t)=e^{-t}$ with real coefficients and $n \ge 0$. We set

$$|(k/n+k)_f(it)|^2 = \frac{1+\beta_1 t^2+\ldots+\beta_k t^{2k}}{1+\alpha_1 t^2+\ldots+\alpha_{n+k} t^{2(n+k)}}$$

If the zeros of the denominator have negative real parts, if $\beta_i \le \alpha_i$ for $i = p+1, \ldots, k$ and if $0 \le \alpha_i$ for $i = k+1,\ldots,n+k$ where p is the integer part of $k/2$, then $(k/n+k)$ is A-acceptable.

Proof : It follows the ideas of the proof given by Crouzeix and Ruamps (20) for Padé approximants.

$(k/n+k)$ is analytic in $Re(t) > 0$ since the zeros of its denominator have negative reals parts. If $n \ge 1$, the second condition is satisfied. By definition of a Padé-type approximant we have

$$(k/n+k)_f(t) = e^{-t} + O(t^{k+1}).$$

Thus

$$|(k/n+k)_f(it)|^2 = 1 + O(t^{k+1})$$

which implies $\alpha_i = \beta_i$ for i = 1, ..., p. Moreover

$$|(k/n+k)_f(it)|^2 = 1 + \frac{(\beta_{p+1}-\alpha_{p+1})t^{2(p+1)}+\ldots+(\beta_{n+k}-\alpha_{n+k})t^{2(n+k)}}{1+\alpha_1 t^2+\ldots+\alpha_{n+k}t^{2(n+k)}}$$

with the convention that $\beta_i = 0$ for $i \geq k+1$. Thus $|(k/n+k)_f(it)|^2 \leq 1$ if $\beta_i \leq \alpha_i$ for i = p+1, ..., n+k.

If n = 0, $0 < \beta_k \leq \alpha_k$ and the second condition is satisfied.∎

In the solution of systems of differential equations it is advantageous from the computational point of view to use rational approximations with only one (multiple) pole. If we want to have convergence the pole must move out to infinity as it was explained in section 3.1.

For these reasons we shall make the choice $x_i = -k^{-1}$ for i = 1,...,k that is

$$(k-1/k)_f(t) = \tilde{w}(t)/(1+t/k)^k$$

From theorem 31 it immediately follows the :

<u>Theorem 34</u> : $\forall\, t \geq 0 \quad \lim_{k\to\infty} (k-1/k)_f(t) = e^{-t}$.

These approximants have been shown to be A-acceptable for $k \leq 4$. So are also the approximants (k/k) with the same denominator for $k \leq 3$ but (4/4) is not A-acceptable.

For application of Padé-type approximants to analytic continuation see (24) and to finite length p-adic arithmetic see (14, 35).

More advanced results and applications will be given in a forthcoming paper (16).

BIBLIOGRAPHY

(1) R. ALT *"Deux théorèmes sur la A-stabilité des schémas de Runge-Kutta simplement implicites"*. RAIRO, R3 (1972) pp 99-104.

(2) R.J. ARMS, A. EDREI, *"The Padé tables and continued fractions generated by totally positive sequences"*. In "Mathematical essays dedicated to A.J. Macintyre", Ohio University Press, Athens, Ohio (1970), pp. 1-21.

(3) G.A. BAKER jr. *"The Padé approximant method and some related generalizations"*. In "The Padé approximant in theoretical physics", G.A. Baker jr. and J.L. Gammel eds., Academic Press, New-York, 1970.

(4) G.A. BAKER jr. *"Essentials of Padé approximants"*. Academic Press, New-York, 1975.

(5) G.A. BAKER jr., P.R. GRAVES-MORRIS, *"Padé approximants, Vols. 1 and 2"*, Encyclopedia of mathematics and its applications, Vols. 13 and 14, Addison Wesley, Reading, Mass., 1981.

(6) C.M. BENDER, S.A. ORSZAG, *"Advanced mathematical methods for scientists and engineers"*. Mc Graw-Hill, 1978.

(7) C. BREZINSKI, *"Computation of Padé approximants and continued fractions"*. J. Comp. Appl. Math., 2(1976) pp. 113-123.

(8) C. BREZINSKI, *"Accélération de la convergence en analyse numérique"* Lecture Notes in Mathematics 584, Springer Verlag, Heidelberg, 1977.

(9) C. BREZINSKI, *"Padé approximants and orthogonal polynomials"* in "Padé and rational approximation", E.B. Saff and R.S. Varga eds., Academic Press, New-York, 1977.

(10) C. BREZINSKI, *"Algorithmes d'accélération de la convergence. Etude numérique"*. Editions Technip, Paris, 1978.

(11) C. BREZINSKI, *"Convergence acceleration of some sequences by the ε-algorithm"*. Numer. Math., 29 (1978) pp. 173-177.

(12) C. BREZINSKI, *"Padé-type approximants for double power series"* J. Indian Math. Soc., 42(1978) pp. 267-282.

(13) C. BREZINSKI, *"Rational approximation to formal power series"* J. Approx. Theory, 25(1979) pp. 295-317.

(14) C. BREZINSKI, *"Padé-type approximation and general orthogonal polynomials"*. ISNM, Vol. 50, Birkhäuser Verlag, Basel, 1980.

(15) C. BREZINSKI, *"The long history of continued fractions and Padé approximants"* in "Padé approximation and its applications.

Amsterdam 1980", M.G. de Bruin and H. Van Rossum eds., Lecture Notes in Mathematics 888, Springer Verlag, Heidelberg, 1981.

(16) C. BREZINSKI, *"Padé approximants : old and new"*. Jahrbuch überblicke Mathematik, 1982.

(17) M.G. DE BRUIN, H. VAN ROSSUM eds. *"Padé approximation and its applications. Amsterdam 1980"*, Lecture Notes in Mathematics 888, Springer Verlag, Heidelberg, 1981.

(18) T. CARLEMAN, *"Sur les équations intégrales singulières à noyau symétrique"*. Uppsala, 1923.

(19) F. CORDELLIER, *"Démonstration algébrique de l'extension de l'identité de Wynn aux tables de Padé non normales"*. in "Padé approximation and its applications", L. Wuytack ed., Lecture Notes in Mathematics 765, Springer Verlag, Heidelberg, 1979.

(20) M. CROUZEIX, F. RUAMPS, *"On rational approximations to the exponential"* RAIRO, R11 (1977) pp. 241-243.

(21) A. DRAUX, *"Approximants of exponential type. General orthogonal polynomials"* in "Padé approximation and its applications. Amsterdam 1980", M.G. de Bruin and H. Van Rossum eds. Lecture Notes in Mathematics 888, Springer Verlag, Heidelberg, 1981.

(22) A. DRAUX, *"Polynômes orthogonaux formels. Applications"*, Thèse, Université de Lille I, 1981.

(23) A. DRAUX,
to appear.

(24) EIERMANN
Thesis, University of Karlsruhe, 1982.

(25) J. GILEWICZ, *"Approximants de Padé"*, Lecture Notes in Mathematics 667, Springer Verlag, Heidelberg, 1979.

(26) J. GILEWICZ, E. LEOPOLD *"Padé approximant inequalities for the functions of the class S"* in "Padé approximation and its applications. Amsterdam 1980", M.G. de Bruin and H. Van Rossum eds., Lecture Notes in Mathematics 888, Springer Verlag, Heidelberg, 1981.

(27) P. HENRICI, *"The quotient-difference algorithm"* NBS Appl. Math. Series, 49 (1958) pp. 23-46.

(28) P. HENRICI, *"Discrete variable methods in ordinary differential equations"*. J. Wiley, New-York, 1962.

(29) P. HENRICI, *"Applied and computational complex analysis"*. J. Wiley, New-York, Vol. 1 (1974) and 2 (1976)

(30) J. VAN ISEGHEM
Thèse 3ème cycle, Université de Lille I, to appear.

(31) A. ISERLES, *"Padé and rational approximations to the exponential and their applications to numerical analysis"*. in "Proceedings of the first French-Polish meeting on Padé approximation and convergence acceleration techniques", J. Gilewicz ed., CPT, CNRS, Marseille, to appear.

(32) W.B. JONES, W.J. THRON *"Continued fractions. Analytic theory and applications"* Encyclopedia of mathematics and its applications, Vol. 11, Addison Wesley, Reading, Mass., 1980.

(33) S. KARLIN, *"Total positivity"* Stanford Univ. Press, Stanford, 1968.

(34) E. LEOPOLD, *"Approximants de Padé pour les fonctions de classe S et localisation des zéros de certains polynômes"*. Thèse 3e cycle, Université de Provence, 1982.

(35) R.A. LEWIS, *"p-adic number systems for error-free computation"* Ph. D., University of Tennessee, Knoxville, 1979.

(36) I.M. LONGMAN, *"Computation of the Padé table"* Intern. J. Comp. Math., 3B (1971) pp. 53-64.

(37) I.M. LONGMAN, M. SHARIR, *"Laplace transform inversion of rational functions"*. Geophys. J. Astr. Soc., 25 (1971) pp. 299-305.

(38) D.S. LUBINSKY *"Exceptional sets of Padé approximants"*. Ph. D. Thesis, University of Witwatersrand, Johannesburg, 1980.

(39) Al. MAGNUS, *"Rate of convergence of sequences of Padé-type approximants and pole detection in the complex plane"*. in "Padé approximation and its applications. Amsterdam 1980", M.G. de Bruin and H. Van Rossum eds., Lecture Notes in Mathematics 888, Springer Verlag, Heidelberg, 1981.

(40) R. DE MONTESSUS DE BALLORE, *"Sur les fractions continues algébriques"*. Bull. Soc. Math. France, 30 (1902) pp. 28-36.

(41) H. PADE, *"Sur la représentation approchée d'une fonction par*

des fractions rationnelles". Ann. Ec. Norm. Sup., 9 (1892) Supp. pp. 1-92.

(42) O. PERRON, *"Die Lehre von dem Kettenbrüchen"*. Chelsea, New-York, 1950.

(43) M. PINDOR, *"A simplified algorithm for calculating the Padé table derived from Baker and Longman schemes"*, J. Comp. Appl. Math., 2(1976) pp. 255-257.

(44) H. VAN ROSSUM, *"A theory of orthogonal polynomials based on the Padé table"*, Thesis, University of Utrecht, Van Gorcum, Assen, 1953.

(45) H. RUTISHAUSER, *"Der Quotienten-Differenzen Algorithmus* Birkhäuser Verlag, Basel, 1957.

(46) E.B. SAFF, R.S. VARGA, *"On the zeros and poles of Padé approximants to e^z"*, Numer. Math., 25 (1976) pp 1-14.

(47) E.B. SAFF, R.S. VARGA eds. *"Padé and rational approximation"* Academic Press, New-York, 1977.

(48) E. STIEFEL, *"Kernel polynomials in linear algebra and their numerical applications"*, NBS Appl. Math. Series, 49 (1958) pp. 1-22.

(49) T.J. STIELTJES, *"Recherches sur les fractions continues"* Ann. Fac. Sci. Toulouse, 8 (1894) pp. 1-122.

(50) H.S. WALL, *"The analytic theory of continued fractions"* Van Nostrand, New York, 1948.

(51) P.J.S. WATSON, *"Algorithms for differentiation and integration"*, in "Padé approximants and their applications", P.R. Graves-Morris ed., Academic Press, New-York, 1973.

(52) J. WIMP, *"Toeplitz arrays, linear sequence transformations and orthogonal polynomials"* Numer. Math., 23 (1974) pp 1-18.

(53) J. WIMP, *"Sequence transformations and their applications"* Academic Press, New York, 1981.

(54) L. WUYTACK ed. *"Padé approximation and its applications"* Lecture Notes in Mathematics 765, Springer Verlag, Heidelberg, 1979.

(55) P. WYNN, *"Upon systems of recursions which obtain among the quotients of the Padé table"*. Numer. Math., 8(1966) pp. 264-269.

(56) P. WYNN, "*A general system of orthogonal polynomials*" Quart. J. Math. Oxford, (2) 18 (1967) pp. 81-96.

NUMERICAL METHODS IN CONFORMAL MAPPING

Dieter Gaier

University of Gießen
Federal Republic of Germany

INTRODUCTION

Complex analysis and in particular conformal mapping (CM) is still an important tool in applied mathematics. Sometimes one is interested in the determination of the mapping function itself, sometimes one looks for global quantities only, like the module of a quadrilateral. The use of high speed computers allows us to solve such mapping problems to considerable accuracy.

We shall first point out some recent practical problems, in which CM was used successfully, and then describe two areas within CM more thoroughly, namely the Theodorsen integral equation method and the application of variational principles. The latter will be applied to the CM of simply and doubly connected domains.

§1. SOME APPLICATIONS OF CONFORMAL MAPPING

It is well known that CM has a wide range of applications. Whenever problems lead to the harmonic equation $\Delta u = 0$ or the biharmonic equation $\Delta\Delta u = 0$, with the appropriate boundary conditions, one might try CM to reduce the problem to a simpler geometric configuration. We mention here a few examples which we met recently.

Given a region G with Jordan boundary $\partial G = C$, and four distinguished points P_1, P_2, P_3, P_4 on C (in positive orientation), we say a quadrilateral V is

H. Werner et al. (eds.), Computational Aspects of Complex Analysis, 51–78.

given with opposite sides P_1P_2 and P_3P_4 . We can map G conformally onto a rectangle $R:(0,a)\times(0,b)$ such that P_1,P_2,P_3,P_4 map into $0,a,a+ib,ib$. The module

$$m(V) = \frac{a}{b}$$

is then a uniquely determined quantity, and it is conformally invariant, i.e., if V_1 is the conformal image of V , then $m(V_1)=m(V)$.

First we mention a method of measuring the sheet resistance of semiconductors (see Versnel [1979]) which goes back to Van Der Pauw. Mathematically one has to determine the module of a quadrilateral as described in Fig. 1a).

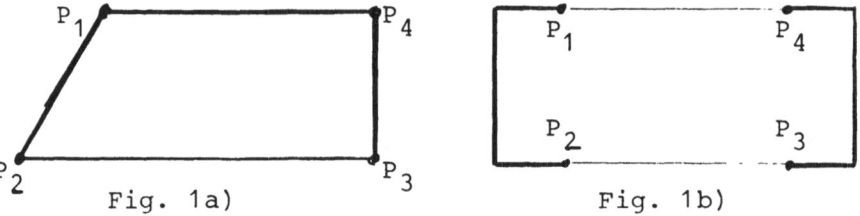

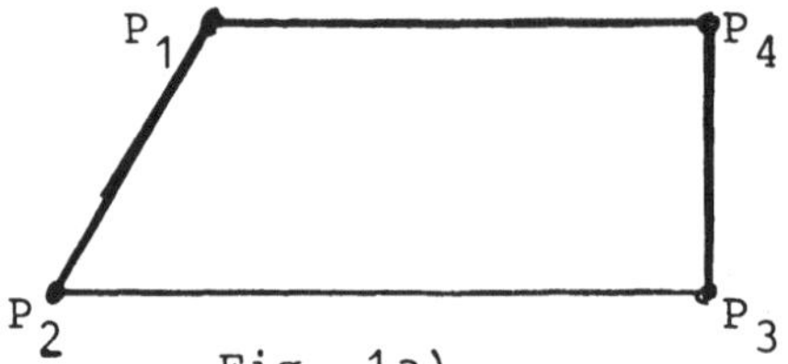

Fig. 1a)

Fig. 1b)

Similarly a method of measuring the diffusion coefficients of solid materials as described by Larsen [1981] leads to finding the conformal module of a quadrilateral of the form in Fig. 1b).

Other applications are in electromagnetic field theory as given in Reppe [1979]. Here we have to deal with quadrilaterals of the form of Fig. 2:

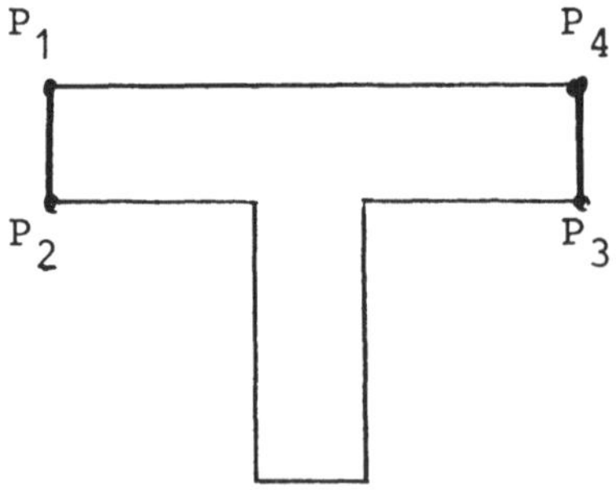

Fig. 2a)

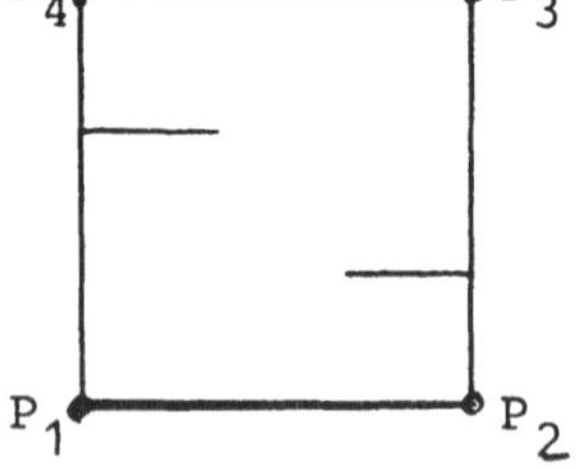

Fig. 2b)

Fig. 2b) shows a metallic square plate with slits, and the question is by how much the electric resistance increases by the introduction of the slits.

Finally there are many applications of CM in aerodynamics. James [1977] treats the lift and velocity distribution on an airfoil close to the ground (short takeoff and landing: STOL) and around a two-airfoil system. Here the module of a ringshaped domain plays an important role. In Prosnak [1977] the flow around a system of N airfoils is studied, the resulting N-fold connected domain outside the N airfoils is conformally mapped onto the exterior of N circles by a suitable rational approximation.

These examples clearly show the need for study of CM methods. In what follows we give a survey of two main areas which have been particularly popular and in which there are recent developments to report.

§ 2. THEODORSEN'S INTEGRAL EQUATION, OLD AND NEW

Although here the mapping problem is transformed to a singular, nonlinear integral equation, it is one of the most popular methods of determining the CM of a simply connected domain G which is starshaped with respect to $w=0$.
More precisely, assume that $\partial G=C$ is a Jordan curve in the w-plane which can be represented in polar coordinates (ρ,Θ), $\rho=\rho(\Theta)$. We want to find the conformal map $w=f(z)$ from $\mathbb{D}=\{z:|z|<1\}$ onto G which is normalized by $f(0)=0, f'(0)>0$. Actually not f will be determined but the boundary correspondence: f has a continuous and bijective extension to $\overline{\mathbb{D}}$, so that

$$(2.1) \qquad f(e^{i\varphi}) = \rho(\Theta(\varphi))e^{i\Theta(\varphi)} \quad ,$$

and it is $\Theta(\varphi)=\arg f(e^{i\varphi})$ that we shall determine.

We shall proceed according to the following flow diagram:

A. Continuous Problem — (1) Integral equation
- (2) existence of solution
- (3) uniqueness of solution
- (4) solution by iteration

discretization error estimates

B. Discrete Problem — (1) Vector equation
- (2) existence of solution
- (3) uniqueness of solution
- (4) solution by iteration

Most of the recent contributions are to B4, but we shall present the whole story.

A. The Continuous Problem

A1. The integral equation (T)
If $F(z)=u(z)+iv(z)$ is regular in $\mathbb{D}$ and continuous in $\overline{\mathbb{D}}$, the functions u and v are related on $\partial\mathbb{D}$ by

$$(2.2) \qquad v(e^{i\varphi}) = v(0) + \frac{1}{2\pi}\int_0^{2\pi} u(e^{it}) \operatorname{ctg} \frac{\varphi-t}{2}\, dt$$

where the integral exists as a Cauchy principal value. We apply this to $F(z) = \log \frac{f(z)}{z}$, where f is our normalized mapping function, and observe (2.1) to obtain the Theodorsen integral equation

$$(T) \qquad \Theta(\varphi) = \varphi + \frac{1}{2\pi}\int_0^{2\pi} \log \rho(\Theta(t)) \operatorname{ctg} \frac{\varphi-t}{2}\, dt$$

or equivalently

$$(T) \qquad \Theta(\varphi) = \varphi + K[\log \rho(\Theta(\cdot))](\varphi)$$

where K is the familiar conjugate operator from

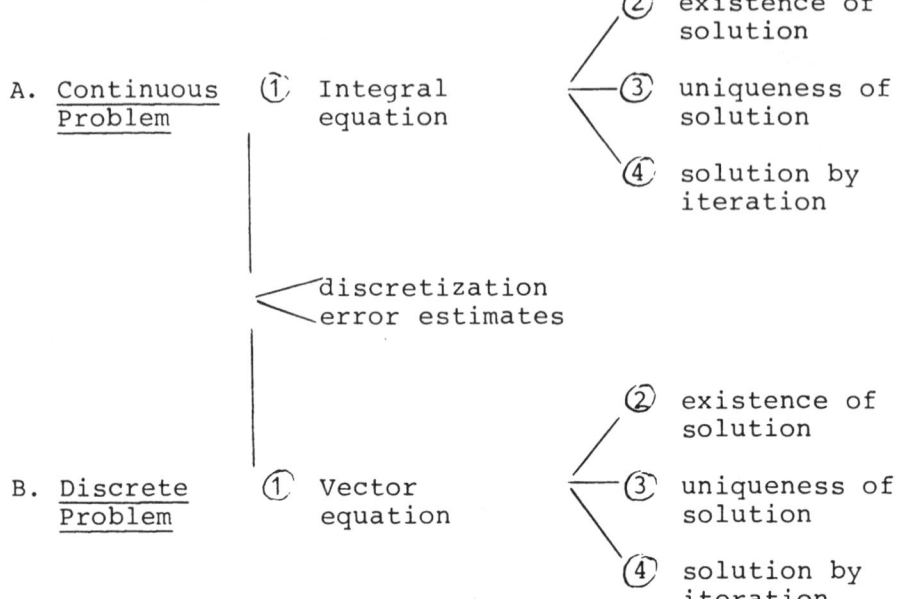

Fourier series. Our aim is to obtain the function $\Theta(\varphi)$ from this nonlinear singular integral equation or to approximate it by a suitable discretization procedure.

Quite frequently we shall impose an additional restriction on ∂G. We say that $C: \rho=\rho(\Theta)$ satisfies an ε-condition for some $\varepsilon>0$ provided that

(i) $\rho(\Theta)$ is absolutely continuous on $[0,2\pi]$, and

(ii) $\left|\frac{\rho'(\Theta)}{\rho(\Theta)}\right| \leq \varepsilon$ for almost all $\Theta \in [0,2\pi]$.

Geometrically, this means that the angle $\kappa(\Theta)$ between the radius vector and the normal to C at $(\Theta,\rho(\Theta))$ satisfies $|\operatorname{tg} \kappa(\Theta)| \leq \varepsilon$ almost everywhere. If ε is small, C will be "nearly circular".

A2. Existence of solution of (T)
This matter is most easily settled by observing that the CM f exists by Riemann's mapping theorem and has a continuous extension to $\bar{\mathbb{D}}$, so $\Theta(\varphi)$ defined by (2.1) is a continuous solution of (T).

A3. Uniqueness of solution of (T)
Here we claim
<u>Theorem 1</u>. <u>The integral equation</u> (T) <u>has exactly one solution</u> $\Theta(\varphi)$ <u>such that</u> $\Theta(\varphi)-\varphi$ <u>is continuous and 2π-periodic on</u> $\mathbb{R}$.

<u>Proof</u>. Let $\Theta(\varphi)$ be one such solution. We form

$$u(e^{i\varphi}) = \log \rho(\Theta(\varphi)) \quad \text{and} \quad v(e^{i\varphi}) = \Theta(\varphi) - \varphi ,$$

two functions continuous on $|z|=1$ which have harmonic extensions u,v, to $\bar{\mathbb{D}}$. These are conjugate because of (T) so that $u+iv$ is regular in $\mathbb{D}$, continuous in $\bar{\mathbb{D}}$. Hence also

$$f(z) := z \cdot \exp\{u(z) + iv(z)\}$$

is regular in $\mathbb{D}$, continuous in $\bar{\mathbb{D}}$. For $z=e^{i\varphi}$

$$|f(e^{i\varphi})| = \rho(\Theta(\varphi)) \quad \text{and} \quad \arg f(e^{i\varphi}) = \varphi+v(e^{i\varphi}) = \Theta(\varphi)$$

so that $f(e^{i\varphi}) \in C = \partial G$. Now as φ increases from 0 to 2π, the point $f(e^{i\varphi})$ traverses C once in the positive direction (monotonically or not), and the argument principle shows that for any $z_0 \in \operatorname{int} C$ $f(z)-z_0$ has exactly one zero in $\mathbb{D}$. That is, f is the normalized CM from $\mathbb{D}$ to G and hence $\arg f(e^{i\varphi})=\Theta(\varphi)$ is the boundary correspondence function.

A4. Solution of (T) by iteration
The most natural way to attack (T) analytically is by iteration:

$$\Theta_{n+1}(\varphi) = \varphi + K[\log \rho(\Theta_n(\cdot))](\varphi) \quad (n=0,1,2,\ldots) \tag{2.3}$$

with a suitable starting function Θ_0 which often will be $\Theta_0(\varphi)=\varphi$. Actually, Θ_n has to be in a certain function class so that Θ_{n+1} can be defined by (2.3). In view of the repeated application of the operator K in (2.3), Theodorsen's method has also been called the <u>method of successive conjugates</u>.

Now convergence of $\{\Theta_n\}$ in the space $L^2(0,2\pi)$ can easily be proved provided C satisfies an ε-condition with some $\varepsilon<1$. Namely, by (2.3) and (T)

$$\Theta_{n+1}(\varphi) - \Theta(\varphi) = K[\log \rho(\Theta_n(\cdot)) - \log \rho(\Theta(\cdot))](\varphi) .$$

Furthermore K is norm-decreasing in L^2: $||Kf||\le||f||$ for all $f\in L^2$. This gives us

$$||\Theta_{n+1} - \Theta|| \le ||\log \rho(\Theta_n(\cdot)) - \log \rho(\Theta(\cdot))|| .$$

But because of the ε-condition we have

$$|\log \rho(x_1)-\log \rho(x_2)| = \left|\int_{x_1}^{x_2} \frac{\rho'(t)}{\rho(t)}\,dt\right| \le \varepsilon|x_1-x_2|$$

and so

$$||\Theta_{n+1} - \Theta|| \le \varepsilon||\Theta_n - \Theta|| ,$$

which gives us convergence of the sequence $\{\Theta_n\}$ in L^2.

A slightly more delicate argument is needed to prove the uniform convergence of $\{\Theta_n\}$ on $[0,2\pi]$. One uses the fact that the norms $||\Theta_n' - 1||$ are bounded and a lemma due to Warschawski to conclude back to $|\Theta_n-\Theta|$ itself; see Gaier [1964] for details. This gives us

Theorem 2. Suppose $\Theta_o(\varphi)=\varphi$, and that C satisfies an ε-condition with $\varepsilon<1$. Then the iterations Θ_n defined by (2.3) converge uniformly to Θ, and moreover we have

$$|\Theta_n(\varphi) - \Theta(\varphi)| \le K(\varepsilon)\cdot\varepsilon^{n/2} \qquad (n = 1,2,\dots)\ .$$

It is not known whether $\{\Theta_n\}$ may diverge if the ε-condition with $\varepsilon<1$ is not fulfilled.

B. The Discrete Problem

In the following we descirbe the standard discretization of the integral equation (T) and point out other possibilities. The new problem will be a non-linear system of equations (T_d), and we study its solution by various iteration methods.

B1. The vector equation (T_d)

The discretization of Theodorsen's integral equation calls for an approximation of the conjugate operator K defined by

$$K[f](\varphi) := \frac{1}{2\pi}\int_o^{2\pi} f(t)\ \text{ctg}\ \frac{\varphi-t}{2}\,dt\ .$$

The most common method is due to Wittich (1941). First f is replaced by its trigonometric interpolation polynomial T of degree N corresponding to the nodes $t_k = k\frac{\pi}{N}$ $(k=0,\pm1,\pm2,\dots)$ where N is an integer. The conjugate $K[T]$ of T is easy to obtain, and its values at the points t_k serve as approximation of $K[f](t_k)$. If the input vector $\{f(t_k)\}(k = -N+1,-N+2,\dots,N)$ is denoted by x, the vector $\{K[T](t_k)\}(k = -N+1,-N+2,\dots,N)$ is given by a linear transformation:

$$(2.4)\qquad y = Wx\ .$$

The $2N\times 2N$ matrix W is called the Wittich matrix; its entries are

$$a_{ik} = \begin{cases} 0 & \text{if } i-k \text{ is even} \\ \frac{1}{N}\ \text{ctg}\ \frac{(i-k)\pi}{2N} & \text{if } i-k \text{ is odd}\ . \end{cases}$$

It is a checker-board matrix with zeros on the main diagonal, cyclic, and its eigenvalues are 0(2 fold) i(N-1 fold), -i(N-1 fold). Like K it has a norm reducing property:

$$(2.5) \qquad ||Wx|| \le ||x||$$

for the Euclidean norm of x.

We mention in passing that various estimates for the error

$$\max_k |K[f](t_k) - K[T](t_k)|$$

have been given by Gaier [1974], Henrici [1981] and Braß [1981]. In particular, Braß showed that in the class of functions f which have a bounded k-th derivative ($k \ge 1$) the Wittich method is optimal in a certain sense compared to all other quadrature methods with t_k as nodes.

Other methods of approximation of the operator K have been proposed. We mention Gutknecht [1979] who uses spline functions to approximate f and the survey by Gabdulhaev [1980] of work that has been done in the Sovjet Union.

We now apply the Wittich method to the integral equation (T). If x is a 2N-vector which we expect as an approximation to the vector $\{\Theta(t_k)\}$, the discretization of (T) will be

$$(T_d) \qquad x = a + W \log \rho(x),$$

where $a = \{t_k\} = \{k \cdot \frac{\pi}{N}\}$, $k = -N+1, \ldots, N$. Error estimates for $\max_k |\Theta(t_k) - x_k|$ are known provided our curve C satisfies an ε-condition with some $\varepsilon < 1$; see Gaier [1964], p. 92 ff.

B2. Existence of solution of (T_d)

We now focus our attention to the solution of the system (T_d). First of all, (T_d) <u>has always at least one solution.</u>

For if $A = \max_{[0,2\pi]} |\log \rho(t)|$ we get for the right hand side of (T_d)

$$||a + W \log \rho(x)|| \le ||a|| + 1 \cdot A\sqrt{2N}$$

which is a constant independent of x. That is the ball $||x|| \leq R$ in $\mathbb{R}^{2N}$ is mapped into itself, provided R is large enough. Brouwer's theorem then guarantees at least one fixed point of the equation (T_d). This proof is due to Hübner (1973) and Gutknecht [1977].

B3. Uniqueness of solution of (T_d)

It would be nice if (T_d) would always have exactly one solution corresponding to the solution of (T) (Theorem 1). Unfortunately this is true only under additional assumptions on C.

Theorem 3. a) The solution of (T_d) is unique if C satisfies an ε-condition for some $\varepsilon<1$.
b) This is not true in general for curves satisfying a 1-condition.

Proof. a) If x and y satisfy (T_d), then

$$x - y = W[\log \rho(x) - \log \rho(y)] \quad \text{and}$$

$$||x-y|| \leq ||\log \rho(x) - \log \rho(y)||$$

because of (2.5).

In view of the ε-condition

$$|\log \rho(x_k) - \log \rho(y_k)| \leq \varepsilon |x_k - y_k|$$

and so $||x-y|| \leq \varepsilon ||x-y||$. If $\varepsilon<1$, this gives $x=y$.

b) Here we follow the construction in Hübner [1982]. Let N be even, for example, and let our curve C be defined such that $\log \rho(t)$ is a zig-zag-function:

$$\log \rho(t) = \frac{1}{N} \sin N t$$

$$\text{for} \quad t = k \cdot \frac{\pi}{2N}, \quad k = 0, \pm 1, \pm 2, \ldots,$$

linear in between.

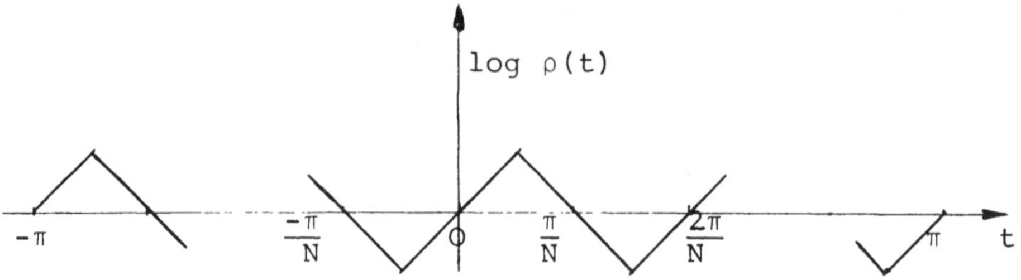

Fig. 3

Notice that C depends on N! Obviously C satisfies a 1-condition, and it is also clear that (T_d) is satisfied by $x=a$, since $\log \rho$ is zero at the components of a.
However, there is an infinity of solutions of the form $x=a+\tau y$ for a suitable vector y and for small $|\tau|$. In order that this is so, we require

$$a + \tau y = a + W \log \rho(a + \tau y) ;$$

the vector $\log \rho$ has components $\log \rho(t_k+\tau y_k) = =(-1)^k \tau y_k$ for small $|\tau|$. We therefore require

$$\tau y = \tau W\{(-1)^k y_k\} \quad \text{or} \quad y = WDy$$

with $D = \text{diag}(-1,+1,-1,+1,\ldots)$. Now it is not difficult to see that $y = (1,-1,-1,1,\ldots)'$ is an eigenvector of WD to the eigenvalue 1, and so all vectors $x = a + \tau y$ solve (T_d) for small $|\tau|$ and this eigenvector y.

We have pointed out that the region G for which (T_d) has many solutions depends on N. The question remains open whether for a given G with sufficiently smooth boundary (T_d) has a unique solution provided N is large enough.

Finally Hübner [1979] has considered the special case that the region G is symmetric with respect to $\mathbb{R}$. Then (T_d) can be reduced to a smaller system $(\hat{T}_d)$, and this has at most one solution with components in $[0,\pi]$, provided that ρ is absolutely continuous and monotonically increasing in $[0,\pi]$.

B4. Solution of (T_d) by iteration
The classical way to solve (T_d) is by Jacobi iteration

$$(2.6) \qquad x_{n+1} = a + W \log \rho(x_n) \qquad (n = 0,1,2\ldots) ,$$

with an arbitrary x_0 which often will be $x_0=a$. If C satisfies an ε-condition for some $\varepsilon<1$, the right hand side of (T_d) defines a contraction mapping and one obtains in a familiar way
Theorem 4. If C satisfies an ε-condition for some $\varepsilon<1$, the sequence $\{x_n\}$ defined by (2.6) converges to the unique solution x of (T_d) , and we have

$$||x_n-x|| \le \frac{\varepsilon^n}{1-\varepsilon} ||x_1-x_0|| \qquad (n = 1,2,\ldots) .$$

In the last ten years many advances have been made in two directions:

(i) Fast evaluation of Wx where W is the Wittich matrix.
(ii) Study of other methods of iteration, different from (2.6).

Concerning (i), a significant advance was made by Henrici who proposed to use FFT to evaluate Wx . Since W is a $2N \times 2N$-matrix, ordinarily we need $O(N^2)$ operations to compute Wx . However, Henrici observed that W was obtained by trigonometric interpolation, conjugation, and evaluation at the nodes t_k . Since conjugation of a trigonometric polynomial is a trivial operation, there remain two real FFT , and Wx can be computed with $O(N \log N)$ operations. First experiments were done by Henrici's student Lundwall-Skaar [1975]; later work by Henrici [1976] and [1979] and Gutknecht [1979].
The idea of using FFT in this context was apparently in the air. Simultaneously with Henrici, D.C. Ives (working in aerospace industry) proposed a "fast conformal mapping" [1975] by applying FFT to solve the Theodorsen integral equation.

Concerning (ii), various iteration methods have been proposed and studied by Niethammer [1966], Gutknecht [1981] and Hübner [1979], [1982]. By these works the range of effectiveness of the Theodorsen method was extended considerably.

Let us first look at the JOR-iterations

(2.7) $$x_{n+1} = (1-\omega)x_n + \omega[a + W \log \rho(x_n)] \qquad \omega \in \mathbb{R} ;$$

the special case $\omega=1$ gives (2.6).

Hübner [1982] showed that in general we cannot expect any improvement by this method. For, given $\varepsilon>1$ and $N>1$, there is a region G whose boundary C satisfies an ε-condition, for which $x=a$ solves (T_d) but is a point of repulsion for each JOR method. In addition, if $\varepsilon<1$, the spectral radius of the Jacobi matrix associated with (2.7) is least for $\omega=1$, i.e. for the ordinary Jacobi iterations. In this example, (2.6) converges faster than (2.7) for $\omega \neq 1$.
He also shows that for a curve satisfying a 1-condition, $\{x_n\}$ from (2.6) may diverge!

But the JOR method (2.7) is advantageous under additional assumptions. Let

$$D(x) = \operatorname{diag}\{ \frac{\rho'}{\rho}(x_{-N+1}), \frac{\rho'}{\rho}(x_{-N+2}), \dots, \frac{\rho'}{\rho}(x_N)\}$$

where we assume the derivatives to exist where they are needed. Then $WD(x)$ is the Jacobian of the right hand side of (T_d), and the position of its eigenvalues is decisive for the convergence of (2.7) for suitable factors ω.

Gutknecht [1981] assumes that $C=\partial G$ is symmetric with respect to $\mathbb{R}$ and, in addition, symmetric with respect to 0; i.e. C is mapped onto itself by a rotation around $w=0$ through an angle $\frac{2\pi}{\nu}$, $\nu \geq 2$. It is further assumed that ρ increases (or decreases) monotonically in the interval $(0, \frac{\pi}{\nu})$.

Then it is shown that there is a solution x of (T_d), having the same symmetries, and the eigenvalues of $WD(x)$ belonging to symmetric eigenvectors are purely imaginary. With these facts it follows easily that given a curve C satisfying an ε-condition and the symmetry assumptions mentioned above, the JOR - iterations (2.7) converge locally at x provided $0<\omega<2/(1+\varepsilon^2)$. The optimal underrelaxation factor is $(1+\varepsilon^2)^{-1}$. For large ε, the spectral radius of the local iteration matrix is close to 1 but <1 for an ω close to 0.

We now return to the solution of (T_d) for general C. Following Niethammer [1966] we apply a reordering in the computation of $z=Wy$,

$$(2.8)\quad \begin{pmatrix} z_1 \\ z_2 \\ \vdots \\ \vdots \\ z_{2N} \end{pmatrix} = \begin{pmatrix} 0 & a_1 & 0 & a_2 & 0 & \dots & a_N \\ a_N & 0 & a_1 & 0 & a_2 & \dots & 0 \\ 0 & a_N & 0 & a_1 & 0 & \dots & \\ & \dots & & & \dots & & \\ & & & & & & \end{pmatrix} \begin{pmatrix} y_1 \\ y_2 \\ \vdots \\ \vdots \\ y_{2N} \end{pmatrix},$$

by writing

$$Py = \begin{pmatrix} y_1 \\ y_3 \\ \vdots \\ y_2 \\ y_4 \\ \vdots \end{pmatrix} = \begin{pmatrix} y' \\ \\ y'' \end{pmatrix}$$

for N-vectors y', y'' and a permutation matrix P. $z=Wy$ is now equivalent to $Pz = \begin{pmatrix} 0 & A \\ B & 0 \end{pmatrix} Py$ so that

$$PWP^T = \begin{pmatrix} 0 & A \\ B & 0 \end{pmatrix}$$

for $N \times N$ matrices A and B; because of the formula after (2.4) we have $A=-B^T$. Altogether $z=Wy$ is equivalent to the system

$$(2.9)\quad z' = Ay'', \quad z'' = -A^T y',$$

with the cyclic $N \times N$ matrix

$$A = \begin{pmatrix} a_1 & a_2 & \cdots & a_N \\ a_N & a_1 & \cdots & a_{N-1} \\ \cdots & & \cdots & \\ a_2 & a_3 & \cdots & a_1 \end{pmatrix} .$$

For our applications it is essential that multiplication of A by an N-vector can be carried out fast by using FFT ; see Gutknecht [1979] and Hübner [1979].

Our $2N \times 2N$ system (T_d) is now equivalent to two $N \times N$ systems

$$(2.10) \quad x' = a' + A \log \rho(x'') , \qquad x'' = a'' - A^T \log \rho(x') ,$$

and the Gauß-Seidel method is according to the formulas

$$(2.11) \quad x'_{n+1} = a' + A \log \rho(x''_n) , \qquad x''_{n+1} = a'' - A^T \log \rho(x'_n) \quad (n = 0,1,2,\ldots) ;$$

in other words we compute the following N-vectors:

$$x'_0 \to x''_1 \to x'_2 \to x''_3 \to x'_4 \to \cdots .$$

One step of (2.11) needs about the same number of operations as one step of (2.6). However, if C satisfies an ε-condition with some $\varepsilon<1$, Niethammer showed that (2.11) converges twice as fast as the Jacobi iteration (2.6); the error is $O(\varepsilon^{2n})$.

Finally we remark that (2.11) can be generalized to the SOR method:

$$(2.12) \quad \begin{aligned} x'_{n+1} &= (1 - \omega)x'_n + \omega[a' + A \log \rho(x''_n)] \\ x''_{n+1} &= (1 - \omega)x''_n + \omega[a'' - A^T \log \rho(x'_n)] \end{aligned} \qquad (n = 0,1,2,\ldots) .$$

Its convergence has been investigated by Gutknecht [1981] and Hübner [1979] in a way similar to the JOR method, and for domains with symmetries there are definite advantages of SOR over Gauß-Seidel.

Of the remaining iteration methods to solve (T_d) we mention Gutknecht's investigations of the Chebyshev semi-iterative method and the second order Euler methods, and Hübner's studies [1979] of two-parametric overrelaxation. Considering all the work that has been done towards the solution of (T_d) , it is fair to claim that probably no nonlinear system of equations has received more attention than (T_d) .

§3. APPLICATION OF VARIATIONAL PRINCIPLES

As pointed out in § 1, in many applications of CM certain domain functionals are wanted rather than the CM of the domain itself. These functionals are the module of a quadrilateral, the module of a ring domain, and the capacity of a Jordan curve. They can be characterized by extremum principles, and consequently bounds can be computed by using test functions. Approximations of the CM are obtained as by-product of the method. We report on error estimates and on numerical experiments.

Finally, we mention the advances in the Bergman kernel method by the introduction of singular functions.

A. The principles of Dirichlet and Gauß-Thomson

Let C be a Jordan curve with four distinguished points P_1, P_2, P_3, P_4 in positive orientation, so that a quadrilateral V is given. Let f_o be the CM of $G = \text{int } C$ onto a rectangle $R : (0,a) \times (0,b)$ such that P_1, P_2, P_3, P_4 map into $0, a, a+ib, ib$. Then

$$m(V) = \frac{a}{b}$$

is the conformal module of the quadrilateral V .

Let G be a bounded ring domain bounded by the Jordan curves Γ_o (interior curve) and Γ_1 (exterior curve). Let f_o be a CM of G onto the concentric circular ring $\{w: 1<|w|<M\}$. Then $M=M(G)$ is the module of the ring domain G .

Finally, let Γ be a Jordan curve and f_o be the CM of $\text{ext } \Gamma$ onto $\{w: |w|>r\}$ normalized at ∞ by $f_o(z) = z+a_o+\frac{a_1}{z}+\ldots$. Then

$$\operatorname{cap} \Gamma = r$$

is the capacity of the curve Γ .

We first show how these domain functionals can be characterized by extremal principles.

A1. The Dirichlet principle gives $m(V)$ and $M(G)$
Let $\Gamma_o = P_1P_2$ and $\Gamma_1 = P_3P_4$ be opposite sides of V , let K be the class of real valued functions u continuous in $\bar{G}$ with $u = 0$ on Γ_o , $u = 1$ on Γ_1 , for which u is in the Sobolev space $W_1(G)$, and let

$$D[u] = \iint_G (u_x^2 + u_y^2)\, db$$

be its Dirichlet integral. Then

$$(3.1) \qquad m(V) = \min\{D[u] : u \in K\} = D[u_o]$$

where $u_o = \operatorname{Im} f_o/b$. For, Dirichlet's principle tells us that it is sufficient to consider functions u harmonic in G . Their Dirichlet integral remains invariant under the CM by f_o , so $D_G[u] = D_R[\tilde{u}]$ with $\tilde{u} = u \circ f_o^{-1}$. $\tilde{u}$ has boundary values 0 on $0 \ldots a$ and 1 on $ib \ldots a+ib$. Therefore

$$1 = [\tilde{u}(x,b) - \tilde{u}(x,0)]^2 =$$

$$= \left(\int_0^b \tilde{u}_y(x,t)\,dt\right)^2 \le b \int_0^b \tilde{u}_y^2(x,t)\,dt \, .$$

Integration with respect to x gives

$$a \le b \iint_R \tilde{u}_y^2(x,y)\,db \le b\, D_R[\tilde{u}]$$

so $D_G[u] \ge \frac{a}{b}$ for all admissible u .

Furthermore $D_G[u_o] = D_R[\frac{y}{b}] = \frac{1}{b^2}\, ab = m(V)$.

Any test function $u \in K$ gives, by (3.1), an upper bound for $m(V)$. Considering the conjugate quadrilateral V' with opposite sides P_2P_3 and P_4P_1 , we get upper bounds for $m(V') = [m(V)]^{-1}$ and therefore lower bounds for $m(V)$. This method of determining $m(V)$ was studied in Gaier [1972].

If we have a ring domain G with $\partial G = \Gamma_0 \cup \Gamma_1$, and if K denotes the class of functions u continuous in $\overline{G}$ with $u = 0$ on Γ_0 , $u=1$ on Γ_1 , for which $u \in W_1(G)$, we have similarly

$$(3.2) \qquad \frac{2\pi}{\log M(G)} = \min\{D[u] : u \in K\} = D[u_0]$$

where $u_0 = [\log M(G)]^{-1} \log |f_0|$. In this case we only get lower bounds for $M(G)$. This method of approximating $M(G)$ was introduced by Opfer [1967] and [1969].

A2. The Gauß-Thomson principle gives cap Γ and $M(G)$
Now let Γ be a rectifiable Jordan curve, and for every $g \in L^1(\Gamma)$ put

$$Q(g) = \int_\Gamma g(z)\, |dz|$$

(total charge of g) and

$$E[g] = \iint_{\Gamma\Gamma} \log \frac{1}{|z_1-z_2|}\, g(z_1)g(z_2)\,|dz_1|\,|dz_2| \quad ,$$

the energy of the logarithmic potential

$$(3.3) \qquad V_g(z) = \int_\Gamma \log \frac{1}{|z-z_1|}\, g(z_1)\,|dz_1| \qquad (z \notin \Gamma)$$

associated with g . In particular, if f_0 is the normalized CM of ext Γ onto $\{w : |w|>r\}$, we study

$$g_0(z) = \frac{1}{2\pi}\frac{d}{ds}\,\Theta(z) \quad \text{where} \quad \Theta(z) = \arg f_0(z) \ , \ z\in\Gamma.$$

Here $g_0 \geq 0$ a.e. on Γ and $Q(g_0)=1$, and furthermore

$$(3.4) \qquad V_{g_0}(z) = \log \frac{1}{r} \quad \text{a.e. on } \Gamma \ ;$$

see Gaier [1976], p.124. Integrating over Γ , we get for every $g \in L^1(\Gamma)$

$$(3.5)\qquad \iint_{\Gamma\Gamma} \log \frac{1}{|z_1-z_2|}\, g_o(z_1)g(z_2)\,|dz_1|\,|dz_2| =$$

$$= \log \frac{1}{r}\cdot Q(g)\ ,$$

in particular $E[g_o] = \log \frac{1}{r}$. From this follows

$$E[g-g_o] = E[g] + E[g_o] - 2\log\frac{1}{r}\cdot Q(g)$$

and therefore for all $g \in L^1(\Gamma)$ with $Q(g)=1$

$$E[g] = E[g-g_o] + \log\frac{1}{r}\ .$$

Because of the well known fact (M.Riesz) that $E[g] \geq 0$ whenever $Q(g)=0$ we get in

$$(3.6)\qquad \log\frac{1}{r} = \min\{E[g] : g \in L^1(\Gamma)\ ,\ Q(g) = 1\} = E[g_o]$$

our desired characterization of $r = \text{cap}\,\Gamma$; see Pólya-Szegö [1951], p.60 and Weisel [1979], p.95. Every test function gives a lower bound for r .

In a similar way we get a characterization of $M(G)$ for a ring domain. If $\partial G = \Gamma_o \cup \Gamma_1$ we consider integrable mass distributions g with $\int_{\Gamma_o} g = 1$ and $\int_{\Gamma_1} g = -1$. In particular,

$$g_o(z) = \begin{cases} \dfrac{1}{2\pi}\dfrac{d}{ds}\,\Theta(z) & z \text{ on } \Gamma_o \\[2ex] -\dfrac{1}{2\pi}\dfrac{d}{ds}\,\Theta(z) & z \text{ on } \Gamma_1 \end{cases}$$

is admitted; here $\Theta(z) = \arg f_o(z)$ where f_o is a CM of G onto $\{w : 1<|w|<M\}$ which carries Γ_o onto $|w|=1$. Proceeding as above one obtains

$$(3.7)\qquad \log M = \min\{E[g] : \int_{\Gamma_o} g = -\int_{\Gamma_1} g = 1\} = E[g_o]\ .$$

In the language of mathematical physics this means that the energy of the gradient field of the logarithmic potential associated with g is minimal for $g=g_o$ where all distributions g are admitted whose charge is ± 1 on Γ_o and Γ_1 .

In the special case that $\Gamma_o = \{z : |z| = 1\}$ the extremal principle (3.7) can be modified and $\log M$ is given by an extremal problem on Γ_1 alone ; see Tsuji [1959], p.95 ff. and Weisel [1979], p.128 ff. For integrable distributions g on Γ_1 we consider

$$\tilde{E}[g] = \int_{\Gamma_1}\int_{\Gamma_1} \log \left|\frac{1-\bar{z}_1 z_2}{z_1 - z_2}\right| g(z_1)g(z_2)\,|dz_1|\,|dz_2| \quad ,$$

and then

$$(3.8) \qquad \log M = \min\{\tilde{E}[g] : \int_{\Gamma_1} g = 1\} \quad .$$

(This is also true if Γ_1 is the inner boundary of G .)

Both principles (3.7) and (3.8) are dual to (3.2) in that they give upper bounds for $M(G)$.

B. Numerical approximation, error estimates

The extremal principles (3.1), (3.2), (3.6), (3.7) and (3.8) will now be applied numerically. For (3.1) and (3.2) we need two-dimensional test functions, for the others onedimensional test functions. The first were used in Gaier [1972] for the approximation of the modules of various quadrilaterals V , whereas Gauß-Thomson was applied by Weisel [1979] following a suggestion by Kühnau (1976). Finite elements were also used in the dissertation of Bosshard [1980].

So far only linear or bilinear splines have been used although splines of higher order would present no essential difficulties (see for example Schwarz [1980], p.67 ff.). Since the exact solutions of the extremal problems are known - in terms of conformal mappings - their behavior at corners of ∂G is known. The use of singular elements is advised and Weisel has successfully applied this in some cases.

Error estimates are given in Weisel [1979] and also in Bosshard [1980]. We report some of them. Suppose that ∂G is a polygon and that G has been divided into a regular mesh $(h/\underline{h} \le \sigma$ fixed); see for example the L-shaped region.

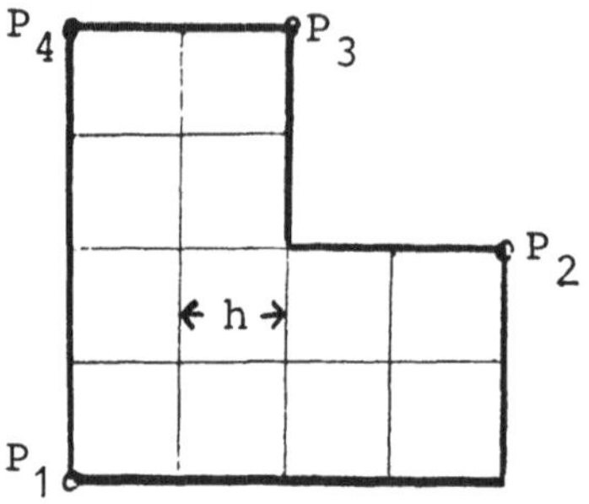

Exact module 0.585 080

$= (4\sqrt{3} - 6)^{1/2}$.

Fig. 4

Let m_h be the minimum of $D[u]$ if $u \in K$ and u is a linear or bilinear spline to mesh size h . Then Weisel ([1979], p.77) proves that

$$0 \le m_h - m(V) \le O(h^{\beta}) .$$

Here β depends on the geometry of the quadrilateral V . Let $\pi\alpha$ be the greatest interior angle of ∂G , and $\pi\omega$ the greatest interior angle at P_1, P_2, P_3, P_4 ; then $\beta = \min(\frac{2}{\alpha}, \frac{1}{\omega})$. In the example of Fig. 4 we would have $\alpha = \frac{3}{2}$, $\omega = \frac{1}{2}$, and $\beta = \frac{4}{3}$. Numerical evidence suggests that the exponent β is best possible.

Similar estimates are possible for the Gauß-Thomson principle. Assume that ∂G is smooth, that the tangent angle $\tau(s)$ has an n-th derivative in Lip α , and that splines of order n are used in (3.6), with mesh size h . Then (Weisel [1979], p.100)

$$0 \le r - r_h \le O(h^{2(n+\alpha)}) .$$

If ∂G is piecewise analytic, as is often the case, weaker estimates hold, but they can be improved if singular functions are added to the Ansatz. If z_j is a corner of ∂G with interior angle α_j , these are of the form

$$s_j(z) = |z-z_j|^{\alpha_j^{-1}-1} ,$$

and their adjunction brings the error down to $O(h^2)$. For example, the module of the ring domain of Fig. 5 was computed.

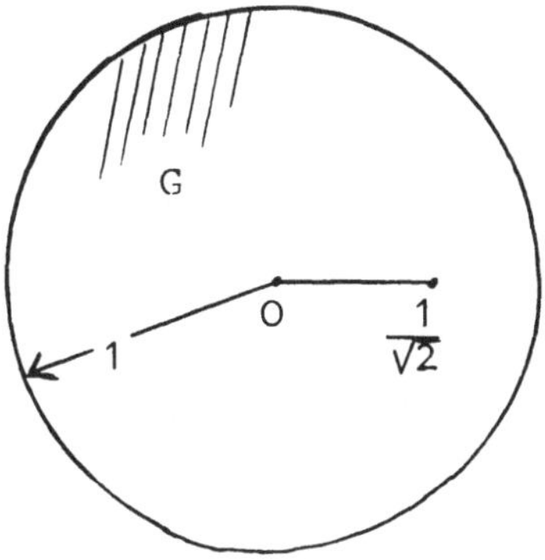

Exact module:

$$M(G) = e^{\pi/2} = 4.810\,477$$

Fig. 5

Dividing $(0, \frac{1}{\sqrt{2}})$ into n intervals of equal length and applying (3.8) to corresponding step functions, Weisel obtained $M_n = 4{,}87042$ for $n = 20$, whereas the addition of just two singular functions $s^{-1/2}$ and $(\frac{1}{\sqrt{2}} - s)^{-1/2}$ gave 4,81051 with only two subintervals.

C. Approximation of quasiconformal mappings

Our methods can be generalized to give approximations of quasiconformal mappings and their modules. This is of physical importance in the study of electrostatic fields in inhomogeneous media, for example. Here we deal with homeomorphisms $w = w(z) = u(x,y)+iv(x,y)$ which satisfy

$$pu_x = v_y , \quad pu_y = -v_x , \quad (x,y) \in G .$$

The mapping is called p-analytic. We restrict ourselves to a piecewise C^1-function p with $0<c\le p(z)\le C<\infty$ in G .
For p=1 we get a conformal map.

Now each quadrilateral V can again be mapped by a p-analytic function f_o onto a rectangle $(0,a)\times(0,b)$, and the domain functional

$$m_p(V) = \frac{a}{b}$$

is called the p-module of V. It can be characterized by an extremal property involving a modified Dirichlet integral:

$$m_p(V) = \\ = \min\{\iint_G \frac{1}{p}(u_x^2 + u_y^2)\,db : \begin{array}{l} u = 0 \text{ on } \Gamma_0 \\ u = 1 \text{ on } \Gamma_1 \end{array},\ u \in W^1(G)\} ,$$

the minimum being attained by $u = \mathrm{Im} f_o/b$. Using test functions, or searching for the minimum in certain subclasses, one obtains upper bounds for $m_p(V)$ and approximations for the imaginary part of the mapping function. Likewise the use of the Dirichlet integral $\iint_G p(u_x^2+u_y^2)\,db$ leads to lower bounds for $m_p(V)$ and an approximation of the real part of the mapping function.

Weisel [1980] hat done all that and has also given error estimates similar to the ones mentioned in section B. They are of the form

$$0 \le m_{p,h}(V) - m_p(V) = O(h^\gamma)$$

where now γ depends on the geometry of V and on the discontinuities of p.

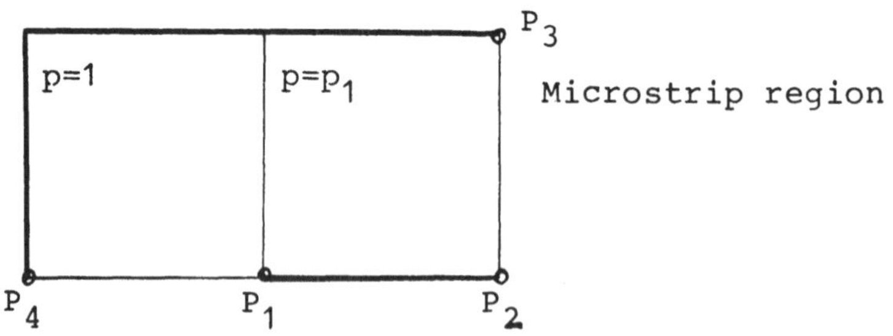

Fig. 6

In the case of the "microstrip region" one obtains $\gamma = \frac{4}{\pi}$ arc tg $\sqrt{p_1}$ as convergence rate. The p-module $m_p(V)$ is here ~ 10.5718 if $p_1 = 10$ is chosen.

In an analogous way a p-module $M_p(G)$ for ring domains G can be introduced and characterized by a weighted Dirichlet integral, but the homeomorphism $w(z)$ mapping G onto a circular ring is now such that $\log w$ is locally p-analytic in G.

D. Advances in the Bergman kernel method

If one speaks of variational methods in CM, the Bergman kernel method (BKM) must be mentioned, too. It has the advantage that it gives an approximation of the mapping function in closed form, i.e.

$$f(z) \approx \sum_{k=1}^{N} a_k U_k(z) .$$

More precisely, let G be a Jordan domain in the z-plane with $0 \in G$, and let f be the CM of G onto $\{w: |w|<R\}$ normalized by

$$f(0) = 0 , \quad f'(0) = 1 .$$

R is called the conformal radius of G at 0. [If Γ is the reflection of ∂G in the unit circle, we get $R \cdot \mathrm{cap}\, \Gamma = 1$, so that the determination of R and of $\mathrm{cap}\, \Gamma$ are equivalent problems.]

Let $L^2(G)$ be the set of all functions u regular in G with $\iint_G |u|^2 db < \infty$ and let

$$K_1 = \{u : u \in L^2(G) , u(0) = 1\} ,$$

$$K_0 = \{u : u \in L^2(G) , u(0) = 0\} .$$

Then it is well known that the extremal problem

$$(3.9) \quad ||u|| = \min! \qquad \text{for } u \in K_1$$

has a unique solution u_0 with the properties

$$(3.10) \quad u_0(z) = f'(z) \quad \text{and} \quad ||u_0||^2 = \pi R^2 .$$

This function u_0 is characterized by the orthogonality condition

$$(u_0, v) = 0 \quad \text{for all} \quad v \in K_0 \ .$$

To solve (3.9) practically one looks for the minimum in an N-dimensional subspace

$$K_1^{(N)} = \{ \sum_{k=1}^{N} a_k \, u_k(z) \}$$

where $u_1 \in K_1$ and $u_k \in K_0$ for $k>1$. The coefficients a_k of the optimal function φ_N in $K_1^{(N)}$ can be found by solving a linear $(N-1)\times(N-1)$-system which is Hermitian, and we get

$$f_N(z) = \int_0^z \varphi_N(t)\,dt \quad \text{and} \quad R_N = ||\varphi_N|| \,/\, \sqrt{\pi}$$

as approximations to f and R ; notice that $R_N \geq R$.

As mentioned above, this gives immediately an approximation of f in closed form:

$$f_N(z) = \sum_{k=1}^{N} a_k \int_0^z u_k(t)\,dt \ ,$$

and $f_N(z) \Rightarrow f(z)$ for $N \to \infty$ in every compact subset of G whenever $\{u_k\}$ forms a basis in $L^2(G)$.

The choice of the basis is indeed very decisive. If $u_k(z) = z^{k-1}$ $(k=1,2,\ldots)$ the f_N are (Bieberbach) polynomials of degree N and convergence to f is fast more or less only if ∂G is an analytic Jordan curve. For this reason the method based on (3.9) and (3.10) was neglected until 1977, when Papamichael and his students introduced singular functions into the basis [1978], [1981]. These reflect the singularities of the mapping function on ∂G and in the complement of $\overline{G}$. When ∂G is a polygon, the location of poles off $\overline{G}$ is given by the reflection principle.

Many practical examples have been carried out. We merely mention the case of a rectangle with sides 2 and 12. Using $N=13$ monomial terms one gets the mapping function with an error of about 10^{-2} , whereas

N=10 terms including two terms with poles at ± 2i and ± 12 suffice to give the CM of G onto a disc with an error of $< 2.10^{-6}$.

Although no theoretical error estimates are available yet the numerical experiments have been very successful, and the introduction of singular functions in the BKM has certainly made this method much more attractive. Concerning the use of singular functions in the CM of doubly connected domains see Eidel [1979] and Papamichael - Kokkinos [1982]. Moduls M(G) for various regions G are obtained to considerable accuracy, for example for square hole in disc, square hole in square, or circular hole in square.

BIBLIOGRAPHY

Bosshard, F. (1980): Die Konstruktion konformer Abbildungen mit der Methode der finiten Elemente. Dissertation Universität Zürich.

Braß,H. (1981): Zur numerischen Berechnung konjugierter Funktionen. Preprint Universität Braunschweig.

Eidel, W. (1979): Konforme Abbildung mehrfach zusammenhängender Gebiete durch Lösung von Variationsproblemen. Diplomarbeit Gießen.

Gabdulhaev, B.G. (1980): Finite-dimensional approximations of singular integrals and direct methods for solving singular integral and integro-differential equations (Russian). See: Math. Reviews 81, p. 5022.

Gaier, D. (1964): Konstruktive Methoden der konformen Abbildung. Springer, Berlin,Göttingen,Heidelberg.

Gaier, D. (1972): Ermittlung des konformen Moduls von Vierecken mit Differenzenmethoden. Numer. Math. 19, 179-194.

Gaier, D. (1974): Ableitungsfreie Abschätzungen bei trigonometrischer Interpolation und Konjugierten-Bestimmung. Computing 12, 145-148.

Gaier, D. (1976): Integralgleichungen erster Art und konforme Abbildung. Math. Z. 147, 113-129.

Gaier, D., Hübner, O. (1976): Schnelle Auswertung von Ax bei Matrizen A zyklischer Bauart, Toeplitz- und Hankel-Matrizen. Mitt. Math. Sem. Gießen 121, 27-38.

Gutknecht, M.H. (1977): Existence of a solution to the discrete Theodorsen equation for conformal mappings. Math. Comp. 31, 478-480.

Gutknecht, M.H. (1979): Fast algorithms for the conjugate periodic function. Computing 22, 79-91.

Gutknecht, M.H. (1979): The evaluation of the conjugate function of a periodic spline on a uniform mesh. Preprint No. 79-05, ETH Zürich.

Gutknecht, M.H. (1981): Solving Theodorsen's integral equation for conformal maps with the fast Fourier transform and various non-linear iterative methods. Numer. Math. 36, 405-429.

Gutknecht, M.H. (1981): Numerical experiments on solving Theodorsen's integral equation for conformal maps. SIAM J.Sci.Stat.Comp.

Henrici, P. (1976): Einige Methoden der schnellen Fouriertransformation. In: Moderne Methoden der numerischen Mathematik. ISNM 32, 111-124.

Henrici, P. (1979): Fast Fourier methods in computational complex analysis. SIAM Rev. 21, 481-527.

Henrici, P. (1981): Zur ableitungsfreien Abschätzung des Fehlers bei trigonometrischer Interpolation und Konjugiertenbestimmung. Preprint, ETH Zürich.

Hübner, O. (1979): Zur Numerik der Theodorsenschen Integralgleichung in der konformen Abbildung. Mitt. Math.Sem. Gießen 140, 1-32.

Hübner, O. (1982): Über die Anzahl der Lösungen der diskreten Theodorsen-Gleichung. Numer. Math. 39, 195-204.

Ives, D.C. (1976): A modern look at conformal mapping including multiply connected regions. AIAA J. 14, 1006-1011.

James, R.M. (1977): The theory and design of two-airfoil lifting systems. Comput. Methods Appl. Mech. Engrg. 10, 13-43.

Larsen, J. (1981): On the measuring of diffusion coefficients of solid materials. Z. Angew. Math. Phys. 32, 229-232.

Levin, D., Papamichael, N., Sideridis, A. (1978): The Bergman kernel method for the numerical conformal mapping of simply connected domains. J.Inst. Math. Appl. 22, 171-187.

Lundwall-Skaar, C. (1975): Konforme Abbildungen mit Fast Fourier Transformationen. Diplomarbeit ETH Zürich.

Niethammer, W. (1966): Iterationsverfahren bei der konformen Abbildung. Computing 1, 146-153.

Opfer, G. (1967): Untere, beliebig verbesserbare Schranken für den Modul eines zweifach zusammenhängenden Gebietes mit Hilfe von Differenzenverfahren. Dissertation Hamburg.

Opfer, G. (1969): Die Bestimmung des Moduls zweifach zusammenhängender Gebiete mit Hilfe von Differenzenverfahren. Arch.Rat.Mech.Anal. 32, 281-297.

Papamichael, N., Kokkinos, C.A. (1981): Two numerical methods for the conformal mapping of simply-connected domains. Comput.Methods Appl.Mech.Engrg. 28, 285-307.

Papamichael, N., Kokkinos, C.A. (1981): Numerical conformal mapping of exterior domains. Technical Report 08/81, Brunel University.

Papamichael, N., Kokkinos, C.A. (1982): The use of singular functions for the approximate conformal mapping of doubly-connected domains. Technical Report 01/82, Brunel University.

Pólya, G., Szegö, G. (1951): Isoperimetric inequalities in mathematical physics. Princeton.

Prosnak, W.J. (1977): Conformal representation of arbitrary multiconnected airfoil sections. Bull. Acad. Polon.Sci.Sér.Sci.Tech. 25, 591-602.

Reppe, K. (1979): Berechnung von Magnetfeldern mit Hilfe der konformen Abbildung. Siemens Forsch.- u. Entwickl.Ber. 8, 190-195.

Schwarz, H.R. (1980): Methode der finiten Elemente. Teubner, Stuttgart.

Tsuji, M. (1959): Potential theory in modern function theory. Tokyo.

Versnel, W. (1979): Analysis of the Greek cross, a Van Der Pauw structure with finite contacts. Solid-State Electronics 22, 911-914.

Weisel, J. (1979): Lösung singulärer Variationsprobleme durch die Verfahren von Ritz und Galerkin mit finiten Elementen - Anwendungen in der konformen Abbildung. Mitt.Math.Sem. Gießen 138, 1-150.

Weisel, J. (1980): Numerische Ermittlung quasikonformer Abbildungen mit finiten Elementen. Numer.Math. 35, 201-222.

Weisel, J. (1981): Approximation quasikonformer Abbildungen mehrfach zusammenhängender Gebiete durch finite Elemente. Z.Angew.Math.Phys. 32, 34-44.

ON COMPLEX RATIONAL APPROXIMATION
PART I: THE CHARACTERIZATION PROBLEM

Martin H. Gutknecht

ETH-Zentrum, CH-8092 Zürich

Abstract. We review results on the characterization of best or local best complex rational approximations. Special attention is paid to uniqueness and strong uniqueness. Two particularly simple sufficient conditions, alternation and circularity, are treated as examples.

1. INTRODUCTION

Uniform complex rational approximation is a classical area of analysis but not of numerical analysis. The interest has centered on questions of the possibility of approximation ("Is the set of rational functions dense in some set of given functions?") and of the degree of approximation ("How fast does the error of the best approximation out of $\mathcal{R}_{mn}$ decrease when $m \to \infty$ or $n \to \infty$ or $m,n \to \infty$?"). Outstanding contributions have been made, e.g., by Runge, Walsh, Mergelyan, Vitushkin, and Alice Roth. There are excellent surveys of these results, see Walsh [33,34], Korevaar [19], Gaier [14].

However, with regard to the construction of the best approximation (BA) $r^* \in \mathcal{R}_{mn}$ of some given f continuous on a

H. Werner et al. (eds.), Computational Aspects of Complex Analysis, 79–101.

compact set $Z \subset \mathbb{C}$, there are many questions - some of them unsettled - that are more important. Walsh showed long ago that a BA exists if Z has no isolated points [33, §12.2] and that it may be non-unique [32; 33, §12.4] (in contrast to polynomial approximation). But only a few years ago it became known that non-uniqueness is quite common in the seemingly simple problem of complex rational approximation of a real function on a real interval (Saff and Varga [28,29]), and that there may even be a continuum of BAs in this case (Ruttan [26]). Related to this is the lack of local strong uniqueness (Newman and Shapiro [22], Cline [10], Gutknecht [15], Williams [35]) and certain anomalies discovered by Wulbert [38] and further investigated by Ruttan [27]. Some of these results are cited in Section 3. Before, we review some consequences of basic results in "general" Chebyshev approximation theory. Later, in Sections 4 and 5 we discuss and extend two particularly simple sufficient conditions for BAs: the existence of an alternation set of sufficient length (Klotz [18]) and circularity of the error curve (Klotz [18], Trefethen [30,31]).

The computation of best rational approximations is not discussed here. But in Part II we will describe the theoretical background of an extension of the recently proposed Carathéodory-Fejér method which for sufficiently smooth functions gives approximations that are typically very close to best.

Notation: For any $Z \subset \overline{\mathbb{C}} := \mathbb{C} \cup \{\infty\}$, $\overline{Z}$ is the closure of Z, ∂Z is boundary, and Z^c the complement with respect to $\overline{\mathbb{C}}$. The symbol $\setminus$ denotes the difference of sets. The following spaces of complex functions are mainly used:

$C(Z)$:= space of continuous functions defined on Z,

$A(Z) := \{f \in C(Z) ; f$ holomorphic in the interior of $Z\}$,

$\mathcal{P}_m$:= space of polynomials of degree $\leq m$,

$$R_{mn} := \{p/q\,;\; p \in P_m\,,\; q \in P_n\,,\; q(z) \not\equiv 0\}\,,$$
$$R_{mn}(Z) := R_{mn} \cap C(Z)\,.$$

$\| f \|$ is always the supremum norm of f; the domain Z over which the supremum is taken will be apparent from the context. $M(f)$ is the corresponding set of extremal points:

$$M(f) := \{z \in Z\,;\; |f(z)| = \| f \|\}\,.$$

The error of the BA of $f \in C(Z)$ from $R_{mn}(Z)$ is denoted by $E_{mn}(f)$. In particular, $E_m(f) := E_{m0}(f)$.

2. WHAT CAN WE LEARN FROM GENERAL CHEBYSHEV APPROXIMATION THEORY?

Chebyshev approximation theory grew out of real polynomial Chebyshev approximation on a real interval. The further we withdraw from this particular problem, the weaker the results are. Its application to complex rational approximation is limited for various reasons: First of all, the problem is highly nonlinear and deviates considerably from some better understood nonlinear problems (such as real rational approximation). Second, one of the main features, analyticity, is not taken into account by the "ordinary" nonlinear theory. However even in the polynomial case, which is an example of approximation from a complex (linear) Haar space, there are essential differences from the corresponding real problem, in particular, the lack of a characterization similar to the alternation criterion and the lack of strong uniqueness (or, strong unicity). Nevertheless, we want to review in this section some well known general results, applied to complex polynomial and rational approximation.

Let $Z \subset \mathbb{C}$ be compact and $f \in C(Z)$. For any $u, v \in C(Z)$, $u \neq 0$, any closed $Z_o \subset Z$, and any $V \subset C(Z)$ let

$$\mu(u,v;Z_o) := \max_{z\in Z_o} \operatorname{Re}\{\overline{u(z)}\,v(z)\} ,$$

$$\gamma(u;V) := \inf_{\substack{v\in V\\ v\neq 0}} \frac{\mu(u,v;M(u))}{\|u\|\ \|v\|} .$$

For *polynomial approximation* Kolmogorov's criterion [20, Thm 18] and the related inclusion theorem due to Collatz and Schwedt [20, Thm 16] apply; moreover, according to Bartelt and McLaughlin [4, Thm 5], inequality in Kolmogorov's criterion means *strong uniqueness* of the BA $p \in P_m$, i.e. there exists $\tilde{\gamma} > 0$ such that

$$\|f-u\| \geqq \|f-p\| + \tilde{\gamma}\,\|u-p\| , \quad \forall u \in P_m . \tag{2.1}$$

We summarize these results in

THEOREM 2.1. (i) p *is the BA of* f *from* P_m *iff* $\gamma(f-p\,;P_m) \geqq 0$ *or* $f = p$.

(ii) p *is a strongly unique BA of* f *from* P_m *iff* $\gamma(f-p\,;P_m) > 0$ *or* $f = p$.

(iii) If $Z_o \subset Z$, $Z_o \neq \emptyset$ *closed, and* $p \in P_m$ *are such that* $f-p$ *does not vanish on* Z_o *and* $\mu(f-p\,,v\,;Z_o) \geqq 0$, $\forall v \in P_m$, *then*

$$\|f-p\| \geqq E_m(f) \geqq \min_{z\in Z_o} |f(z)-p(z)| .$$

$\tilde{\gamma} = \gamma(f-p\,;P_m)$ is in fact the greatest constant satisfying (2.1). In the case $\gamma(f-p\,;P_m) > 0$ this has been shown in [4, Thm 5]; in general it follows from the fact that $\gamma(f-p\,;\{v\})$ is the directional derivative at p of the error functional $p \mapsto \|f-p\|$ in the direction v; this can be shown by arguments similar to those in the usual proof of Kolmogorov's criterion.

Unlike in real polynomial approximation (or, generally, in real Haar spaces), where strong uniqueness always holds [22, Thm 4], Newman and Shapiro obtained in the complex case

$$\| u-p \| \leqq \beta_1(\| f-u \| - \| f-p \|)^{1/2} + \beta_2(\| f-u \| - \| f-p \|), \quad \forall u \in P_m \tag{2.2}$$

(with $\beta_1, \beta_2 > 0$) [22, Thm 4'], or, for $u \to p$,

$$\| f-u \| \geqq \| f-p \| + \beta \| u-p \|^2 + o(\| u-p \|^2) \tag{2.3}$$

(with $\beta > 0$). Although Newman and Shapiro were certainly aware of a counterexample to (2.1), the first published one seems due to Cline [10]. Further examples and a discussion of some consequences are given in [15] and below (Thms 3.1, 4.2, 5.5). In particular, *if* $Z \subset \mathbb{R}$ *and* p *is the real BA to a real-valued* $f \in C(Z)$, *then* p *is not strongly unique in* P_m (by Thm 1 in [15], cf. Thm 3.1 below). So, failure of strong uniqueness seems to be quite common. However, Blatt [6] showed that *strong uniqueness of the BA to* $f \in C(Z)$ *out of* P_m *is a generic property of* $C(Z)$ (i.e. the set of functions with strongly unique BA is open and dense in $C(Z)$) *if and only if* Z *has at most* m *isolated points*. Examples of failure are not limited to $Z \subset \mathbb{R}$ of course: e.g., the best constant approximating $f(z) = z^3 - 2z$ on the unit disk is $p = 0$, and $M(f-p) = \{i, -i\}$; perturbation of p by a small real constant obviously leads to a non-strong increase in $\| f-p \|$ (cf. Thm 4.2 below).

In practice, strong uniqueness is important since it implies the Lipschitz continuity of the BA [9, p. 82]. Likewise, (2.2) *implies the Hölder continuity with exponent* $1/2$ *of the BA* (Cline [10]): *If* $p, \tilde{p} \in P_m$ *are the BAs of* f *and* $\tilde{f}$, *respectively,*

$$\| p-\tilde{p} \| \leqq \sqrt{2}\,\beta_1 \| f-\tilde{f} \|^{1/2} + 2\beta_2 \| f-\tilde{f} \| . \tag{2.4}$$

(Proof: $\| \tilde{f} - \tilde{p} \| \leqq \| \tilde{f} - p \| \leqq \| \tilde{f} - f \| + \| f - p \|$, hence $\| f - \tilde{p} \| - \| f - p \| \leqq 2 \| \tilde{f} - f \|$. Now, apply (2.2) with $u = \tilde{p}$.) It is easy to see that the exponent $1/2$ is best.

In *nonlinear approximation* Kolmogorov's criterion branches into a sufficient global condition and a necessary local condition. The former is due to Meinardus and Schwedt [20, Thm 86; 21, Satz 2]. It is easy to derive likewise conditions that ensure uniqueness or even strong uniqueness. In accordance with (2.1), a BA $r \in R_{mn}$ of f is called *strongly unique* if for some $\tilde{\gamma} > 0$

$$\| f - s \| \geqq \| f - r \| + \tilde{\gamma} \| s - r \| , \quad \forall s \in R_{mn}(Z) \tag{2.5}$$

A local best approximation (LBA) is called *(locally) strongly unique* if (2.5) holds only for those $s \in R_{mn}(Z)$ in some neighborhood of r.

THEOREM 2.2. Let $r \in R_{mn}(Z)$, $f \neq r$.

(i) If $\gamma(f - r; \{r\} - R_{mn}(Z)) \geqq 0$, *then* r *is a BA of* f *from* $R_{mn}(Z)$.

(ii) If

$$\mu(f-r , r-s ; M(f-r)) > 0 , \quad \forall s \in R_{mn}(Z), \quad s \neq r , \tag{2.6}$$

r *is the unique BA to* f *out of* $R_{mn}(Z)$.

(iii) If $\gamma(f - r ; \{r\} - R_{mn}(Z)) > 0$, *the BA* r *is strongly unique.*

(iv) If $\gamma(f - r ; \{r\} - U(r)) > 0$ *for some neighborhood* $U(r) \subset R_{mn}(Z)$ *of* r, r *is a strongly unique LBA.*

(There is a related inclusion theorem [20, Thm 85; 21, Satz 1] also.)

Proof of part (iii): Given $s \in R_{mn}(Z)$, choose $z \in M(f-r)$ such that $\mu_s := \mu(f-r, r-s; M(f-r)) = \mathrm{Re}\{\overline{(f-r)(z)}(r-s)(z)\}$. For this z

$$\begin{aligned} |f(z)-s(z)|^2 &= |f(z)-r(z)|^2 + 2\,\mathrm{Re}\{[\overline{f(z)-r(z)}][r(z)-s(z)]\} \\ &\quad + |r(z)-s(z)|^2 \\ &\geqq \|f-r\|^2 + 2\mu_s + (\mu_s/\|f-r\|)^2 \\ &\geqq [\|f-r\| + \tilde{\gamma}\|r-s\|]^2 \end{aligned}$$

if we let $\tilde{\gamma} := \gamma(f-r;\{r\} - R_{mn}(Z))$ and note that $\mu_s/\|f-r\| \geqq \tilde{\gamma}\|r-s\|$. □

Unfortunately, checking the conditions of Theorem 2.2 is seldom possible in practice (see §§4 and 5 for two exceptions.) In contrast, the local Kolmogorov condition is very useful both for recognizing and constructing stationary points of the error functional (see [16] and Refs. given there). In its original form it is restricted to Fréchet differentiable families of approximants, but, unfortunately, $R_{mn}(Z)$ is not Fréchet differentiable at any *degenerate* r, i.e. any $r \in R_{m-1,n-1}$. If $r = p/q$ is non-degenerate, $qP_m - pP_n = P_{m-n}$, and

$$T_r := \frac{q\,P_m - p\,P_n}{q^2} = \frac{P_{m+n}}{q^2} \tag{2.7}$$

is the tangent space of R_{mn} at r. Then the local Kolmogorov condition [21, Satz 8] and the related criterion for local strong uniqueness (given by Wulbert[36] for real families) apply, cf.[15]:

THEOREM 2.3. (i) If $r \in R_{mn}(Z) \setminus R_{m-1,n-1}$ *is an LBA of* $f \neq r$, $\gamma(f-r; T_r) \geqq 0$.

(ii) $r \in R_{mn}(Z) \setminus R_{m-1,n-1}$ *is a strongly unique LBA of* f *iff* $\gamma(f-r; T_r) > 0$ *or* $f = r$.

According to Braess [8, Lemma 2.1] one can replace T_r in (i) by the *tangent cone* C_r if r is degenerate. Let $r = p/q \in R_{m-d,n-d}$, $u \in P_m$, $v \in P_n$ such that $qu \neq pv$, $w \in P_d$ with $w(z) \neq 0$ on Z, $t \in \mathbb{R}$, and

$$s_t := \frac{pw + tu}{qw + tv}, \qquad s'_o := \frac{qu - pv}{q^2 w} .$$

Then, as $t \to 0$, $s_t \to r$ and

$$\| s_t - r - s'_o t \| = \left\| s'_o t \frac{vt}{qw + vt} \right\| = o(t) .$$

Hence, $\pm s'_o$ are tangential directions and belong to C_r. Since part (i) of Theorem 2.3 holds for C_r, it holds a fortiori for any subset of C_r; the same applies to one direction of part (ii) (note that $qP_m - pP_n = P_{m+n-d}$ if $r \notin R_{m-d-1,n-d-1}$):

THEOREM 2.4. (i) If $r \in R_{m-d,n-d} \setminus R_{m-d-1,n-d-1}$ *is an LBA of* f *from* R_{mn}, *then* $\gamma(f - r ; C_r^o) \geqq 0$, *where*

$$C_r^o := \left\{ \frac{\tilde{p}}{q^2 w} ; \tilde{p} \in P_{m+n-d}, w \in P_d, w(z) \neq 0 \text{ on } Z \right\}.$$

(ii) If, in addition, r *is (locally) strongly unique,* $\gamma(f - r ; C_r^o) > 0$.

This result extends those of Williams [35], who only considered directions with $w(z) \equiv 1$. (Note that his definition of an LBA is different from ours.)

The fact that $\gamma(f - r ; V)$ only depends on $\arg(f - r)$ and $M(f - r)$, but not on $\| f - r \|$ or on f itself, made Rivlin and Shapiro [24] introduce extremal signatures as a generalization of Collatz's H-sets. However, as we will see in the next section, extremal signatures are in general not appropriate for the discussion of complex rational approximation. Nevertheless, we want to cite some results that are essentially results on the cardinality of certain extremal signatures. We assume that $r \in R_{mn}$ is a non-degenerate LBA [strongly unique LBA] to $f \notin R_{mn}$ and define a *primitive (local)[strongly] extremal point set* to be any subset $Z_o \subset M(f-r)$ with the property that r is a [strongly unique] LBA to f on Z_o, but not on any proper subset of Z_o. Results summarized in [15], when put together with Theorem 2.3 and the fact that T_r in (2.7) is an $(m+n+1)$-dimensional complex Haar subspace, yield

THEOREM 2.5. *Let* $r \in R_{mn}(Z) \setminus R_{m-1,n-1}$ *be an LBA to* $f \notin R_{mn}$. *Then* $m+n+2 \leqq |Z_o| \leqq 2m+2n+3$ *for every primitive extremal point set* Z_o, *and* $|Z_o| \leqq 2m+2n+2$ *if* r *is not a strongly unique LBA. If* r *is a strongly unique LBA, then* $2m+2n+3 \leqq |Z_o'| \leqq 4(m+n+1)$ *for every primitive strongly extremal point set* Z_o'. *Moreover, all these bounds are in general best possible.*

($|Z_o|$ denotes the cardinality of Z_o.)

Note that if $|M(f-r)| \leqq 2m+2n+2$ or even $|Z| \leqq 2m+2n+2$, r cannot be strongly unique unless $r = f$. (This makes a result of Rivlin [23, p. 84f] easily understandable.)

3. SOME EXAMPLES OF MISBEHAVIOR

Walsh's example of non-uniqueness [32; 33, §12.4] had a flavor of peculiarity: He considered $f(z) = z + 1/z$ in a crescent-shaped region symmetric about the unit circle and showed that there must be at least two BAs $r_1^*, r_2^* \in R_{11}$ related by $r_2^*(z) = r_1^*(1/z)$. It was a surprise when Saff and Varga (re)discovered that non-uniqueness holds whenever in certain real rational approximation problems with degenerate BA complex rational functions are allowed to compete ([28,29], see also their reference to previous work of Gončar and Lungu).

For $Z \subset \mathbb{R}$ let us denote in this section by $C'(Z)$ the subset of real-valued functions in $C(Z)$, let $\mathcal{R}'_{mn}(Z) := \mathcal{R}_{mn} \cap C'(Z)$, and define $\mathcal{R}'_{mn}$ likewise. Let $I \subset \mathbb{R}$ be an interval. Saff and Varga [29, Prop. 2] proved: *If* $r \in \mathcal{R}'_{nn}(I) \setminus \mathcal{R}_{n-1,n-1}$ *is the BA to* $f \in C'(I)$ *out of* $\mathcal{R}'_{n+k,n+k}$, *where* $k \geqq n+1$, *but not best out of* $\mathcal{R}'_{2k,2k}$, *then* r *is not best out of* $\mathcal{R}_{n+k,n+k}$; *hence, if* r^* *is a BA out of* $\mathcal{R}_{n+k,n+k}$, $\overline{r^*}$ *is another one.* For example, this occurs if an even nonconstant $f \in C'[-1,1]$ that is monotone on $[0,1]$ is approximated out of $\mathcal{R}_{11}$ [29]. Bos and Williams [7, Thm 7] noticed that then r^* and $\overline{r^*}$ do not satisfy the condition of Meinardus and Schwedt (Thm 2.2 (i) above) if $\operatorname{Im} r^*(z) \neq 0$ on $M(f - r^*)$. On the other hand, Saff and Varga [29, Cor. 1] showed that $r \in \mathcal{R}'_{nn}(I) \setminus \mathcal{R}_{n-1,n-1}$ *which is best in* $\mathcal{R}'_{2n+2k,2n+2k}$ $(k \geqq 0)$ *is also best in* $\mathcal{R}_{n+k,n+k}$. Ruttan [26] presented an example of an even $f \in C'[-1,1]$ with a continuum of BAs. A simpler example with a continuum of BAs on a discrete set of points is due to Williams [35, Ex. 2]. The question by how much the error of the complex BA from $\mathcal{R}_{nn}$ is at most smaller than the one of the real BA is in general still open [29, 5, 13]. However, in a forthcoming paper Trefethen and the author show that for $\mathcal{R}_{mn}$ the infimum of the quotient of these two errors is zero if $n \geqq m+3$. In [39] we present examples of analytic functions whose BA on the *disk* D is not unique. The number of distinct BAs can be arbitrarily large, and, again, the BA to a real analytic function can be better than the best real rational approximation, which, on the other hand, is in general not unique.

In most published examples of non-uniqueness on a real interval the BA is degenerate [5,7,26,28,29,35]. In [15, Ex. 2] we had a discrete non-degenerate example. Here is one on [-1,1]:

Example: Let $f(z) = 1 + e(z)$, where $e(-1) = e(1) = -\eta$, $e(1/2) = \eta \neq 0$, and $e(z)$ is piecewise linear in between. $r^*(z) :\equiv 1$ is the BA out of $\mathcal{R}'_{01}[-1,1]$. Consider approximations of the form

$$r(z) := \frac{1+i\alpha}{1+i\beta z}$$

with α,β real. For $\alpha,\beta \to 0$,

$$f(z)-r(z) = e(z)+r^*(z)-r(z)$$
$$= e(z)-i\alpha+i\beta z-\alpha\beta z+\beta^2 z^2+O(\alpha\beta^2)+O(\beta^3) ,$$

$$|f(z)-r(z)|^2 \sim |e(z)|^2+2e(z)\beta z(\beta z-\alpha)+(\beta z-\alpha)^2 .$$

For $z \in M(f-r^*) = \{-1,1/2,1\}$ and $\eta = 1$, $\beta = 3\alpha/2$ straightforward calculations yield

$$|f(z)-r(z)|^2 \leqq 1-\frac{5}{16}\alpha^2+O(\alpha^3) , \forall z \in M(f-r^*) ,$$

and with, say, 5/16 replaced by 1/4 this holds in a neighborhood of $M(f-r^*)$, so that by standard arguments $\|f-r\| < \|f-r^*\|$ for sufficiently small $|\alpha|$. Hence r^* is not best in R_{01} and consequently, by symmetry, there are at least two BAs in R_{01}.

In the above example r^* is a *saddle point*, i.e. a point with $\gamma(f-r^*, C_{r^*}) \geqq 0$ (a so-called *critical point* [8]) which is not an LBA. This follows from the following extension of Theorem 1 in [15]:

THEOREM 3.1. Assume $Z \subset \mathbb{R}$, $f \in C'(Z)$, *and that* $r \in R'_{mn}(Z) \setminus R_{m-1,n-1}$ *is BA to* f *out of* $R'_{mn}(Z)$. *Then* $\gamma(f-r, C_r) = \gamma(f-r, T_r) = 0$, *i.e.* r^* *is a critical point which is not a strongly unique LBA.*

Remarks. (i) Note that a fortiori r is not a strongly unique LBA with respect to any $R_{\tilde{m}\tilde{n}}$ with $\tilde{m} \geqq m$, $\tilde{n} \geqq n$. However, even if r is best out of $R'_{\tilde{m}\tilde{n}}(Z)$, we do not claim that it is a critical point in $\underset{\sim}{R}_{\underset{\sim}{m}\underset{\sim}{n}}$. Counterexamples are given in [39] and [27, Thm 2.1].

(ii) Theorem 3.1 is easily generalized to similar nonlinear problems studied by Dunham [11].

Proof. Since r is not degenerate $C_r = T_r$. If $v \in T_r$, $\mathrm{Re}\, v \in T'_r$, where T'_r is the tangent space of $R'_{mn}(Z)$, cf. (2.7). Hence,

$$\gamma(f-r,T_r) = \frac{1}{\| f-r \|} \inf_{\substack{v\in T_r \\ v\neq 0}} \max_{z\in M(f-r)} \left\{ \frac{f(z)-r(z)}{\| v \|} \mathrm{Re}\, v(z) \right\} \geqq 0 ,$$

where we have used that $\gamma(f-r, T'_r) \geqq 0$ since r is best in $R'_{mn}(Z)$. (Even $\gamma(f-r, T'_r) > 0$, see Cheney [9, p. 165].) For $v := i/q^2 \in T_r$ obviously $\mathrm{Re}\, v(z) \equiv 0$, so that actually $\gamma(f-r, T_r) = 0$. □

In the situation of Theorem 3.1 there are of course also cases where r is best in R_{mn} , see, e.g., [7, Thm 6]. Hence, one cannot decide by linearization (i.e. by replacing R_{mn} by $\{r\}+T_r$) whether r is locally best or a saddle point. Wulbert [38] and Ruttan [27] investigated the situation thoroughly. Wulbert, assuming r non-degenerate, replaced T_r by a non-linear approximation V_r of R_{mn} at $r \in R'_{mn}(I)$ with the property that r is an LBA to $f \in C'(I)$ if and only if 0 is a BA to $f-r$ from V_r [38, Thm 10]. Here are some of his interesting results; note that $\lambda f + (1-\lambda)r - r = \lambda(f-r)$:

THEOREM 3.2 [*38, Prop. 12, Thm 14, Prop. 23*].
(i) Assume $f \in C'(I)$ *and* $r \in R'_{mn}(I) \setminus R_{m-1,n-1}$. *Then* r *is a BA to* f *from* $R'_{mn}(I)$ *if and only if for all sufficiently small* $\lambda > 0$, r *is the unique BA to* $\lambda f + (1-\lambda)r$ *from* R_{mn} .

(ii) For any $m,n > 0$ *there is* $f \in C'(I)$ *and* $r \in \mathcal{R}'_{mn}(I) \setminus \mathcal{R}_{m-1,n-1}$ *such that* r *is the unique global BA of* f *out of* $\mathcal{R}_{mn}$*, but for sufficiently large* λ, r *is not an LBA to* $\lambda f + (1-\lambda) r$.

(iii) For any $n > 0$ *there is* $f \in C'(I)$ *and* $r \in \mathcal{R}'_{nn}(I)$ *such that* r *is the unique global BA of* f *out of* $\mathcal{R}_{nn}$*, but* $-r$ *is not an LBA of* $f - 2r$ *(though the error function is the same).*

Part (i) might be illustrated by choosing η small enough in our example; then $r^*(z)$ becomes best. Part (ii) makes clear that there is no chance to characterize BAs by extremal signatures (introduced in [24]). By (iii) there cannot be a characterization of BAs or LBAs based on T_r (or C_r) and $f - r$ only [38]. Statement (iii) holds likewise for approximation from $\mathrm{Re}\,\mathcal{R}_{mn}$ [37].

Ruttan [27] extended some of these results to degenerate approximants and added new ones. In particular, he also improved the theorem of Saff and Varga stated above: *If* $r \in \mathcal{R}'_{mn}(I) \setminus \mathcal{R}_{m-1,n-1}$ *is best out of* $\mathcal{R}'_{m+k,n+k}$*, where* $k \geq 1$*, but not best out of* $\mathcal{R}'_{m+2k,n+2k}$*, then it is not locally best in* $\mathcal{R}_{m+k,n+k}$ [27, *Thm 2.1*].

4. SUFFICIENT CONDITIONS FOR BEST RATIONAL APPROXIMATIONS: ALTERNATION

In this and the next section we state some simple sufficient conditions assuming that the error curve is either circular with sufficiently large winding number or alternates between $\pm$ const a sufficient number of times. Less general conditions of this type have been given, e.g., by Klotz [18] and Trefethen [30,31]. Most examples of explicitly known or algebraically computable BAs in complex *and* real rational approximation can be understood on this basis, cf. Part II, §5, and [17].

In this section approximation on the unit circle ∂D is considered only. In view of the just mentioned real examples we choose approximants out of the space

$$z^{\ell}R_{mn}(\partial D) := \{z \mapsto z^{\ell}r(z) \,;\, r \in R_{mn}(\partial D)\} \,, \tag{4.1}$$

where $\ell \in \mathbb{Z}$.

We say that $f \in C(\partial D)$ *alternates* on $A := \{e^{i\omega_j}\}_{j=1}^{2\kappa} \subset \partial D$ if f takes the values $\pm \|f\|$ alternatively at the 2κ points $e^{i\omega_j}$, $\omega_1 < \omega_2 < \ldots < \omega_{2\kappa} < \omega_1 + 2\pi$. A is called a *set of alternation points*, and 2κ is its *length*.

The following result extends Thm. 3 of Klotz [18]. Our proof, which yields uniqueness also, is nevertheless much simpler.

THEOREM 4.1. Suppose $f \in C(\partial D)$, $r \in z^{\ell}R_{mn}(\partial D)$ *and* $w_o \in \partial D$ *are such that* $w_o(f-r)$ *has a set* A *of alternation points whose length is* 2κ, *where*

$$\kappa := \max\{m+n+\ell,\ 2n-\ell\} + k + 1 \tag{4.2}$$

with some $k \geqq 0$. *Then* r *is the BA to* f *on* D *out of* $z^{\ell}R_{m+k,n+k}(\partial D)$, *and the BA is unique.*

Proof. We may assume $f \neq r$. According to Theorem 2.2 (ii) adapted to the space (4.1) we have to show that for no $s \in z^{\ell}R_{m+k,n+k}(\partial D)$, $s \neq r$, does one have

$$\mu(f-r\,,\,r-s\,;\,A) \leqq 0 \tag{4.3}$$

(Note that $A \subset M(f-r)$, and thus $\mu(f-r\,,\,r-s\,;\,M(f-r)) \geqq \mu(f-r\,,\,r-s\,;\,A)$.) We denote the elements of A as above, and assume a negative error $w_o(f-r)$ at $e^{i\omega_1}$. Then (4.3) may be written

$$\max_{1 \leq j \leq 2\kappa} (-1)^j \operatorname{Re} \hat{s}(e^{i\omega_j}) \leq 0 , \tag{4.4}$$

where $\hat{s}(z) := w_o[r(z) - s(z)]$, i.e. $\hat{s} = \hat{p}/\hat{q}$ with $\hat{p} \in z^{\ell} P_{m+n+k}$, $\hat{q} \in P_{2n+k}$, $\hat{q}(z) \neq 0$ on ∂D .

We assume to the contrary that there is $s \neq r$ and hence $\hat{s} \neq 0$ such that (4.3) and (4.4) hold for s and $\hat{s}$, respectively. Now,

$$\begin{aligned} \operatorname{Re} \hat{s}(e^{i\omega}) &= \frac{1}{2} \frac{\hat{p}(e^{i\omega})\overline{\hat{q}(e^{i\omega})} + \overline{\hat{p}(e^{i\omega})}\hat{q}(e^{i\omega})}{|\hat{q}(e^{i\omega})|^2} \\ &= \frac{\alpha(\omega)}{|\hat{q}(e^{i\omega})|^2} , \end{aligned} \tag{4.5}$$

where α is a real trigonometric polynomial of degree at most $\kappa - 1$. According to (4.4), α has at least 2κ zeros (multiplicity included) in $[\omega_1, \omega_1 + 2\pi)$, hence $\alpha(\omega) \equiv 0$. Therefore,

$$\begin{aligned} \hat{s}(e^{i\omega}) &= i \operatorname{Im} \hat{s}(e^{i\omega}) \\ &= \frac{1}{2} \frac{\hat{p}(e^{i\omega})\overline{\hat{q}(e^{i\omega})} - \overline{\hat{p}(e^{i\omega})}q(e^{i\omega})}{|\hat{q}(e^{i\omega})|^2} \\ &= \frac{i\beta(\omega)}{|\hat{q}(e^{i\omega})|^2} , \end{aligned} \tag{4.6}$$

where β is also a real trigonometric polynomial of degree at most $\kappa - 1$. Now,

$$\begin{aligned} \| f - s \| &\geq \max_{1 \leq j \leq 2\kappa} |w_o(f - s)(e^{i\omega_j})| \\ &= \max_{1 \leq j \leq 2\kappa} |(w_o f - w_o r + \hat{s})(e^{i\omega_j})| , \end{aligned}$$

and since $w_o(f - r)(e^{i\omega_j}) = \pm \| f - r \|$ and $\hat{s}(e^{i\omega_j}) \in i\mathbb{R}$, obviously $\|f - s\| > \|f - r\|$ unless $\hat{s}(e^{i\omega_j}) = 0 \ \forall j$, i.e. $\beta(\omega_j) = 0 \ \forall j$, which implies $\beta(\omega) \equiv 0$, $\hat{s} = 0$, $s = r$. □

Remark. If $r = 0$, Theorem 4.1 holds with $\kappa := \max\{m+\ell, n-\ell\}+k+1$, and one may choose $\ell = m = n = 0$, so that $\kappa = k+1$.

For certain values of k, ℓ, m, and n we can easily see from Theorem 2.4 (ii) that strong uniqueness does not hold if $A = M(f-r)$:

THEOREM 4.2. Under the assumptions of Theorem 4.1, if $M(f-r) = A$ *and either* $0 \leqq \ell \leqq k$ *and* $m+k \geqq n$ *or* $\ell < 0$ *and* $m+k \geqq n-\ell$ *, then* r *is not locally strongly unique with respect to* $z^\ell R_{m+k,n+k}$ *. If* $\ell \leqq 0$ *and* $m \geqq n-\ell$ *,* r *is not even locally strongly unique in* $z^\ell R_{mn}$ *.*

Proof. Adapted to the space (4.1) and our notation here, the cone $C_r^o \subset C_r$ in Theorem 2.4 consists of the functions

$$z \mapsto z^\ell \frac{\tilde{p}(z)}{w(z)[q(z)]^2}$$

with $\tilde{p} \in P_{m+n+k}$, $w \in P_k$, $w(z) \neq 0$ on ∂D if $r \notin R_{m-1,n-1}$. By choosing $\tilde{p}(z) = i w_o q^2(z)$, $w(z) = z^\ell$ if $\ell \geqq 0$ and $\tilde{p}(z) = i w_o z^{-\ell} q^2(z)$, $w(z) \equiv 1$ if $\ell < 0$, we get $z \mapsto \hat{s}(z) :\equiv i\bar{w}_o \in C_r^o$,

$$\mu(f-r, \hat{s}; M(f-r)) = \pm \max_{1 \leqq j \leqq 2\kappa} (-1)^j \mathrm{Re}\{w_o \| f-r \| i\bar{w}_o\} = 0 .$$

Hence $\gamma(f-r; C_r^o) \leqq 0$, and, by Theorem 2.4 (ii), r cannot be locally strongly unique. (The case $\ell \leqq 0$, $m \geqq n-\ell$ could also be settled by applying the general form [15, p. 210] of one of the results in Theorem 2.5.) Finally, if we restrict approximations to $z^\ell R_{mn}$, $\tilde{p} \in P_{m+n}$ and $w(z) \equiv 1$ in (4.8), but $\hat{s}(z) \equiv i\bar{w}_o$ is still admissable if $\ell \leqq 0$ and $m \geqq n-\ell$. The assertions remain true if r is degenerate in R_{mn}. □

Remark. Some of Al'per's examples [3,25] of functions whose partial sums of their Maclaurin series are polynomial best approximations are of the type discussed. According to Theorem 4.2 they are not strongly unique in contrast to a remark at the end of [23].

5. SUFFICIENT CONDITIONS FOR BEST RATIONAL APPROXIMATIONS: CIRCULARITY

The following result ([30, Thm 1] and, for D, [18, Thm 4]) is a simple application of Rouché's theorem:

THEOREM 5.1. Let Ω *be a Jordan region,* $f \in A(\overline{\Omega})$, $r \in R_{mn}(\overline{\Omega})$, *and suppose the error curve* $(f-r)(\partial\Omega)$ *is a circle about* 0 *with winding number at least* $m+n+1$. *Then* r *is a BA to* f *out of* $R_{mn}(\overline{\Omega})$.

If approximation on $\partial\Omega$ is considered only, it suffices to assume $f \in C(\partial\Omega)$ since only the topological aspect of Rouché's theorem actually matters (cf. Rivlin [23, p. 80] who states this theorem for P_m). Moreover, Trefethen [31, Lemma 2.3] noticed that the result remains true (on ∂D) for certain extended rational functions. We first consider a corresponding extension to Jordan curves. (One could go further and consider boundaries of simply connected regions with compact complement and bounded boundary rotation.)

Notation: Let Ω be a Jordan region in the ζ-plane with positively oriented rectifiable boundary $\partial\Omega$. Let Ω^c be the complement of Ω with respect to the extended plane, and let $\psi \in A(\Omega^c)$ be the homeomorphism of D^c onto Ω^c that is conformal in $\overline{D}^c$ and normalized by $\psi(\infty) = \infty$, $\psi'(\infty) > 0$. The inverse map is called ϕ . For any $f \in C(\partial\Omega)$ with $0 \notin f(\partial\Omega)$

the winding number with respect to 0 is denoted by $\iota(f)$. Furthermore, for $m, n \in \mathbb{Z}$, $n \geqq 0$, we define $\tilde{R}_{mn}(\bar{\Omega})$ to be the space of functions that are meromorphic in Ω^c, have a total of $\nu \leqq n$ poles in $\Omega^c \setminus \{\infty\}$, are at most of order $O(\zeta^{m-\nu})$ at ∞, and are continuous in some neighborhood of $\partial\Omega$ relative to Ω^c. (In order to avoid technical difficulties we require this continuity here; if one wanted to approximate functions $f \in L_\infty(\partial\Omega)$ that are not continuous, one would have to replace it by boundedness as in the theory of Adamjan, Arov, and Krein [1] for ∂D. One still has non-tangential limits nearly everywhere on $\partial\Omega$ and a number of other properties, cf. Duren [11, Ch. 10].)

Under our assumptions, if $\zeta_o \in \Omega$ and if we let

$$H(\Omega^c) := \{h \in A(\Omega^c);\ h(\infty) = 0\},$$

we have

$$\tilde{R}_{mn}(\bar{\Omega}) = (\zeta - \zeta_o)^{m-n+1}\left(R_{n-1,n}(\bar{\Omega}) + H(\Omega^c)\right), \tag{5.1}$$

i.e. $r \in \tilde{R}_{mn}(\bar{\Omega})$ is of the form

$$\tilde{r}(\zeta) = (\zeta - \zeta_o)^{m-n+1}\left(r(\zeta) + h(\zeta)\right), \tag{5.1'}$$

with $r \in R_{n-1,n}(\bar{\Omega})$ and $h \in H(\Omega^c)$. The error of the BA of some $f \in C(\partial\Omega)$ out of $\tilde{R}_{mn}(\bar{\Omega})$ is denoted by $\tilde{E}_{mn}(f;\bar{\Omega})$.

Here is our version of Trefethen's result [31, Lemma 2.3]; its usefulness will become apparent in Part II:

THEOREM 5.2. Let $f \in C(\partial\Omega)$ *and* $\tilde{r} \in \tilde{R}_{mn}(\bar{\Omega})$ *be such that* $0 \notin (f - \tilde{r})(\partial\Omega)$ *and* $\iota(f - \tilde{r}) = m + n + k + 1$ *for some* $k \geqq 0$. *Then*

$$\min_{\zeta \in \partial\Omega} |f(\zeta) - \tilde{r}(\zeta)| \leqq \tilde{E}_{m+k,n+k}(f;\bar{\Omega}) \leqq \|f - \tilde{r}\|.$$

In particular, if the error curve is a circle about the origin, $\tilde{r}$ *is a BA of* f *out of* $\tilde{R}_{m+k,n+k}(\bar{\Omega})$. *If* $\tilde{r} = 0$, *the assertions hold for arbitrary* n *with* $k := \iota(f - \tilde{r}) - m - 1$, if $k \geqq 0$.

Proof. Suppose to the contrary that for some $\tilde{s} \in \tilde{R}_{m+k,n+k}(\bar{\Omega})$

$$\|f - \tilde{s}\| < \min_{\zeta \in \partial\Omega} |f(\zeta) - \tilde{r}(\zeta)|.$$

Then $\iota(f - \tilde{r}) = \iota(\tilde{s} - \tilde{r})$ [2, p. 464]. Since $\tilde{s} - \tilde{r} \in \tilde{R}_{m+n+k,2n+k}(\bar{\Omega})$, $\tilde{s} - \tilde{r}$ has at most $2n + k + (m + n + k - 2n - k) = m + n + k$ poles in $\bar{\Omega}^c$, so that $\iota(\tilde{s} - \tilde{r}) \leqq m + n + k$ in contrast to our assumption on $\iota(f - \tilde{r})$. The case $\tilde{r} = 0$ is treated the same way. □

Let us state another, fairly trivial generalization of Theorem 5.1 that is nevertheless sufficient to understand and extend several published examples of best approximations (cf. [17]). It is proved the same way as Theorem 5.2, or it is readily derived from the latter. The case $r = 0$ is left to the reader.

THEOREM 5.3. Let $f \in C(\partial\Omega)$ *and* $r \in (\zeta - \zeta_o)^{\ell} R_{mn}(\partial\Omega)$ *be such that* $(f - r)(\partial\Omega)$ *is a circle about the origin and either* $\iota(f - r) = m + n + \ell + k + 1$ *or* $\iota(f - r) = -2n + \ell - k - 1$ *for some* $k \geqq 0$. *Then* r *is on* $\partial\Omega$ *a BA to* f *out of* $(\zeta - \zeta_o)^{\ell} R_{m+k,n+k}$.

In $R_{mn}(\bar{D})$ circularity is also easily seen to be sufficient for uniqueness and local strong uniqueness if $f \in \tilde{R}_{MN}(\bar{D})$. In order to prove this we need generalizations of two further results due to Trefethen [30, Thm 2 and Lemma 5].

LEMMA 5.4. Let $s = u/v \in R_{mn}(\partial\Omega) \setminus R_{m,n-1}$ *and* $k \geqq 0$. *Assume that either* s *has a zero* $\zeta' \in \bar{\Omega}^c \setminus \{\infty\}$ *or* $m + k < n$. *Then, for any* $p \in P_k$ *the image* $ps(\Omega^c)$ *covers every point of a disk of radius* $\alpha\|ps\|$ *about* 0, *where* $\alpha > 0$ *depends on* Ω, s, *and* k *only, but not on* p.

Proof. Assume, say, $1 \notin ps(\Omega^c)$. Then $(1 - pu/v)(\zeta) \neq 0$ in Ω^c. Assume v monic, and u, v mutually prime. Let $g :=$ g.c.d.(p, v), g monic. Then $h := (v - pu)/g \in P_\mu$, where $\mu \leqq \max\{n, m+k\}$. Also, $h(\zeta) \neq 0$ in Ω^c since the zeros of v/g are poles of s and neither zeros of u nor zeros of p/g.

If $\zeta' \in \overline{\Omega}^c$ is a finite zero of s, $u(\zeta') = 0$ and $h(\zeta') = v(\zeta')/g(\zeta')$. Consequently, h is of the form

$$h(\zeta) = \frac{v(\zeta')}{g(\zeta')} \prod_{j=1}^{\mu} \frac{\zeta - \zeta_j}{\zeta' - \zeta_j} \tag{5.2}$$

with $\zeta_j \in \Omega$. Now, let $\alpha_1 := \max\{|\zeta|;\ \zeta \in \partial\Omega\}$, $\alpha_2 := \text{dist}(\zeta', \partial\Omega)$, and, if w ranges among all monic divisors of v, let $\alpha_3 := \max |w(\zeta')|$, $\alpha_4 := \min\{\min |w(\zeta)|;\ \zeta \in \partial\Omega\}$. Note that $0 < \alpha_j < \infty$, $j = 1,\ldots,4$. By (5.2), $\|h\| \leqq \alpha_3(2\alpha_1/\alpha_2)^\mu$ (on $\partial\Omega$), and, since $ps = 1 - (g/v)h$, we get

$$\|ps\| \leqq \alpha_5 := 1 + (\alpha_3/\alpha_4)(2\alpha_1/\alpha_2)^\mu .$$

Thus the lemma holds with $\alpha := 1/\alpha_5$ in this case.

If $m + k < n$, h is monic and thus

$$h(\zeta) = \prod_{j=1}^{\mu} (\zeta - \zeta_j)$$

with $\zeta_j \in \Omega$. Hence, $\|ps\| \leqq 1 + (2\alpha_1)^\mu/\alpha_4$. □

THEOREM 5.5. Let $\Omega := D$ *be the unit disk and* $\zeta_0 := 0$, $z := \zeta - \zeta_0$ *in Theorem 5.3, and assume that* f *or* $z \mapsto f(1/z)$ *lies in* $\tilde{R}_{MN}(\overline{D})$ *for some* M *and* N. *Then* f *is rational,* r *is the unique BA of* f *out of* $z^\ell R_{m+k,n+k}(\partial D)$, *and* r *is locally strongly unique in* $z^\ell R_{mn}(\partial D)$ *if* $r \notin z^\ell R_{m-1,n-1}$. *(In particular, if* $n = 0$, r *is (globally) strongly unique in* P_m.*)*

Proof. By the reflection principle, $f - r$ can be extended to a function meromorphic in $\overline{\mathbb{C}}$, hence $f - r$ and f are rational, cf. [30, Thm 2]. Since $(f - r)/\|f - r\|$ is unimodular on ∂D, it is of the form $(f - r)(z) = \eta z^j w(z)/w^*(z)$, where w^* denotes the reciprocal polynomial of w. On ∂D, $\overline{(f - r)(z)} = \overline{\eta} z^{-j} w^*(z)/w(z)$.

Let $s = u/v \in z^\ell R_{m+k,n+k}(\partial D)$ be arbitrary, with $v \notin P_{n+k-1}$ (u, v not necessarily prime), let $r = p/q$, and assume $\iota(f-r) = m+n+k+\ell+1$ first. Then w^* has at least $m+n+k+\ell+1-j$ zeros in $\overline{D}^c$, $\tilde{s} := \bar{\eta} z^{\ell-j} w^*/(wqv)$ has at least $\nu := m-n+\ell+1-j$ of them, and $\tilde{s}\tilde{p} = O(z^{\nu-1})$ as $z \to \infty$ for any $\tilde{p} \in P_{\tilde{k}}$, $\tilde{k} := m+n+k$. Consequently, we may apply Lemma 5.4 to $\tilde{s}$ and $\tilde{k}$ (if $\nu > 0$, ζ' exists, otherwise the alternative assumption holds); it follows that $\overline{(f-r)(z)}(r-s)(z) > 0$ for some $z \in \partial D$, i.e. (2.6) holds, hence r is unique.

Similarly, if we apply Lemma 5.4 to $\tilde{s} := \eta z^{\ell-j} w^*/(wq^2)$ and $\tilde{k} := m+n$, it follows that $\max \mathrm{Re}\{\tilde{s}(z)\tilde{p}(z)\} \geqq \alpha \|\tilde{s}\tilde{p}\| = \alpha|\eta| \|\tilde{p}/q^2\|$, where α is independent of $\tilde{p}$. Since $T_r = \{\zeta^{\ell}\tilde{p}/q^2 ; \tilde{p} \in P_{m+n}\}$, this implies $\gamma(f-r, T_r) \geqq \alpha$ if $r \notin \zeta^{\ell} R_{m-1,n-1}$.

The case $\iota(f-r) = -2n+\ell-k-1$ can be treated by an inversion $z \mapsto 1/z$. □

Remark. It follows from Theorem 5.2 above and from Theorem 7.3 in Part II that the uniqueness statement in Theorem 5.5 still holds for $\Omega \neq D$. In fact, r is uniquely best either out of $\tilde{R}_{m+k+\ell,n+k}(\overline{\Omega})$ or out of a similar space obtained after the inversion $\zeta \mapsto 1/(\zeta-\zeta_0)$. In particular, r may be computed by the CF method (without truncation) of Part II.

5. REFERENCES

1. Adamjan, V.M., Arov, D.Z., Krein, M.G.: Analytic properties of Schmidt pairs for a Hankel operator and the generalized Schur-Takagi problem. Math. USSR Sbornik *15*, 31-73 (1971).
2. Alexandroff, P., Hopf, H.: Topologie. Chelsea Publ. Co., Bronx, N.Y., 1972.
3. Al'per, S.Ja.: Asymptotic values of best approximation of analytic functions in a complex domain (Russian). Uspehi Mat. Nauk *14*, 131-134 (1959).
4. Bartelt, M.W., McLaughlin, H.W.: Characterization of strong unicity in approximation theory. J. Approx. Theory *9*, 255-266 (1973).
5. Bennett, C., Rudnick, K., Vaaler, J.D.: Best uniform approximation by linear fractional transformations. J. Approx. Theory *25*, 204-224 (1979).
6. Blatt, H.-P.: On strong uniqueness in linear and complex Chebyshev approximation. Approximation Theory III (Ed. E.W. Cheney), Academic Press, New York, 1980; pp. 221-222.

7. Bos, L., Williams J.: Some examples of best uniform complex approximations. Numer. Anal. Report 32, Dept. of Mathematics, Univ. of Manchester, 1978.
8. Braess, D.: Kritische Punkte bei der nichtlinearen Tschebyscheff-Approximation. Math. Z. *132*, 327-341 (1973).
9. Cheney, E.W.: Introduction to Approximation Theory. McGraw-Hill, New York, 1966.
10. Cline, A.K.: Lipschitz conditions on uniform approximation operators. J. Approx. Theory *8*, 160-172 (1973).
11. Dunham, C.B.: Failure of complex strong uniqueness. Approximation Theory III (Ed. E.W. Cheney), Academic Press, New York, 1980; pp. 367-370.
12. Duren, W.L.: Theory of H^p spaces. Academic Press, New York, 1970.
13. Ellacott, S.W.: A note on a problem of Saff and Varga concerning the degree of complex approximation to real valued functions. Bull. (New Series) Amer. Math. Soc. *6*, 218-220 (1982).
14. Gaier, D.: Vorlesungen über Approximation im Komplexen. Birkhäuser, Basel, 1980.
15. Gutknecht, M.H.: Non-strong uniqueness in real and complex Chebyshev approximation. J. Approx. Theory *23*, 204-213 (1978).
16. Gutknecht, M.H.: Ein Abstiegsverfahren für nicht-diskrete Tschebyscheff-Approximationsprobleme. Numerische Methoden der Approximationstheorie, Bd. 4 (Hrsg. L. Collatz, G. Meinardus, H. Werner), Birkhäuser, Basel, 1978; pp. 155-171.
17. Gutknecht, M.H.: A survey of explicitly or algebraically solvable Chebyshev approximation problems. Forthcoming.
18. Klötz, V.: Gewisse rationale Tschebyscheff-Approximationen in der komplexen Ebene. J. Approx. Theory *19*, 51-60 (1977).
19. Korevaar, J.: Polynomial and rational approximation in the complex domain. Aspects of Contemporary Complex Analysis (Eds. D.A. Brannan, J.G. Clunie), Academic Press, 1980; pp. 251-292.
20. Meinardus, G.: Approximation of Functions: Theory and Numerical Methods. Springer, Berlin, 1967.
21. Meinardus, G., Schwedt, D.: Nicht-lineare Approximationen. Arch. Rat. Mech. Anal. *17*, 297-326 (1964).
22. Newman, D.J., Shapiro, H.S.: Some theorems on Čebyšev approximation. Duke Math. J. *30*, 673-681 (1963).
23. Rivlin, T.J.: Best uniform approximation by polynomials in the complex plane. Approximation Theory III (Ed. E.W. Cheney), Academic Press, New York, 1980; pp. 75-86.
24. Rivlin, T.J., Shapiro, H.S.: A unified approach to certain problems of approximation and minimization. J. SIAM *9*, 670-699 (1961).
25. Rivlin, T.J., Weiss, B.: Some best polynomial approximations in the plane. Duke Math. J. *35*, 475-482 (1968).
26. Ruttan, A.: On the cardinality of a set of best complex rational approximations to a real function. Padé and Rational

Approximation (Eds. E.B. Saff, R.S. Varga), Academic Press, New York, 1977; pp. 303-319.
27. Ruttan, A.: The length of the alternation set as a factor in determining when a best real rational approximation is also a best complex rational approximation. J. Approx. Theory *31*, 230-243 (1981).
28. Saff, E.B., Varga, R.S.: Nonuniqueness of best approximating complex rational functions. Bull. Amer. Math. Soc. *83*, 375-377 (1977).
29. Saff, E.B., Varga, R.S.: Nonuniqueness of best complex rational approximations to real functions on real intervals. J. Approx. Theory *23*, 78-85 (1978).
30. Trefethen, L.N.: Near-circularity of the error curve in complex Chebyshev approximation. J. Approx. Theory *31*, 344-367 (1981).
31. Trefethen, L.N.: Rational Chebyshev approximation on the unit disk. Numer. Math. *37*, 297-320 (1981).
32. Walsh, J.L.: On approximation to an analytic function by rational functions of best approximation. Math. Z. *38*, 163-176 (1934).
33. Walsh, J.L.: Interpolation and Approximation by Rational Functions in the Complex Domain. Amer. Math. Soc., Providence, R.I., 1935 (5th ed., 1969).
34. Walsh, J.L.: Approximation by rational functions: open problems. J. Approx. Theory *3*, 236-242 (1970).
35. Williams, J.: Characterization and computation of rational Chebyshev approximation in the complex plane. SIAM J. Numer. Anal. *16*, 819-827 (1979).
36. Wulbert, D.E.: Uniqueness and differential characterization of approximations from manifolds of functions. Amer. J. Math. *93*, 350-366 (1971).
37. Wulbert, D.E.: The rational approximation of real functions. Amer. J. Math. *100*, 1281-1315 (1978).
38. Wulbert, D.E.: On the characterization of complex rational approximations. Illinois J. Math. *24*, 140-155 (1980).
39. Gutknecht, M.H., Trefethen, L.N.: Nonuniqueness of rational Chebyshev approximations on the unit disk. Preprint, Aug. 1982.

ON COMPLEX RATIONAL APPROXIMATION
PART II: THE CARATHÉODORY-FEJÉR METHOD

Martin H. Gutknecht
ETH-Zentrum, CH-8092 Zürich

Abstract. We give a complete constructive treatment of the following extended rational Chebyshev approximation problem: A function that is meromorphic and except for a finite number of poles continuous on and outside the unit circle is to be approximated on the circle by a function of the same type with fewer poles. This problem can be solved via the singular value decomposition of a finite matrix. It is the key to the CF method for near-best Chebyshev approximation of rational functions analytic on a disk by functions of the same type but lower order. This method can also be applied to, for example, real rational approximation on an interval and complex rational approximation on Jordan regions.

1. INTRODUCTION

The Carathéodory-Fejér method (or, briefly, CF method) is both a numerical method for computing near-best polynomial or rational complex Chebyshev approximations on the unit disk D and a constructive tool for proving asymptotic results on the true best approximation (BA) r^*, in particular results on the error and the near-circularity of the error curve [20,36,37]. Initially, Trefethen [36] used the classical Carathéodory-Fejér theorem [5,28] (hence the name) for approximating a high degree poly-

H. Werner et al. (eds.), Computational Aspects of Complex Analysis, 103–132.

nomial (thought of as a computable representation of an analytic function) by a lower degree polynomial. Later he recognized that this was in fact a special case of a general theoretical method due to Adamjan, Arov, and Krein [3] for computing the BA of an arbitrary complex valued $f \in L_\infty(\partial D)$ by a function $\tilde{r}$ from a space $\tilde{R}_{mn}$ containing $R_{mn}(\bar{D})$. Here, as in Part I, $R_{mn}(\bar{D})$ denotes the space of complex rational functions with numerator degree at most m, denominator degree at most n, and no poles on $\bar{D}$. $\tilde{R}_{mn}$ is defined by

$$\tilde{R}_{mn} := z^{m-n+1}(R_{n-1,n}(\bar{D}) + H) \tag{1.1}$$

where H is the space of functions analytic and bounded in $\bar{D}^c = \{z ; 1 < |z| \leqq \infty\}$ and vanishing at ∞. Hence, $\tilde{r} \in \tilde{R}_{mn}$ can be written in the form

$$\tilde{r}(z) = \frac{1}{q(z)} \sum_{j=-\infty}^{m} \tilde{a}_j z^j , \tag{1.2}$$

where $q \in P_n$ has no zero on $\bar{D}$, and the series converges for $|z| > 1$ and is bounded on bounded subsets of $\bar{D}^c$. However, as in Part I, we restrict ourselves here to a smaller space by requiring that the functions are continuous on ∂D, so that $H := H(D^c)$ and $\tilde{R}_{mn} := \tilde{R}_{mn}(\bar{D})$ in the notation of §5 in Part I. (This restriction is not essential, however.)

Adamjan, Arov, and Krein showed that the mentioned Chebyshev approximation problem can be reduced to an algebraic problem, namely the computation of the singular value decomposition (or, the Schmidt series) of an infinite Hankel matrix [12]. If f is a polynomial, the Hankel matrix becomes finite and a full understanding of the approximation problem (in $\tilde{R}_{mn}$) can be achieved from Takagi's classical, but not widely known, extension of the Carathéodory-Fejér theorem [31,32]; in particular, we have shown that degenerate BAs always occur in square blocks in the CF table [15]. Previously, Trefethen [37] was able to generalize most of his asymptotic results of [36] to rational approximation on D, and in his experiments the rational CF approximation r^{cf}, obtained by projecting $\tilde{r} \in \tilde{R}_{mn}$ into $R_{mn}(\bar{D})$, turned out to be exceedingly close to the BA if f had a sufficiently rapidly converging Maclaurin series. Often the method is still very efficient if f is far from having this property, as, e.g., in digital filter approximation problems [18]. To motivate our interest in the CF method we cite Trefethen's main result:

THEOREM 1.1 [37, *Thms.* 6.2, 6.3]. *Let* f *be analytic at* 0, *and assume the (m,n)-Padé approximant* r^p *of* f *has* n *finite poles, and its Maclaurin series agrees with* f *exactly through the term of degree* $m+n$. *Let* $\|\cdot\|_\varepsilon$ *be the supremum norm on the disk* $D_\varepsilon := \{z\,;\,|z| \leqq \varepsilon\}$. *Then, as* $\varepsilon \downarrow 0$,

$$\| r^{cf} - r^* \|_\varepsilon = \| f - r^* \|_\varepsilon O(\varepsilon^{m+n+2}) = O(\varepsilon^{2m+2n+3}), \quad (1.3)$$

and for both $r := r^{cf}$ *and* $r := r^*$

$$\begin{aligned} \| f - r \|_\varepsilon - \min_{|z|=\varepsilon} |(f-r)(z)| &= \| f - r \|_\varepsilon O(\varepsilon^{m+n+2}) \\ &= O(\varepsilon^{2m+2n+3}) \ . \end{aligned} \quad (1.4)$$

(In contrast to (1.3), $\| r^p - r^* \|_\varepsilon = \| f - r^* \|_\varepsilon O(\varepsilon) = O(\varepsilon^{m+n+2})$ for the (m,n)-Padé fraction.)

Here we should mention that there are various definitions for r^{cf} depending on the truncation of $\tilde{r}$ (i.e. the projection of $\tilde{r}$ into $R_{mn}(\bar{D})$) [15,37]. Theorem 2.1 was originally proved for the simplest projection defined by deletion of all negative powers in the series in (1.2).

There are versions of the CF method for approximations on other domains than the disk. On one hand, by applying a splitting technique similar to the one used by Gragg and others [13,14] for the definition of Laurent-Padé fractions, we obtained a method for rational approximation on the unit circle (Laurent-CF), and, as a special case, one for ordinary real rational approximation on an interval (Chebyshev-CF) [15,19,38]. For the latter case, Trefethen and the author [19,38] presented asymptotic results for small intervals showing that Chebyshev-CF approximation is far superior to Chebyshev-Padé approximation. However, we also became aware that Chebyshev-CF approximation is not entirely new. Special cases of it and similar less accurate methods appear in the work of, e.g., Bernstein, Achieser, Talbot, Clenshaw, Meinardus, Darlington, Lam and D. Elliott, Hollenhorst, and G.H. Elliott [8] (see [15,38] for the other references). But none of them had the tools to treat the rational case completely (not even under the assumption $m \geqq n$), and only the last two clearly displayed the connection of the Chebyshev-CF method with the CF method on the disk, the polynomial case of which they seem to have discovered independently.

On the other hand, the basic idea of CF approximation can also be used to economize Faber series in order to compute near-best polynomial and rational approximations on fairly general simply-connected domains in the plane (Faber-CF approximation) [7; 6; 36, §8]. We present a new approach to and an extension of this method in Section 7.

First, we consider yet another extension: If f is rational, the infinite Hankel matrix whose singular value decomposition is needed in the method of Adamjan, Arov, and Krein [3] has finite rank and can be replaced by a finite matrix, so that a numerical treatment becomes possible. There is great interest in system theory for applying this method to the model reduction problem, and it has been investigated by Bultheel and Dewilde [4], Kung [21], and Silverman and Bettayeb [29]. They call it approximation in the Hankel norm. In his remarkable paper [21], Kung was able to establish the basic results independently of (though often in analogy to) the operator theoretic treatment of Adamjan, Arov, and Krein [3]. We do the same here using a few times Kung's arguments, but often simpler ones. (Some are adapted from Takagi [31,32] or Trefethen [37].) While Kung restricted himself to the approximation of $f \in \mathcal{R}_{N-1,N}(\overline{D})$ with real coefficients by $\tilde{r} \in \tilde{\mathcal{R}}_{n-1,n}$, we generally approximate $f \in \tilde{\mathcal{R}}_{MN}$ by $\tilde{r} \in \tilde{\mathcal{R}}_{mn}$. In addition we allow for a weight function. Among other new results we obtain an extension of our earlier block structure theorem for the CF table [15]. (The earlier version reappears as special case $N = 0$.)

Again there are applications of this method to approximation on other domains, in particular on a real interval, and to various related questions in approximation theory, e.g., algebraically computable best rational approximations [1,17,22,33,34]. Also related is the convergence of relative error curves [24,26]. Moreover, we should mention that there is a close connection with the Nevanlinna-Pick problem [3,11,28].

Finally, let us summarize some of our further notation: $\mathcal{P}_m$ is the space of (complex) polynomials of degree at most m, $\mathcal{P}_m^*$ is the subset of polynomials that are self-reciprocal in $\mathcal{P}_m$, i.e.

$$\mathcal{P}_m^* \;:=\; \{p \in \mathcal{P}_m \,;\, p(z) = z^m \overline{p}(1/z)\ (\forall z)\} \;.$$

($\overline{p}$ is the polynomial whose coefficients are complex conjugate to those of p.) Also, $\tilde{\mathcal{P}}_m := \tilde{\mathcal{R}}_{m0}$. The space of functions analytic in D and continuous in $\overline{D}$ is denoted by $A(\overline{D})$.

2. KUNG'S REDUCTION OF THE INFINITE SINGULAR VALUE PROBLEM TO A FINITE ONE

In this and the following two sections we give a complete characterization of and an algorithmic approach to the solution of the following

PROBLEM A. Given $m, n \in \mathbb{Z}$, $n \geq 0$, $f \in \tilde{R}_{MN}$ *and* $g \in \tilde{P}_L$, g *nonvanishing in* $\{z ; 1 \leq |z| < \infty\}$, *find* $\tilde{r} \in \tilde{R}_{mn}$ *such that (in the supremum norm on* ∂D) $\| gf - g\tilde{r} \|$ *is minimal. (We assume in the sequel that* L,M,N *are smallest possible.)*

Remark. One can handle fairly general weight functions: If $g_0 \in C(\partial D)$, $g_0 > 0$ is a given weight function for which the conjugate periodic function $t \mapsto \hat{g}_1(e^{it})$ of $t \mapsto g_1(e^{it}) := \log g_0(e^{-it})$ is continuous also, then

$$g_2(e^{it}) := \exp[g_1(e^{it}) + i\hat{g}_1(e^{it})]$$

defines a function that can be extended from ∂D to D such that $g_2 \in A(\overline{D})$ and $g(z) \neq 0$ on $\overline{D}$. (g_2 can be computed with FFTs, cf. [18, §4].) Then, $g_3(z) := g_2(1/z)$ satisfies

$$g_3 \in \tilde{P}_0, \quad g_3(z) \neq 0 \ \forall z \in D^c, \quad |g_3(z)| = g_0(z) \ \forall z \in \partial D .$$

Note also that for any given rational weight function an equivalent admissible one is obtained by reflecting zeros and poles outside ∂D at ∂D.

Let us first note that we may reduce problem A to one with uniform weight function $g(z) \equiv 1$: Given g as above, $gf \in \tilde{R}_{L+M,N}$ and $g\tilde{r} \in \tilde{R}_{L+m,n}$ if and only if $f \in \tilde{R}_{MN}$ and $\tilde{r} \in \tilde{R}_{mn}$. In particular, multiplication of f and $\tilde{r}$ by a power of z leaves the modulus of the error function invariant. Therefore, we may assume $M = N$. Finally, we may assume $m < M = N$ or $n < N$ since $\tilde{r} = f$ otherwise. Hence problem A reduces to

PROBLEM A'. Given $m,n \in \mathbb{Z}$, $n \geq 0$, $f \in \tilde{R}_{NN} \setminus (\tilde{R}_{N-1,N} \cup \tilde{R}_{N,N-1})$, *where* $N > \min\{m,n\}$, *find* $\tilde{r} \in \tilde{R}_{mn}$ *such that* $\| f - \tilde{r} \|$ *is minimal.*

In some annulus $1 < |z| < \rho_f$ the function f has a Laurent expansion

$$f(z) = \sum_{k=-\infty}^{\infty} f_k z^k .$$

For any $j \in \mathbb{Z}$ we define the infinite Hankel matrix

$$\underset{\sim}{H}_j := \begin{pmatrix} f_{j+1} & f_{j+2} & \cdots \\ f_{j+2} & & \\ \vdots & & \end{pmatrix}$$

whose singular values (or Schmidt numbers) are denoted by $\sigma_0^{(j)} \geq \sigma_1^{(j)} \geq \ldots$ [12]. Since $\tilde{P}_{m-n} + \tilde{R}_{mn} = \tilde{R}_{mn}$, we might reduce our problem to one where $f_k = 0$ for $k \leq m-n$, but we need not make explicit use of this. By Kronecker's theorem [10] $\underset{\sim}{H}_j$ has finite rank $(\forall j)$, cf. Lemma 2.2 below. The general theory of Adamjan, Arov, and Krein [3] now implies

THEOREM 2.1 [3]. *Problem A' has a unique solution* $\tilde{r}^*$, *namely*

$$\tilde{r}^*(z) := f(z) - \sigma z^{\ell+1} \frac{\grave{u}(z)}{\acute{u}(1/z)} ; \tag{2.1}$$

here $\ell := m-n$, $\sigma := \sigma_n^{(\ell)}$, $\grave{u}(z) := \sum_{k=0}^{\infty} \grave{u}_k z^k$, $\acute{u}(z) := \sum_{k=0}^{\infty} \acute{u}_k z^k$, *and* $\underset{\sim}{\grave{u}} := (\grave{u}_0, \grave{u}_1, \ldots)^T$, $\underset{\sim}{\acute{u}} := (\acute{u}_0, \acute{u}_1, \ldots)^T$ *is* <u>*any*</u> *pair of singular vectors of* $\underset{\sim}{H}_\ell$ *corresponding to* σ :

$$\underset{\sim}{H}_\ell \underset{\sim}{\acute{u}} = \sigma \underset{\sim}{\grave{u}} , \quad \underset{\sim}{H}_\ell^H \underset{\sim}{\grave{u}} = \sigma \underset{\sim}{\acute{u}} . \tag{2.2}$$

$|(f - \tilde{r}^*)(z)|$ *has constant modulus* σ *on* ∂D .

Trefethen [37] noted that due to the (complex) symmetry of $\underset{\sim}{H}_\ell$ there are bases of the singular spaces such that $\acute{\underset{\sim}{u}}$ is the complex conjugate of $\grave{\underset{\sim}{u}}$. Hence, we may replace (2.1) and (2.2) by

$$\tilde{r}^*(z) \quad := \quad f(z) - \sigma z^{\ell+1} \frac{u(z)}{\bar{u}(1/z)} \quad , \qquad (2.1')$$

$$\underset{\sim}{H}_\ell \bar{\underset{\sim}{u}} = \sigma \underset{\sim}{u} \quad . \qquad (2.2')$$

It is our aim to give a relatively simple proof of Theorem 2.1. (Concerning uniqueness we only show that all solutions of (2.2') corresponding to a fixed σ yield the same $\tilde{r}^*$, cf. Lemma 2.3 and Thm. 3.1 below.) We also want to explore the structure of the solutions of (2.2) and (2.2') in order to determine the actual 'degrees' m and n of $\tilde{r}^*$.

The first step consists in showing that u is a rational function with known denominator (as has been noticed before by several authors [4,21,29]) and that the numerator may be computed by solving a finite dimensional singular value problem due to Kung [21].

It is clear from the operator theoretic background [12] that $\underset{\sim}{u} \in \ell_2$, hence $u \in H_2$ (the Hardy space with index 2). By considering the Fourier series of f and u, (2.2') is seen to be the Fourier coefficients relation equivalent to

$$z \mapsto f(z)\bar{u}(1/z) - \sigma z^{\ell+1} u(z) \in \tilde{\mathcal{P}}_\ell \qquad (2.3)$$

($\tilde{\mathcal{P}}_\ell := \tilde{\mathcal{R}}_{\ell 0}$). By assumption and by (1.2),

$$f = \frac{\tilde{a}}{b} \quad \text{with} \quad \tilde{a} \in \tilde{\mathcal{P}}_N \setminus \tilde{\mathcal{P}}_{N-1} , \quad b \in \mathcal{P}_N \setminus \mathcal{P}_{N-1} , \qquad (2.4)$$

where b has all zeros outside ∂D. Multiplication of (2.3) by $b(z)$ yields

$$z \mapsto \tilde{a}(z)\bar{u}(1/z) - \sigma z^{\ell+1} u(z) b(z) \in \tilde{P}_{\ell+N} . \tag{2.5}$$

Here the first term is in $\tilde{P}_N$. Therefore, the second is in $\tilde{P}_{\max\{N,N+\ell\}}$, and, since $ub \in H_2$, ub is a polynomial if $\sigma > 0$:

$$v := ub \in P_K ,$$
$$\text{where} \quad K := N - 1 + \ell^- , \quad \ell^- := \max\{-\ell,0\} . \tag{2.6}$$

Consequently, u is a rational function with poles at most at the zeros of b and at ∞ . Multiplying (2.5) by $\bar{b}(1/z)/b(z)$ and inserting (2.6) leads to

$$z \mapsto \frac{1}{b(z)} [\tilde{a}(z)\bar{v}(1/z) - \sigma z^{\ell+1} \bar{b}(1/z) v(z)] \in \tilde{P}_\ell ,$$

or, in view of $\ell^+ := \max\{\ell,0\} = \ell + \ell^- = \ell + 1 + K - N$, to

$$\tilde{d} : z \mapsto \tilde{d}(z) := \frac{\tilde{a}(z)}{b(z)} v^*(z) - \sigma z^{\ell^+} \frac{b^*(z)}{b(z)} v(z) \in \tilde{P}_{\ell+K} , \tag{2.7}$$

where

$$v^*(z) := z^{-K} \bar{v}(1/z) , \quad b^*(z) := z^{-N} \bar{b}(1/z)$$

are the reciprocal polynomials of v and b in P_K and P_N , respectively. Under the additional assumption $f_k = 0$ for $k \leq \ell$, it is easy to verify that

$$\tilde{d} \in z^\ell P_K \quad \text{if} \quad \ell \geq 0 , \qquad \tilde{d} \in z^{\ell+1} P_{K-1} \quad \text{if} \quad \ell < 0 . \tag{2.8}$$

It is also readily checked that the condition imposed by (2.7) on v and σ is sufficient for $u = v/b$ and σ to satisfy (2.3). In particular, linearly independent solutions v of (2.7) correspond to linearly independent solutions u of (2.3), and vice versa. On the other hand, by inserting the Laurent series (2.0) of f and the Maclaurin series

$$\frac{b^*(z)}{b(z)} = \beta_0 + \beta_1 z + \beta_2 z^2 + \dots \tag{2.9}$$

of b^*/b (analytic on $\overline{D}$) into (2.7), the condition $\tilde{d} \in \tilde{P}_{\ell+K}$ takes the form of *Kung's relations* (cf. [21], where the case $\ell = 0$ is treated for real f):

$$\begin{pmatrix} f_{\ell+1} & f_{\ell+2} & \cdots & f_{\ell+K+1} \\ f_{\ell+2} & & \cdot^{\cdot^{\cdot}} & \vdots \\ \vdots & \cdot^{\cdot^{\cdot}} & & \vdots \\ f_{\ell+K+1} & \cdots & \cdots & f_{\ell+2K+1} \end{pmatrix} \overline{\underset{\sim}{v}} = \sigma \begin{pmatrix} \beta_N & \beta_{N-1} & \cdots & \beta_{N-K} \\ \beta_{N+1} & \ddots & & \vdots \\ \vdots & & \ddots & \vdots \\ \beta_{N+K} & \cdots & \cdots & \beta_N \end{pmatrix} \underset{\sim}{v}, \tag{2.10}$$

$$\begin{pmatrix} f_{\ell+K+2} & f_{\ell+K+3} & \cdots & f_{\ell+2K+2} \\ f_{\ell+K+3} & & \cdot^{\cdot^{\cdot}} & \vdots \\ \vdots & \cdot^{\cdot^{\cdot}} & & \vdots \\ & & & \end{pmatrix} \overline{\underset{\sim}{v}} = \sigma \begin{pmatrix} \beta_{N+K+1} & \cdots & \cdots & \beta_{N+1} \\ \beta_{N+K+2} & \ddots & & \vdots \\ \vdots & & \ddots & \vdots \\ & & & \end{pmatrix} \underset{\sim}{v} \tag{2.11}$$

(where $\beta_j := 0$ for $j < 0$). The matrices in (2.10) are square and of order $K + 1$; one is of Hankel type and one of Toeplitz type, and we briefly write

$$\underset{\sim}{H}_{\ell,K} \overline{\underset{\sim}{v}} = \sigma \underset{\sim}{T} \underset{\sim}{v} \tag{2.10'}$$

from now on. This is a (particularly simple) *generalized singular value problem* in the sense of Paige and Saunders [23]. In (2.11) the matrices have size $\infty \times (K+1)$, however.

The coefficients $f_{\ell+j}$, $j \geq 1$, of the Hankel matrices in (2.10) and (2.11) are the Taylor coefficients at ∞ of the proper rational function $z \mapsto f_{\ell+1} z^{-1} + f_{\ell+2} z^{-2} + \dots$, which is of exact order $K + 1$ as can be verified as a consequence of $f \in \tilde{R}_{NN}$, $f \notin \tilde{R}_{N,N-1} \cup \tilde{R}_{N-1,N}$. Likewise, the coefficients β_{N-K+j}, $j \geq 0$, of the Toeplitz matrices in (2.10) and (2.11) are Taylor coefficients at ∞ of a proper rational function of exact order $K+1$ having the *same* denominator $b(1/z)$. Here, cancellation is impossible and it follows from a classical result due to Kronecker [10, Ch. XVI] that $\underset{\sim}{H}_{\ell,K}$ *and* $\underset{\sim}{T}$ *are regular* and that each row in (2.11) is a linear combination of the rows in (2.10), so that (2.10) *implies* (2.11), cf. [21]. Hence we get

LEMMA 2.2. The relation $v(z) = u(z)b(z)$ *defines a one-to-one correspondence between the solutions* $(\sigma,\underset{\sim}{u})$ *with* $\sigma > 0$ *of (2.2') and the solutions* $(\sigma,\underset{\sim}{v})$ *of the generalized singular value problem (2.10)* [$\equiv$ *(2.10')*] *and the singular value problem*

$$\underset{\sim}{T}^{-1}\underset{\sim}{H}_{\ell,K}\bar{\underset{\sim}{v}} = \sigma\underset{\sim}{v} \quad . \tag{2.12}$$

Kung's approach is not the only one for reducing the infinite singular problem to a finite one. An alternative was proposed by Young [39].

We now establish the uniqueness of $\tilde{r}^*$ defined by (2.1'):

LEMMA 2.3. For fixed $\sigma = \sigma_n^{(\ell)}$ *all nontrivial solutions* $\underset{\sim}{v}$ *of (2.10) yield the same unimodular function* v/v^* *and hence the same function*

$$\tilde{r}^*(z) \quad := \quad f(z) - \sigma z^{\ell^+} \frac{b^*(z)}{b(z)}\frac{v(z)}{v^*(z)} = \frac{\tilde{d}(z)}{v^*(z)} \quad . \tag{2.13}$$

Proof. (An extension of Kung's proof [21, App. A].) Let v_1, v_2 be two solutions of (2.10), so that (2.7) holds for both. Multiplying the first of these two relations (2.7) by v_2 and subtracting the second one multiplied by v_1 yields

$$\frac{\tilde{a}}{b}[v_1^* v_2 - v_2^* v_1] \in \tilde{P}_{\ell+2K} \tag{2.14}$$

Hence, b having all zeros outside ∂D must be a divisor of $-i(v_1^* v_2 - v_2^* v_1) \in P_{2K}^*$ (i.e. of a polynomial of degree at most $2K$ that is self-reciprocal in P_{2K}). Consequently, bb^* must be a divisor of it:

$$v_1^* v_2 - v_2^* v_1 = ibb^*\hat{w} , \tag{2.15}$$

where $\hat{w} \in P_{2K-2N}^*$ if $K \geqq N$ and $\hat{w} = 0$ otherwise. In the first case, we obtain by inserting (2.15) into (2.14) $\tilde{a}b^*\hat{w} \in \tilde{P}_{\ell+2K}$, so $\hat{w} \in \tilde{P}_{\ell+2K-2N}$. Now, $K \geqq N$ implies $\ell < 0$, $\ell = N - K - 1$. Therefore, $\hat{w} \in P_{2K-2N}^* \cap P_{K-N-1}$, but this implies that $\hat{w} = 0$. Hence, $v_2/v_2^* = v_1/v_1^*$ by (2.15). □

3. THE STRUCTURE OF THE SINGULAR VECTORS

Lemma 2.3 is the first step in our investigation of the structure of v. In the following let $\sigma := \sigma_n^{(\ell)} > 0$ be a fixed common singular value of $\underset{\sim}{H}_{\ell,K}$ and $\underset{\sim}{H}_{\ell}$. Its multiplicity is denoted by $\mu(\geqq 1)$, and, more generally, $\mu_j \geqq 0$ is the multiplicity of σ as a singular value of $\underset{\sim}{H}_j$. Since the solution $\underset{\sim}{u}$ of (2.2') only depends on $\ell = m - n$ and σ, we may adapt n and $m = n + \ell$ such that $\sigma_{n-1}^{(\ell)} > \sigma$:

$$\begin{aligned}\sigma_0^{(\ell)} \geqq \dots \geqq \sigma_{n-1}^{(\ell)} > \sigma_n^{(\ell)} = \dots = \sigma_{n+\mu-1}^{(\ell)} > \sigma_{n+\mu}^{(\ell)} \geqq \\ \sigma_{n+\mu+1}^{(\ell)} \geqq \dots \geqq \sigma_K^{(\ell)} > \sigma_{K+1}^{(\ell)} = \sigma_{K+2}^{(\ell)} = \dots = 0 .\end{aligned} \tag{3.1}$$

Of course, there are μ linearly independent solutions of (2.2') and corresponding μ linearly independent solutions of (2.10').

In the following argument we will use that the singular values of $\underset{\sim}{H}_j$ interlace with those of $\underset{\sim}{H}_{j-1}$:

$$\sigma_k^{(j-1)} \geqq \sigma_k^{(j)} \geqq \sigma_{k+1}^{(j-1)} \;, \quad j \in \mathbb{Z} \,. \tag{3.2}$$

This can be concluded from the corresponding classical result on eigenvalues of Hermitian operators (of finite rank) [12] since the Hermitian form $\underset{\sim}{u}_{j-1}^{H} \underset{\sim}{H}_{j-1}^{H} \underset{\sim}{H}_{j-1} \underset{\sim}{u}_{j-1}$ reduces to $\underset{\sim}{u}_{j}^{H} \underset{\sim}{H}_{j}^{H} \underset{\sim}{H}_{j} \underset{\sim}{u}_{j}$ if the first component of $\underset{\sim}{u}_{j-1}$ is forced to be zero; cf. [35] also. In particular, (3.2) implies

$$\mu_j - \mu_{j-1} \in \{-1,0,1\} \,. \tag{3.3}$$

Now, let $\gamma \geqq 0$ be defined by

$$u_0 = \ldots = u_{\gamma-1} = 0 \tag{3.4a}$$

for every $\underset{\sim}{u}$ satisfying (2.2'),

$$u_\gamma \neq 0 \quad \text{for some } \underset{\sim}{u} \text{ satisfying (2.2')} \,. \tag{3.4b}$$

If $\gamma > 0$, σ is called an *irregular* singular value of $\underset{\sim}{H}_\ell$, otherwise σ is *regular* [15,32]. It is readily verified that (3.4) implies that $\underset{\sim}{u}$ is a solution of (2.2') if and only if $\underset{\sim}{\tilde{u}} := (u_\gamma \,, u_{\gamma+1} \,, \ldots)^T$ is a solution of

$$\underset{\sim}{H}_{\ell+2\gamma} \underset{\sim}{\overline{\tilde{u}}} = \sigma \underset{\sim}{\tilde{u}} \;. \tag{3.5}$$

Hence, σ is a regular singular value of $\underset{\sim}{H}_{\ell+2\gamma}$ of multiplicity

$$\mu_{\ell+2\gamma} = \mu_\ell = \mu \,. \tag{3.6}$$

Moreover, inserting $u(z) = z^\gamma \tilde{u}(z)$ into (2.3) yields

$$z \mapsto f(z)\overline{\tilde{u}}(1/z) - \sigma z^{\ell+2\gamma} \tilde{u}(z) \in \tilde{P}_{\ell+\gamma} \;, \tag{3.7}$$

which means that $\overline{\tilde{u}}(1/z)$ and $z^\gamma \tilde{u}(z)$ correspond to a pair of singular vectors $\underset{\sim}{\acute{u}} = (\overline{u}_\gamma, \overline{u}_{\gamma+1}, \ldots)^T$, $\underset{\sim}{\grave{u}} = (0,\ldots,0,u_\gamma,u_{\gamma+1},\ldots)^T$ of $\underset{\sim}{H}_{\ell+\gamma}$. If $\gamma > 0$, one can conclude that the function in

(3.7) is in general *not* in $\tilde{P}_{\ell+\gamma-1}$, because otherwise $\overline{\tilde{u}}(1/z)$ and $z^{\gamma+1}\tilde{u}(z)$ would always correspond to a singular vector pair of $\underset{\sim}{H}_{\ell+\gamma-1}$, and thus $z^{-\gamma+1}\overline{\tilde{u}}(1/z)$ and $z^{\gamma+1}\tilde{u}(z)$ would correspond to a pair of $\underset{\sim}{H}_{\ell}$, in contrast to (3.4a), which holds for all singular vectors of $\underset{\sim}{H}_{\ell}$ belonging to σ since it holds for bases of the two singular spaces.

The set of vectors $\underset{\sim}{\tilde{u}}$ constructed as above from the set of solutions $\underset{\sim}{u}$ of (2.2') of course span the μ-dimensional left singular space of $H_{\ell+2\gamma}$ belonging to σ. In view of $u = v/b$ and Lemma 2.3, they all yield the same unimodular function

$$\frac{v(z)}{v^*(z)} = z^{N-K+2\gamma} \frac{\tilde{u}(z)}{\overline{\tilde{u}}(1/z)} \frac{b(z)}{b^*(z)} . \tag{3.8}$$

Hence, for each $\tilde{u}$ there must be a self-reciprocal factor $s \in P^*_{\mu-1}$ that cancels, and vice versa, i.e.

$$u(z) = z^{\gamma}s(z)u^{o}(z) , \qquad v(z) = z^{\gamma}s(z)v^{o}(z) , \tag{3.9}$$

with fixed $v^{o} = u^{o}b$ and arbitrary $s \in P^*_{\mu-1}$, cf. [32]. As a consequence of (3.4b) $u^{o}(0) \neq 0$ and, since $b(0) \neq 0$, $v^{o}(0) \neq 0$ also.

Division of (3.7) by $s(z) = z^{\mu-1}\overline{s}(1/z)$ leads to

$$z \mapsto f(z)\overline{u^{o}}(1/z) - \sigma z^{\ell+2\gamma+\mu-1}u^{o}(z) \in \tilde{P}_{\ell+\gamma} \subset \tilde{P}_{\ell+2\gamma+\mu-1} \tag{3.10}$$

showing that $\underset{\sim}{u}^{o}$ is a singular vector of $\underset{\sim}{H}_{\ell+2\gamma+\mu-1}$. As we have seen before, (3.7) and thus (3.10) does in general not hold for $\tilde{P}_{\ell+\gamma-1}$ if $\gamma > 0$. Generally, we let $\delta \geqq 0$ be the greatest integer such that (3.10) holds for $\tilde{P}_{\ell+\gamma-\delta}$:

$$z \mapsto f(z)\overline{u^{o}}(1/z) - \sigma z^{\ell+2\gamma+\mu-1}u^{o}(z) \in \tilde{P}_{\ell+\gamma-\delta} \setminus \tilde{P}_{\ell+\gamma-\delta-1} . \tag{3.11}$$

Note that

$$\gamma \geqq 0 , \qquad \delta \geqq 0 , \qquad \gamma\delta = 0 . \tag{3.12}$$

Multiplication of (3.11) by $\overline{s}(1/z) = z^{1-j}s(z)$, where $s \in P^*_{j-1}$, is arbitrary, $1 \leqq j \leqq \gamma+\delta+\mu$, yields

$$z \mapsto f(z)\overline{s}(1/z)\overline{u^o}(1/z) - \sigma z^{\ell+2\gamma+\mu-j} s(z)u^o(z) \in \tilde{P}_{\ell+\gamma-\delta} \subset \tilde{P}_{\ell+2\gamma+\mu-j} , \tag{3.13}$$

which means that su^o corresponds to a singular vector of $\underset{\sim}{H}_{\ell+2\gamma+\mu-j}$. Since $s \in P^*_j$ is arbitrary, it follows in view of (3.3) and (3.6) that

$$\mu_{\ell+2\gamma+\mu-j} = j , \quad 1 \leqq j \leqq \gamma+\delta+\mu , \tag{3.14}$$

and that σ is regular in this range of the index. In fact, it is also easy to see that whenever σ is regular for $\underset{\sim}{H}_i$, then $\mu_{i+1} = \mu_i - 1$, since the step from $\underset{\sim}{H}_i$ to $\underset{\sim}{H}_{i+1}$ means forcing the first component of the right singular vector $\acute{\underset{\sim}{u}}$ to be zero [32]. It follows that (3.14) is also true for $j = 0$, and that v^o and v^{o*} are mutually prime (cf. the derivation of (3.10)).

If $\gamma > 0$, $\mu_{\ell+j} \leqq \mu+j$ by (3.3), but $\mu_{\ell+\gamma} = \gamma+\mu$ by (2.29) since $\delta = 0$. Hence,

$$\mu_{\ell+j} = \mu+j , \quad 0 \leqq j \leqq \gamma . \tag{3.15}$$

What remains is to investigate the behavior of μ_j for $j < \ell$ if $\gamma > 0$ and for $j < \ell-\delta$ if $\gamma = 0$ $(\delta \geqq 0)$. Assume first that σ is a regular singular value of $\underset{\sim}{H}_{\ell-\delta-1}$. Then, according to the foregoing, $\mu_{\ell-\delta-1} = \mu_{\ell-\delta}+1$, and σ is regular for $\underset{\sim}{H}_{\ell-\delta}$ also. Hence, $\gamma = 0$, and $\mu_{\ell-\delta-1} = \mu_{\ell-\delta}+1$ $= \mu+\delta+1$ by (3.14). Writing down the corresponding relation (3.7) and inserting (3.9) with $\gamma = 0$ and $s \in P^*_{\mu+\delta}$ then shows that the function in (3.11) lies in $\tilde{P}_{\ell-\delta-1}$, which is a contradiction of the definition of δ . Consequently, σ cannot be a regular singular value of $\underset{\sim}{H}_{\ell-\delta-1}$, and $\mu_{\ell-\delta-1} = \mu_{\ell-\delta} - 1$ by (3.3) or (3.15) (if $\mu_{\ell-\delta} = 1$). By the same argument it follows that μ_j decreases further until $\mu_j = 0$. Summarizing we get

THEOREM 3.1. Let μ, γ, δ *be defined by (3.1), (3.4), and (3.11). Then the function* $u(z) = \Sigma u_k z^k$ *corresponding to a solution* $\underset{\sim}{u} = (u_0, u_1, \ldots)^T$ *of (2.2') is of the form* $u(z) = v(z)/b(z)$, *where* b *is the "denominator" of* f *(cf. (2.4)) and* $v \in P_K$ $(K := N-1+\max\{-\ell, 0\}$, $\ell := m-n)$ *corresponds likewise to a solution of (2.10') and has the form*

$$v(z) = z^{\gamma} s(z) v^{o}(z) \tag{3.16}$$

with $s \in P^*_{\mu-1}$ *and fixed* $v^o \in P_K$. *On the other hand, every* $s \in P^*_{\mu-1}$ *defines solutions of (2.10') and (2.2') in this way. More generally, all left singular vectors* $\underset{\sim}{\acute{u}}$ *of* $\underset{\sim}{H}_{\ell}$ *are obtained if* $s \neq 0$ *is an arbitrary polynomial of degree at most* $\mu - 1$.

For $|k| \leqq \gamma + \delta + \mu$, $\sigma = \sigma_n^{(\ell)}$ *is a singular value of* $\underset{\sim}{H}_{\ell+\gamma-\delta+k}$ *of multiplicity*

$$\mu_{\ell+\gamma-\delta+k} = \gamma + \delta + \mu - |k|, \quad |k| \leqq \gamma + \delta + \mu . \tag{3.17}$$

The corresponding solutions of $\underset{\sim}{H}_j \overline{\underset{\sim}{u}} = \sigma \underset{\sim}{u}$, $j = \ell + \gamma - \delta + k$, $|k| < \gamma + \delta + \mu$, *all yield the same function*

$$\tilde{r}^*(z) = f(z) - \sigma z^{j+1} \frac{u(z)}{\overline{u}(1/z)} . \tag{3.18}$$

For a fixed index j the last part of the theorem follows from Lemma 2.3. However, (3.13) shows directly that $\tilde{r}^*$ is always the same whenever $k \geqq 0$. Finally, if $k = -\gamma' < 0$, σ is irregular and $u(z)/\overline{u}(1/z) = z^{2\gamma'}\tilde{u}(z)/\overline{\tilde{u}}(1/z)$, where $\underset{\sim}{\tilde{u}}$ is a singular vector of $\underset{\sim}{H}_{\ell+\gamma-\delta+\gamma'}$, cf. (3.5).

4. PROOF OF THEOREM 2.1; THE CF TABLE

In order to finish the proof of Theorem 2.1 we still have to show that $\tilde{r}^* \in \tilde{R}_{mn}$ and that it is best. However, since we know from Theorem 3.1 that $\tilde{r}^*$ results also from the singular value problem for $\underset{\sim}{H}_{\ell+2\gamma+\mu}$, where σ is a simple regular singular value, we may restrict part of our further argument to this case.

However, let us first note that (3.11), when devided by $\bar{u}^o(1/z)$, immediately yields

LEMMA 4.1. $\tilde{r}^*(z) = O(z^{\ell+\gamma-\delta})$ *but not* $O(z^{\ell+\gamma-\delta-1})$ *as* $z \to \infty$.

Assume now that $\mu = 1$, $\gamma = 0$, $\delta \geq 0$, and that $v = v^o$ has exactly n' zeros in D. Then $v^*(z) = z^K \bar{v}(1/z)$ has exactly $K - n'$ zeros in D (v^o and v^{o*} are mutually prime), and $v/v^*(\partial D)$ has winding number $\iota(v/v^*) = 2n' - K$. So, by (2.13),

$$\iota(f - \tilde{r}^*) = \ell^+ + N + 2n' - K = \ell + 2n' + 1 .$$

Moreover, since $\tilde{r}^* = \tilde{d}/v^*$, $\tilde{r}^*$ has exactly n' poles outside ∂D and, by Lemma 4.1, $\tilde{r}^* \in \tilde{R}_{m'n'}$ exactly, where $m' := n' + \ell - \delta$. So, $k = \iota(f - \tilde{r}) - (2n' + \ell - \delta + 1) = \delta$ in Theorem 5.2 of Part I, i.e. $\tilde{r}^*$ is BA out of $\tilde{R}_{m'+\delta,n'+\delta}$.

We can perturb f to $\tilde{f}$ such that all singular values $\sigma_j^{(\ell)}$ become simple singular values $\tilde{\sigma}_j^{(\ell)}$. Since $\sigma = \sigma_n^{(\ell)}$ is simple, the corresponding singular space of $\underset{\sim}{T}^{-1} \underset{\sim}{H}_{\ell,K}$ spanned by $\underset{\sim}{v}$ is the limit of the perturbed singular space [30, Thm. 6.4]; in particular, the number of zeros of v in D remains n' as long as the perturbation is small enough. Let $\tilde{r}_n^*$ be the solution of the perturbed problem: $\tilde{r}_n^* \in \tilde{R}_{n'+\ell,n'}$ is BA out of this space according to the above ($\delta = 0$ for the perturbed problem), and $\|\tilde{f} - \tilde{r}_n^*\| = \tilde{\sigma}_n^{(\ell)}$. Likewise, each $\tilde{\sigma}_j^{(\ell)}$, $j = 0,\ldots,K$, is the error of a BA in $\tilde{R}_{j'+\ell,j'}$ for some j' with $0 \leq j' \leq K$. Since all the singular values are distinct, we must have $j = j'$. (The same argument was used by Achieser [2, p. 274] and Trefethen [37, p. 303].) This concludes the proof of Theorem 2.1 (except for the uniqueness statement). Moreover, we get as a consequence of Theorem 3.1:

THEOREM 4.2. Let μ,γ,δ *be defined by (3.1), (3.4), and (3.11). Then the solution* $\tilde{r}^*$ *defined by (2.1') of problem A' satisfies*

$$\tilde{r}^* \in \tilde{R}_{m',n'}, \qquad \tilde{r}^* \notin \tilde{R}_{m'-1,n'} \cup \tilde{R}_{m',n'-1}, \qquad \text{with} \quad m' := m - \delta, \quad n' := n - \gamma, \tag{4.1}$$

and $\tilde{r}^*$ *is BA out of* $\tilde{R}_{m'+\nu,n'+\nu}$, *where* $\nu := \gamma + \delta + \mu - 1$. *The error curve* $(f - \tilde{r}^*)(\partial D)$ *is a circle of radius* $\sigma = \sigma_n^{(m-n)}$ *and winding number*

$$\iota(f - \tilde{r}) = m + n + \mu + 1 \tag{4.2}$$

around 0. *The polynomial* v° *in (3.16) has exactly* n' *zeros in* D.

This result is best displayed in terms of the *CF table* shown in Fig. 1 (where we allow $M \neq N$ again). (4.1) means that the half-plane $n \geq 0$ is partitioned into *square blocks* in each of which $\tilde{r}^*$ and σ are fixed. Note that each descending diagonal of the table corresponds to singular value problems (2.10') and (2.2') with fixed matrices $\underset{\sim}{H}_{\ell,K}$, $\underset{\sim}{T}$ and $\underset{\sim}{H}_{\ell}$, respectively. The size of each square block is $\nu = \gamma + \delta + \mu - 1$. In any such block with $\nu > 0$ the elements on the j-th lower diagonal are

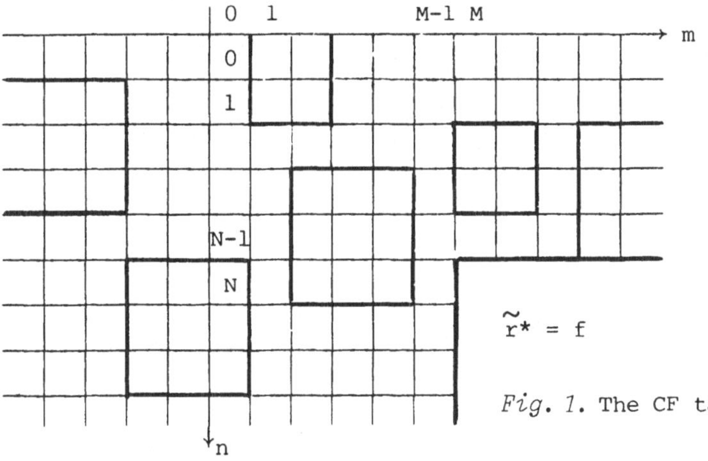

Fig. 1. The CF table.

irregular with $\gamma = j$, and on the j-th upper diagonal $\delta = j$.

Remark: At first sight the fact that n' may be smaller than n seems to contradict a result of Kung [21], who claims that his $(n-1,n)$ approximation has exact degree n . However, Kung's result concerns z times the analytic part of $\tilde{r}^*(z)/z$.

5. ALGEBRAICALLY COMPUTABLE BEST RATIONAL APPROXIMATIONS

As we have seen, for any $f \in \tilde{R}_{MN}$ the BA $\tilde{r}^*$ out of $\tilde{R}_{mn}$ can be computed by solving a finite-dimensional singular value problem. However, the usual aim is to compute approximations out of $R_{mn}(\bar{D})$. The idea of the *CF method* is to project $\tilde{r}^*$ suitably into $R_{mn}(\bar{D})$. As mentioned in §1, this can be done in various ways, e.g., by deleting the negative powers in (1.2) (type 1 CF approximation) or, if $m \geqq n-1$, by deleting the co-analytic part of $\tilde{r}^*$ (type 2 CF approximation), see [15]. The experimental fact is that for $f \in A(\bar{D})$ the resulting CF approximation r^{cf} is often extremely close to $\tilde{r}^*$ and to the BA r^* out of $R_{mn}(\bar{D})$ [36,37].

In this section we are interested in cases where $\tilde{r}^* \in R_{mn}(\bar{D})$ for $f \in A(\bar{D})$ and hence $\tilde{r}^* = r^*$. According to (2.13) we basically need that all poles of v/v^* are outside the unit circle:

LEMMA 5.1. If the square block containing the pair (m,n) *of the CF table has a common boundary point with the infinite block (where* $\tilde{r}^* = f$*), then* $f - \tilde{r}^*$ *is analytic in* $\bar{D}$. *In particular, this holds if either* $m = M-1$ *and* $n \geqq N-1$ *or* $m \geqq M-1$ *and* $n = N-1$.

Proof. Consider the pair (m',n') defined in (4.1) that corresponds to the upper left corner of the block, and redefine $(m,n) := (m',n')$. Then, under our assumption, $K = n + \mu - 1$. By (3.16) the degree of v^o is at most $K - \mu + 1 = n$; so, according to the last statement of Theorem 4.2, it is exactly n, and all n zeros are in D. Hence, from (2.13), $f - \tilde{r}^* \in A(\bar{D})$. □

In view of applications in [17], we formulate the following consequence in terms of the original problem A:

THEOREM 5.2. Assume that $f \in R_{MN}(\bar{D})$, $g \in R_{0L}(D^c)$ *(where* M,N,L *are smallest possible,* $f,g \neq 0$*) and that the pair* (m,n) *lies in the CF table of* gf *in a square block that has a common boundary point with the infinite block. Then the BA* $\tilde{r}^*$ *that minimizes* $\| gf - g\tilde{r} \|$ *(on* ∂D*) among all* $\tilde{r} \in \tilde{R}_{mn}$ *lies in* $R_{m'n'}(\bar{D}) \subset R_{mn}(\bar{D})$, *and hence* $\tilde{r}^*$ *is the BA to* f *out of* $R_{m'+\nu,n'+\nu}(\bar{D})$. *(*$m',n',\nu$ *are as in Theorem 4.2.)*

Proof. Since $g\tilde{r}^*$ is obtained as the BA of gf, cf. §2, $gf - g\tilde{r}^* \in A(\bar{D})$ by Lemma 5.1, and because g is a rational function vanishing only at ∞, $f - \tilde{r}^* \in A(\bar{D})$, hence $\tilde{r} \in A(\bar{D})$. The rest follows from Theorem 4.2 and the inclusion $R_{mn}(\bar{D}) \in \tilde{R}_{mn}$. □

This theorem in conjunction with the transplantation of rational functions from ∂D to $[-1,1]$ (cf. [16]) yields a unifying approach to families of functions with algebraically computable best approximation. Included are those presented by Achieser [1], Mirakyan [22], and Talbot [33,34], see [17]. An alternate approach to the complex problem has been tried by G.H. Elliott [8] with limited success.

Note that the first assertion of Theorem 5.5 in Part I is converse to Theorem 5.2 above.

6. LAURENT - CF APPROXIMATION

If f has a large coanalytic part, it is certainly not appropriate to approximate f on ∂D by elements of $R_{mn}(\bar{D})$ (cf. [25]). Instead we then approximate f by trigonometric rational functions, i.e. out of the space

$$T_{mn} := \{p/q \,;\, p \in T_m \,, q \in T_n \,, q(z) \neq 0 \text{ on } \partial D\} \,, \tag{6.1}$$

where $T_m := z^{-m} P_{2m}$ $(m,n \geq 0)$. Here is a brief summary of the *Laurent-CF method:*

Given $f \in T_{MN}$, $g \in T_{L'L''}$ *with* $g(z) > 0$ *on* ∂D, *and* $m,n \geq 0$, *set* $f^+ := f$, $f^-(z) := f(1/z)$, *and determine* $g^+ \in R_{L'L''}(D^c)$ *and* $g^- := \overline{g^+}$ *by spectral factorization such that* $g(z) = g^+(z)g^-(1/z)$ *and all zeros and poles of* g^+ *lie in* $D \cup \{\infty\}$. *Then solve problem A twice for* $f^{\pm}$, $g^{\pm}$, m, *and* n; *denote the solutions by* $\tilde{r}^{\pm}$, *and let* $e^{\pm} := f^{\pm} - \tilde{r}^{\pm}$,

$$\hat{r}(z) := f(z) - e^{+}(z) - e^{-}(1/z) . \tag{6.2}$$

Finally, project $\hat{r}$ *suitably into* T_{mn} *to obtain* r^{cf} .

(One could handle more general weight functions: If $g \in C(\partial D)$, $g > 0$ is such that the conjugate periodic function of $\log g(e^{-it})$ is continuous, a factorization of the required form with $g^{+} \in \tilde{P}_0$ nonvanishing in D^{c} exists; $g^{+} := \sqrt{g_3}$, where g_3 is defined in §2.)

There are again various reasonable projections, see [15], where the Laurent-CF method for $f \in T_M$ is discussed in detail. If $m \geqq n$, the method can also be understood on the basis of an additive splitting of f (in analogy to Gragg's Laurent-Padé method [13,14]). The choice (6.2) of $\hat{r}$ is not obvious, but it can be shown as in [15,38] that $\hat{r}$ is close to T_{mn} if the Laurent coefficients of $e^{\pm}(z)q^{+}(z)q^{-}(1/z)$ die off fast enough as their index tends to $-\infty$. (Here q^{+} and q^{-} are the denominators of $\tilde{r}^{+}$ and $\tilde{r}^{-}$ in (1.2).) If $f(z) = f(1/z)$, the approximation problem is equivalent to one on $[-1,1]$ (let $x := \frac{1}{2}(z + 1/z)$). If, in addition, f is real-valued on ∂D , then $e^{+} = \overline{e^{+}} = e^{-}$, and the circularity of e^{+} on ∂D with winding number (4.2) is seen to imply the correct number of alternations for r^{cf} being near-best to f with respect to the weighted error $\| \sqrt{g}(f - r^{cf}) \|$ (if the truncation error $\| \hat{r} - r^{cf} \|$ is small); see [15, Thm. 3.1]. Asymptotic error estimates and convincing numerical examples for this *Chebyshev-CF method* are given in [19,38] (case $N = 0$). The case $f(z) = \overline{f}(1/z)$ corresponds to real trigonometric rational approximation (*Fourier-CF method*) [15].

7. FABER - CF APPROXIMATION

The application of the CF method is not restricted to rational approximation on a disk, a circle, or a real interval. Ellacott and the author [6,7] proposed a version of it, called the *Faber-CF method*, that is tailored for fairly general simply connected domains including Jordan regions Ω. (A similar less promising method for $n=0$ was sketched by Trefethen [36, §8].) Basically, assuming $N=0$, $m \geqq n-1$, we applied to the first $M+1$ Faber series coefficients of $F \in A(\overline{\Omega})$ the same process one applies in the CF method to the Maclaurin coefficients. Here this method is presented in a more general framework. In particular, we overcome the restrictions $m \geqq n-1$, $N=0$. For simplicity we assume $F \in C(\partial\Omega)$, but the generalization of the basic facts to the case $F \in L_\infty(\partial\Omega)$ treated for the disk by Adamjan, Arov, and Krein [3] is straightforward. Again, the related singular value problem becomes finite-dimensional if F is itself rational. For simplicity we also restrict ourselves to Jordan regions with rectifiable boundary (cf. the related remark in §5 of Part I, however).

We adopt the notation and the assumptions of §5 in Part I, but we use now capital letters for functions in the ζ-plane. (For clarity, we write now $\tilde{R}_{mn}(\overline{D})$ instead of $\tilde{R}_{mn}$ again.) The following simple lemma is basic:

LEMMA 7.1. The conformal transplantation ψ, *which defines an isometric isomorphism of* $C(\partial D)$ *onto* $C(\partial\Omega)$, *induces also a one-to-one correspondence between* $\tilde{R}_{mn}(\overline{D})$ *and* $\tilde{R}_{mn}(\overline{\Omega})$. *If* $r \in \tilde{R}_{mn}(\overline{D})$, $r \circ \phi \in \tilde{R}_{mn}(\overline{\Omega})$. *If* $R \in \tilde{R}_{mn}(\overline{\Omega})$, $R \circ \psi \in \tilde{R}_{mn}(\overline{D})$. *The poles of* $r \circ \phi$ *(or* R*) in* $\overline{\Omega}^c$ *are the images under* ψ *of those of* r *(or* $R \circ \psi$*) in* $\overline{D}^c$.

Proof. Assume $r \in \tilde{R}_{mn}(\bar{D})$. Trivially, $\| r \| = \| r \circ \phi \|$. If r has exactly $n' \leqq n$ finite poles in D^c, so has $r \circ \phi$ in Ω^c. In particular, since $\phi'(\zeta) \neq 0$ in $\bar{\Omega}^c$, the order of a pole remains invariant under transplantation. The same applies to the behavior at ∞ since $0 < \phi'(\infty) < \infty$. Finally, since r and ϕ are continuous in a relative (exterior) neighborhood of ∂D and $\partial\Omega$, respectively, $r \circ \phi$ is continuous in a relative neighborhood of $\partial\Omega$ also. Hence, $r \circ \phi \in \tilde{R}_{mn}(\bar{\Omega})$. Exactly the same arguments apply to the inverse transplantation. Finally, it is clear that the induced mappings of the spaces are one-to-one and onto. □

COROLLARY 7.2. (i) If $\tilde{r}$ *is a BA of* $f \in C(\partial D)$ *out of* $\tilde{R}_{mn}(\bar{D})$ *on* ∂D, $\tilde{r} \circ \phi$ *is a BA to* $f \circ \phi$ *out of* $\tilde{R}_{mn}(\bar{\Omega})$ *on* $\partial\Omega$, *and* $\| f \circ \phi - \tilde{r} \circ \phi \| = \| f - \tilde{r} \|$.

(ii) If $\tilde{R}$ *is a BA of* $F \in C(\partial\Omega)$ *out of* $\tilde{R}_{mn}(\bar{\Omega})$ *on* $\partial\Omega$, $\tilde{R} \circ \psi$ *is a BA to* $F \circ \psi$ *out of* $\tilde{R}_{mn}(\bar{D})$ *on* ∂D, *and* $\| F \circ \psi - \tilde{R} \circ \psi \| = \| F - \tilde{R} \|$.

The *Faber-CF method* is now immediate: *Given* $F \in C(\partial\Omega)$ *and* $m,n \geqq 0$, *apply the method of Adamjan, Arov, and Krein for* ∂D *(or, in particular, if* $F \in \tilde{R}_{MN}(\bar{\Omega})$ *for some* M *and* N, *its finite-dimensional version implied by Lemma 2.2) to compute the BA* $\tilde{r}^*$ *of* $f := F \circ \psi \in C(\partial D)$ *out of* $\tilde{R}_{mn}(\bar{D})$. *Then project* $\tilde{R}^* := \tilde{r}^* \circ \phi$ *suitably into* $R_{mn}(\bar{\Omega})$.

In view of Lemma 7.1 and Corollary 7.2 it is clear that $\tilde{R}^*$ is best and that the general form of Theorem 2.1 [3] and Theorem 4.2, both with obvious modifications, hold for $\tilde{R}^*$; in particular, the structure of the CF table remains the same:

THEOREM 7.3. (i) Let $F \in C(\partial\Omega)$ *and* $m,n \in \mathbb{Z}$, *with* $n \geqq 0$ *be given. Then* $\tilde{R}^*$, *as defined by the Faber-CF method, is the BA to* F *out of* $\tilde{R}_{mn}(\bar{\Omega})$ *with respect to the supremum norm* *on* $\partial\Omega$. *The error function* $F - \tilde{R}^*$ *has constant modulus equal to the* $(n+1)$*st singular value* $\sigma_n^{(m-n)}$ *of the infinite Hankel matrix* $(f_{m-n+1+i+j})_{i,j=0}^{\infty}$ *whose elements* f_k *are Fourier coefficients of* $t \mapsto f(e^{it})$, $f := F \circ \psi$. *The BA is unique.*

(ii) If, in addition, $F \in \tilde{R}_{MN}(\bar{\Omega})$ *for some (smallest possible)* M *and* N , *the BA out of* $\tilde{R}_{mn}(\bar{\Omega})$ *satisfies*

$$\tilde{R}^* \in \tilde{R}_{m',n'}(\bar{\Omega})$$

$$\textit{with } m' := m - \delta \,, \quad n' := n - \gamma \quad \textit{best possible,} \tag{7.1}$$

$$\iota(F - \tilde{R}^*) = m + n + \mu + 1 \,,$$

and is best out of $\tilde{R}_{m'+\nu,n'+\nu}(\bar{\Omega})$, $\nu := \gamma + \delta + \mu - 1$. $(\gamma, \delta, \mu$ *are defined by (3.1), (3.4), and (3.13).)*

As in problem A (§2) we could allow a *weight function* $G \in \tilde{P}_L(\bar{\Omega})$ with no zeros in Ω^c , since again $G\tilde{R} \in \tilde{R}_{L+m,n}(\bar{\Omega})$ iff $\tilde{R} \in \tilde{R}_{mn}(\bar{\Omega})$. Part (ii) of Theorem 7.3 could also be formulated as a general Carathéodory-Fejér extension theorem (with weight function); (7.1) is the basic result needed, cf. [15, §1] for $N = 0$, $\Omega = D$, and [36, §8] for $N = n = 0$, Ω a Jordan domain.

However, what is a suitable projection from $\tilde{R}_{mn}(\bar{\Omega})$ into $R_{mn}(\bar{\Omega})$? Clearly,

$$R_{mn}(\bar{\Omega}) \subset \tilde{R}_{mn}(\bar{\Omega}) \qquad \text{if} \qquad m,n \geqq 0 \,. \tag{7.2}$$

On the other hand, every $\tilde{R} \in \tilde{R}_{mn}(\bar{\Omega})$ can be split up additively into a function $\tilde{R}_+$ analytic in Ω and a function $\tilde{R}_-$ analytic in $\bar{\Omega}^c$ and vanishing at ∞ . In fact,

$$\tilde{R} \in \tilde{R}_{mn}(\bar{\Omega}) \Rightarrow \tilde{R} = \tilde{R}_+ + \tilde{R}_- \,. \tag{7.3a}$$

where, with $\bar{m} := \max\{m,n-1\}$,

$$\tilde{R}_+ \in \mathcal{R}_{\overline{m}n}(\overline{\Omega}) \ , \qquad \tilde{R}_- \in \tilde{\mathcal{P}}_{-1}(\overline{\Omega}) \ . \tag{7.3b}$$

(Here $\tilde{\mathcal{P}}_m(\overline{\Omega}) := \tilde{\mathcal{R}}_{m0}(\overline{\Omega})$.) For if we add up the principal parts of all Laurent series of $\tilde{R}$ about poles of $\tilde{R}$ in $\overline{\Omega}^c$ and include all terms with nonnegative index of the Laurent series at ∞, we obtain a function $\tilde{R}_+ \in \mathcal{R}_{\overline{m}n}(\overline{\Omega})$, and $\tilde{R} - \tilde{R}_+ \in \tilde{\mathcal{P}}_{-1}(\Omega)$. (Splittings of the form (7.3a) can be defined more generally via the Cauchy integral of $\tilde{R}$, cf. Privalow [27, p. 136].)

In terms of the Faber polynomials ϕ_k [9, p. 47] $\tilde{R}^*_+$ can be written in the form

$$\tilde{R}^*_+ = \frac{1}{Q} \sum_{k=0}^{\overline{m}} a_k^+ \phi_k \ , \qquad Q = \sum_{k=0}^{n} b_k \phi_k \ . \tag{7.4}$$

If $F \in A(\overline{\Omega})$ and $m \geqq n-1$, $\tilde{R}^*_+$ is a reasonable choice of an approximation from $\mathcal{R}_{mn}(\overline{\Omega})$. Generally, we define (in analogy to the terminology for the disk [15]) the *type 2 Faber-CF approximation* $R_2^{cf} \in \mathcal{R}_{mn}(\overline{\Omega})$ by deleting in (7.4) $a_{m+1}^+ \phi_{m+1} + \dots + a_{n-1}^+ \phi_{n-1}$ if $m < n-1$. Note that Q is the polynomial whose zeros are the finite poles of $\tilde{R}^*$ in Ω^c . Hence, $Q\tilde{R}^* \in \tilde{\mathcal{P}}_m(\overline{\Omega})$ and

$$R_1^{cf} := [Q\tilde{R}^*]_+ / Q \in \mathcal{R}_{mn}(\overline{\Omega}) \tag{7.5}$$

is an alternative, the *type 1 Faber-CF approximation.*

Numerically, $\tilde{R}_+$ can be computed via a Fourier analysis of $\tilde{R} \circ \psi$ since the Fourier coefficients with non-negative index of $\tilde{R} \circ \psi$ are the Faber series coefficients of $\tilde{R}_+$. (Likewise, if $F \in A(\overline{\Omega})$, the elements f_k of the Hankel matrix in Theorem 7.3 are the Faber coefficients of F .) The mapping

$$T_f : \tilde{r}_+ = [\tilde{R} \circ \psi]_+ \mapsto \tilde{R}_+ = [\tilde{r} \circ \phi]_+ \tag{7.6}$$

is called *Faber transform* [9, p. 50]. (The plus sign at left means the analytic part with respect to a splitting of type

(7.3a) on ∂D .) Note that Lemma 7.1 together with (7.3) (which holds also for $\Omega = D$, of course) immediately yield

THEOREM 7.4. Let $m,n \geq 0$, $\bar{m} := \max\{m,n-1\}$.

(i) $r \in R_{mn}(\bar{D}) \Rightarrow T_f r \in R_{\bar{m}n}(\bar{\Omega})$, *and the poles of* $T_f r$ *are the images under* ψ *of those of* r .

(ii) $R \in R_{mn}(\bar{\Omega}) \Rightarrow T_f^{-1} R \in R_{\bar{m}n}(\bar{D})$, *and the poles of* $T_f^{-1} R$ *are the images under* ϕ *of those of* R .

Thm. 7.4 is due to Ellacott [6, Thm 1.1]. Another of his results [6, Thm 1.3; 7, Thm 2.3] is readily improved and generalized now:

THEOREM 7.5. Let $F \in C(\partial\Omega)$, $m,n \geq 0$. *Then*

$$\inf_{R \in R_{mn}(\bar{\Omega})} \| F-R \| \geq \tilde{E}_{mn}(F;\bar{\Omega}) = \| F-\tilde{R}^* \| = \sigma_n^{(m-n)} ,$$

where $\sigma_n^{(m-n)}$ *and* $\tilde{R}^*$ *are the same as in Theorem 7.3.*

Proof. Apply the inclusion (7.2) and Theorem 7.3. □

Basically, one can expect the Faber-CF approximation R^{cf} (after truncation) to be good if and only if the error curve of the BA R^* out of $R_{mn}(\bar{\Omega})$ is close to circular, cf. Thm 5.2. Also, as on the disk, it is unlikely that the Faber-CF approximation R^{cf} (after truncation) is close to best unless at least $F \in A(\bar{\Omega})$. Therefore, in analogy to the Laurent-CF approximation on the disk one should in general use a *two-sided Faber-CF approximation* for $F \in C(\partial\Omega)$. Here we need also the conformal mapping Ψ of the exterior of ∂D on Ω , and its inverse Φ . In extension of (6.1) we define $T_{m^+,n^+,m^-,n^-}(\partial\Omega)$ to be the subset of functions $\tilde{r} \in \zeta_0^{n^- - m^-} R_{m^+ + m^-, n^+ + n^-}(\partial\Omega)$ with $\nu^- \leq n^-$ poles in $\Omega \setminus \{\zeta_0\}$, $\nu^+ \leq n^+$ poles in $\bar{\Omega}^c \setminus \{\infty\}$ which are of order $O(\zeta^{m^+ - \nu^+})$ as $\zeta \to \infty$, $O(|\zeta - \zeta_0|^{\nu^- - m^-})$ as $\zeta \to \zeta_0$, where

$\zeta_0 \in \Omega$ is fixed, cf. (I.5.1). The main points of the method are: *Given* $F \in \mathcal{T}_{M^+,N^+,M^-,N^-}(\partial\Omega)$ *and* $m^+,m^-,n^+,n^- \geqq 0$ *, let*

$$f^+ := f \circ \psi \in \tilde{R}_{M^+,N^+}(\bar{D}) \;, \quad f^- := F \circ \Psi \in \tilde{R}_{M^-,N^-}(\bar{D}) \;,$$

and compute the error functions $e^\pm := f^\pm - \tilde{r}^\pm$ *of the BAs of* $f^\pm$ *out of* $\tilde{R}_{m^\pm,n^\pm}(\bar{D})$. (Solution of two problems of type A; weight functions could also be taken into account.) *Then define*

$$\hat{R} := F - e^+ \circ \phi - e^- \circ \Phi$$

and project $\hat{R}$ *suitably into* $\mathcal{T}_{m^+,n^+,m^-,n^-}(\partial\Omega)$.

8. REFERENCES

1. Achieser, N.I.: Ueber ein Tschebyscheffsches Extremumproblem. Math. Ann. *104*, 739-744 (1931).
2. Achieser, N.I.: Theory of Approximation. Frederick Ungar, New York, 1965.
3. Adamjan, V.M., Arov, D.Z., Krein, M.G.: Analytic properties of Schmidt pairs for a Hankel operator and the generalized Schur-Takagi problem. Math. USSR Sbornik *15*, 31-73 (1971).
4. Bultheel, A., Dewilde, P.: On the Adamjan-Arov-Krein approximation, identification and balanced realization of a system. Report TW 48, Kath. Univ. Leuven, 1980; ECCTD Conference, Warsaw, 1980.
5. Carathéodory, C., Fejér, L.: Ueber den Zusammenhang der Extremen von harmonischen Funktionen mit ihren Koeffizienten und über den Picard-Landauschen Satz. Renc. Circ. Mat. Palermo *32*, 218-239 (1911).
6. Ellacott, S.W.: On the Faber transform and efficient numerical rational approximation. Preprint, 1981; revised, 1982.
7. Ellacott, S.W., Gutknecht, M.H.: The polynomial Carathéodory-Fejér approximation method for Jordan regions. Report 82-02, Seminar ang. Math., ETH Zurich, 1982.
8. Elliott, G.H.: The construction of Chebyshev approximations in the plane. Ph. D. thesis, University of London, 1978.
9. Gaier, D.: Vorlesungen über Approximation im Komplexen. Birkhäuser, Basel, 1980.
10. Gantmacher, F.R.: Theory of Matrices. Chelsea Publ. Co., New York, 1959.
11. Genin, Y.V., Kung, S.: A two-variable approach to the model reduction problem with Hankel norm criterion. IEEE Trans.

Circuits Syst. *CAS-28*, 912-924 (1981).
12. Gohberg, I.C., Krein, M.G.: Introduction to the Theory of Linear Nonselfadjoint Operators. Amer. Math. Soc., Providence, R.I., 1969.
13. Gragg, W.B.: Laurent, Fourier, and Chebyshev-Padé tables. Padé and Rational Approximation (Eds. E.B. Saff & R.S. Varga), Academic Press, New York, 1977; pp. 61-72.
14. Gragg, W.B., Johnson, G.D.: The Laurent-Padé table. Information Processing 74, North-Holland, Amsterdam, 1974; pp. 632-637.
15. Gutknecht, M.H.: Rational Carathéodory-Fejér approximation on a disk, a circle, and an interval. Report 82-04, Seminar ang. Math., ETH Zurich, 1982.
16. Gutknecht, M.H.: On the computation of the conjugate trigonometric rational function and on a related splitting problem. Preprint, 1982.
17. Gutknecht, M.H.: A survey of explicitly or algebraically solvable Chebyshev approximation problems. Forthcoming.
18. Gutknecht, M.H., Trefethen, L.N.: Recursive digital filter design by the Carathéodory-Fejér method. Num. Anal. Proj. Manuscript NA-80-01, Stanford University, 1980. Extended version (with J.O. Smith) to appear in IEEE Trans. ASSP.
19. Gutknecht, M.H., Trefethen, L.N.: Real polynomial Chebyshev approximation by the Carathéodory-Fejér method. SIAM J. Numer. Anal. *19*, 358-371 (1982).
20. Henrici, P.: Topics in computational complex analysis: III. The asymptotic behavior of best approximations to analytic functions on the unit disk. This Proceedings.
21. Kung, S: Optimal Hankel-norm model reduction—scalar systems. To appear in IEEE Trans. Automat. Contr.; short version: Proceedings, 1980 Joint Automatic Control Conference, San Francisco, IEEE, New York, 1980.
22. Mirakyan, G.: Sur une nouvelle fonction qui s'écarte le moins possible de zéro. Comm. Soc. Math. Kharkof, sér. 4, t. *12*, 41-48 (1935).
23. Paige, C.C., Saunders, M.A.: Towards a generalized singular value decomposition. SIAM J. Numer. Anal. *18*, 398-405 (1981).
24. Poreda, S.J.: On the behavior of best uniform deviations. J. Approx. Theory *6*, 387-390 (1972).
25. Poreda, S.J.: A characterization of badly approximable functions. Trans. Amer. Math. Soc. *169*, 249-255 (1972).
26. Poreda, S.J.: On the convergence of best uniform deviations. Trans. Amer. Math. Soc. *179*, 49-59 (1973).
27. Privalow, I.I.: Randeigenschaften analytischer Funktionen. Deutscher Verlag der Wissenschaften, Berlin, 1956.
28. Schur, I.: Ueber Potenzreihen, die im Innern des Einheitskreises beschränkt sind. Crelle's J. (J. Reine Angew. Math.) *147*, 205-232 (1917), ibid. *148*, 122-145 (1918).
29. Silverman, L.M., Bettayeb, M.: Optimal approximation of linear systems. Proceedings, 1980 Joint Automatic Conference, San

Francisco, IEEE, New York, 1980.
30. Stewart, G.W.: Error and perturbation bounds for subspaces associated with certain eigenvalue problems. SIAM Review *15*, 727-764 (1973).
31. Takagi, T.: On an algebraic problem related to an analytic theorem of Carathéodory and Fejér and on an allied theorem of Landau. Japan. J. Math. *1*, 83-91 (1924).
32. Takagi, T.: Remarks on an algebraic problem. Japan. J. Math. *2*, 13-17 (1925).
33. Talbot, A.: The Tschebysheffian approximation problem of one rational function by another. Proc. Cambridge Philos. Soc. *60*, 877-890 (1964).
34. Talbot, A.: The uniform approximation of polynomials by polynomials of lower degree. J. Approx. Theory *17*, 254-279 (1976).
35. Thompson, R.C.: Principal submatrices IX: Interlacing inequalities for singular values of submatrices. Linear Algebra Applic. *5*, 1-12 (1972).
36. Trefethen, L.N.: Near-circularity of the error curve in complex Chebyshev approximation. J. Approx. Theory *31*, 344-367 (1981).
37. Trefethen, L.N.: Rational Chebyshev approximation on the unit disk. Numer. Math. *37*, 297-320 (1981).
38. Trefethen, L.N., Gutknecht, M.H.: The Carathéodory-Fejér method for real rational approximation. SIAM J. Numer. Anal. (in print).
39. Young, N.J.: The singular value decomposition of an infinite Hankel matrix. Preprint, 1981.

Acknowledgment. The author is indebted to Lloyd Trefethen for his useful comments on these two papers.

TOPICS IN COMPUTATIONAL COMPLEX ANALYSIS:
I. METHODS OF DESCENT FOR POLYNOMIAL EQUATIONS.

Peter Henrici
ETH-Zentrum, CH-8092 Zürich

Abstract. Methods of descent for solving polynomial equations were introduced by Henrici (1974). The following recent developments concerning such methods are presented: (i) A proof, due to Martin (1976), of a result concerning the non-existence of continuous descent functions, which was stated without proof by Henrici; (ii) A descent function with global uniform convergence, introduced by M. Kneser (1981) in connection with a constructive proof of the Fundamental Theorem of Algebra, which is shown to be identical with Newton's method in the neighborhood of simple zeros.

1. INTRODUCTION

Our interest here is in methods for finding a zero of a given non-constant polynomial p with complex coefficients. Of the many methods that have been proposed, some, like the method of Graeffe and the qd algorithm, have a definitely analytic flavor in the sense that they do not lend themselves to an obvious geometric interpretation. Other methods, such as the use of the principle of the argument, are either directly inspired by geometric ideas, or, although the inspiration is analytic, the algorithm can be visualized.

Here we focus our attention on a class of methods which are derived from a consideration of the graph of the real function $\phi : z \longrightarrow \phi(z) := |p(z)|$ which we

H. Werner et al. (eds.), Computational Aspects of Complex Analysis, 133–147.

visualize as a surface spread over $\mathbb{C}$. By the principle of the minimum, any local minimum of ϕ can occur only at a zero of p . The search for the zeros of p thus is equivalent to a search for the local minima of ϕ . For conducting a systematic search the idea of descent offers itself. Starting from a point z_0 where $\phi(z_0) > 0$, we proceed to a point $z_1 = f(z_0)$ such that $\phi(f(z_0)) < \phi(z_0)$. The function f with domain of definition T is called a descent function for the polynomial p if it has the following properties (see Henrici (1974)):

(i) $f(T) \subset T$

(ii) $p(z) \neq 0 \Longrightarrow |p(f(z))| < |p(z)|$

(iii) $p(z) = 0 \Longrightarrow f(z) = z$

The descent algorithm defined by f consists in constructing the iteration sequence $\{z_n\}$ where $z_0 \in T$ is arbitrary, and where

$$z_{n+1} := f(z_n) \quad , \; n = 0, 1, 2, \ldots \; . \qquad (1.1)$$

By (i), this sequence is well defined, and by (ii), the sequence $\{|p(z_n)|\}$ decreases monotonically as long as $p(z_n) \neq 0$. Naturally one hopes that the sequence $\{z_n\}$ converges to a local minimum of ϕ , and hence to a zero of p .

In this report we point out some difficulties that exist in connection with the existence of continuous descent functions. We then describe in detail a descent function that has recently been proposed by M. Kneser.

2. CONTINUOUS DESCENT FUNCTIONS

The following basic fact was proved by Henrici (1974).

THEOREM 1. Let f be a descent function for the non-constant polynomial p whose domain of definition T is a closed set, and let f be continuous on T . Then the iteration sequence $\{z_n\}$ defined by (1.1) converges to a zero of p for any choice of the starting point $z_0 \in T$.

Unfortunately, the assumption that f be continuous places a severe restriction on the domain of definition T, for it implies that no component of T can contain more than one zero of p. As a consequence, if p has more than one zero, no continuous descent function for p exists whose domain of definition is the whole complex plane. Henrici (1974) supported this fact merely by a plausible intuitive argument (which was characterized as such). We therefore present the proof, due to D. H. Martin (1976), of a more general result.

THEOREM 2. *Suppose p is a non-constant polynomial, and f is a continuous descent function for p, defined on a closed set T containing $r \geq 2$ distinct zeros $w_1, \ldots, w_r$ of p. Then T is the union of disjoint closed sets $T_1, \ldots, T_r$ such that $w_i \in T_i$ $(i = 1, \ldots, r)$, and such that the restriction of f to each T_i is a descent function for p in its own right.*

Proof. Let

$$3\varepsilon := \min_{i \neq j} |w_i - w_j| . \qquad (2.1)$$

Since f is continuous at the w_i and leaves these points fixed there exists δ, $0 < \delta < \varepsilon$, such that if $|z - w_i| < \delta$ for any i, then $|f(z) - w_i| < \varepsilon$ for the same i. By contradiction one easily proves the existence of $\eta > 0$ such that $|p(z)| < \eta$ implies $|z - w_i| < \delta$ for some i. For $i = 1, \ldots, r$, let

$$U_i := \{z \in T : |z - w_i| < \delta \ \& \ |p(z)| < \eta\} .$$

We assert that

$$f(U_i) \subset U_i , \quad i = 1, \ldots, r . \qquad (2.2)$$

Indeed, let $z \in U_i$. Then $|z - w_i| < \delta$, implying $|f(z) - w_i| < \varepsilon$. Furthermore, $|p(z)| < \eta$. By the descent property, $|p(f(z))| \leq |p(z)| < \eta$, hence by the definition of η, $|f(z) - w_j| < \delta < \varepsilon$ for some j. In view of (2.1) we must have $j = i$. Hence $|f(z) - w_i| < \delta$, and it follows that $f(z) \in U_i$.

For $i = 1, \ldots, r$ we now define T_i as the set of all $z_0 \in T$ such that the iteration sequence (1.1) converges to w_i. These sets are obviously disjoint;

by Theorem 1 their union is T. For each i, T_i contains U_i, and thus at least the point w_i. It remains to be shown that T_i is closed. Let $\hat{z}$ be a limit point of T_i which does not belong to T_i. Then $\hat{z}$ belongs to some set T_j such that $j \neq i$. Because the iteration sequence started with $\hat{z}$ stays in T and converges to w_j, there exists n such that

$$f^{[n]}(\hat{z}) := \underbrace{f \circ f \circ \ldots \circ f(\hat{z})}_{n \text{ terms}} \in U_j .$$

By the definition of U_j, $\varepsilon_1 := \delta - |f^{[n]}(\hat{z})| > 0$. Because $f^{[n]}$ as a composition of continuous functions is continuous, there exists $\delta_1 > 0$ such that $z \in T$, $|z - \hat{z}| < \delta_1$ implies

$$|f^{[n]}(z) - f^{[n]}(\hat{z})| < \varepsilon_1 ,$$

hence $f^{[n]}(z) \in U_j$. In view of (2.2) any point $z \in T$ sufficiently close to $\hat{z}$ thus belongs to T_j, contradicting our assumption that there are points arbitrarily close to $\hat{z}$ which belong to T_i. Thus each set T_i is closed, completing the proof of Theorem 2.

By the well-known definition of connectedness, it follows, in particular, that no component of the domain of definition (assumed to be closed) of a continuous descent function can contain more than one zero of p.

We illustrate the difficulties that arise with continuous descent functions by considering the method of <u>steepest descent</u>. If $z = x + iy$ and $p(z) = u + iv$, the gradient of the function

$$(x,y) \longrightarrow \frac{1}{2}|p(z)|^2 = \frac{1}{2}(u^2 + v^2)$$

obviously is $(uu_x + vv_x, uu_y + vv_y)$ or, by the Cauchy-Riemann equations,

$$(uu_x + vv_x, -uv_x + vu_x) .$$

In the complex plane, this vector is represented by the complex number

$$\overline{p'(z)}\, p(z)$$

which if $p'(z) \neq 0$ has the same argument as $p(z)/p'(z)$.

Thus if $\lambda(z) > 0$ is sufficiently small, the function

$$f(z) := z - \lambda(z) \frac{p(z)}{p'(z)}$$

will produce a descent at all points z where $p'(z) \neq 0$. If f is to be continuous also at points z_0 where $p'(z_0) = 0$, $p(z_0) \neq 0$, we necessarily must have

$$\lambda(z) = o(p'(z)), \quad z \to z_0 .$$

However, this implies that $f(z_0) = 0$. That is, no descent will take place at z_0.

To see what happens in a concrete case, consider

$$p(z) = 1 + z^2 .$$

If started at $z = 1$, a suitably restricted Newton's method will produce a sequence $\{z_n\}$ that either converges to $z = 0$ or hits the point $z = 0$ in a finite number of steps. In the first case, we never even get close to a zero of p. In the second case, a decision must be made whether to proceed north or south. For either choice, f will become discontinuous.

It will be interesting to see how Kneser's descent function to be described in §4 deals with such a situation.

3. DISCONTINUOUS DESCENT FUNCTIONS

In order to be realistically applicable, a descent function should be defined on the whole complex plane. We know from Theorem 2 that such a descent function is by necessity discontinuous. As a practical consequence, if a descent function defined everywhere is to be described analytically, such a description by necessity must allow for discontinuities.

The general convergence result of Theorem 1 does not apply to discontinuous descent functions. Indeed, the example at the end of §2 shows that the sequence $\{z_n\}$ produces by a discontinuous descent function may fail to converge to a zero of p. A new condition is necessary to ensure convergence. The simplest such condition is the following:

(iv) For any $z_0 \in T$, the sequence $\{z_n\}$ defined by (1.1) is such that

$$p(z_n) \to 0 . \tag{3.1}$$

A descent function with the property (3.1) will be called <u>range-convergent</u>.

Range convergence by itself does not imply that the sequence $\{z_n\}$ itself converges to a zero of p, but the following is easily established:

THEOREM 3. *Let f be a range-convergent descent function for the non-constant polynomial p, and let f be defined and continuous at all zeros of p. Then for any $z_0 \in T$ the iteration sequence (1.1) converges to a zero of p.*

Proof. From the fact that f is range-convergent it follows as in the proof of Theorem 2 that for any $z_0 \in T$ an element z_m of the iteration sequence will eventually be located in one of the sets U_i described in that proof. Since f is continuous at w_i, it follows that $z_n \in U_i$ for all $n > m$. In view of (3.1) it then follows that $z_n \to w_i$.

Several examples of discontinuous descent functions are described in §6.14 of Henrici (1974). It is easily seen that these functions are range-convergent and continuous at the zeros, and hence produce convergent iteration sequences.

4. KNESER'S DESCENT FUNCTION

Here we describe a descent function which was proposed by M. Kneser (1981) with the purpose of giving a constructive proof of the Fundamental Theorem of Algebra. Kneser's function can also be seen as a tool for actually finding zeros of a polynomial, and as such it has several interesting properties.

The definition of Kneser's function requires a certain amount of preparation. First of all, the given polynomial of degree $n \geq 1$ is assumed to have highest coefficient 1 :

$$p(z) = a_0 + a_1 z + a_2 z^2 + \dots + a_{n-1} z^{n-1} + z^n . \tag{4.1}$$

Let $z \in \mathbb{C}$ be arbitrary. In order to define f at z we write

$$p(z + h) = \sum_{k=0}^{n} b_k h^k . \tag{4.2}$$

Here $b_n = 1$ in view of (4.1). If $b_0 = 0$, then $p(z) = 0$, and we set $f(z) := z$. If $b_0 \neq 0$, we consider the real function

$$\mu(\sigma) := \max_{1 \leqq j \leqq n} \{|b_j| \sigma^j\} , \quad \sigma \geqq 0 . \tag{4.3}$$

It is necessary to establish some properties of the function μ . Since $b_n \neq 0$, the range of indices j considered in (4.3) may be restricted to those for which $b_j \neq 0$. Thus μ is the maximum of a finite set of continuous and strictly increasing functions, and thus is itself <u>continuous and strictly increasing</u>. For each $\sigma \geqq 0$, there exists an index $k = k(\sigma)$ for which

$$\mu(\sigma) = |b_k| \sigma^k . \tag{4.4}$$

If there are several such indices, we denote by $k(\sigma)$ the largest one. We assert that the integer-valued function $k(\sigma)$ is <u>non-decreasing</u>. Indeed, let $\tau > \sigma$ and

$$\ell := k(\sigma) , \quad m := k(\tau) .$$

Then by the definition of μ ,

$$|b_m| \tau^m \geqq |b_\ell| \tau^\ell ,$$

$$|b_\ell| \sigma^\ell \geqq |b_m| \sigma^m ,$$

hence

$$\tau^{m-\ell} \geqq \frac{|b_\ell|}{|b_m|} \geqq \sigma^{m-\ell},$$

thus

$$(\tau/\sigma)^{m-\ell} \geqq 1$$

which in view of $\tau/\sigma > 1$ is possible only if $m \geqq \ell$, as asserted.

Let now τ be fixed as the unique solution of the equation

$$\mu(\tau) = |b_0| . \tag{4.5}$$

Let γ be a real number satisfying $0 < \gamma \leqq \frac{1}{3}$, and for $j = -1, 0, 1, \ldots$ put

$$k_j := k(\gamma^j \tau) .$$

The k_j are integers satisfying

$$1 \leqq k_j \leqq k_{j-1} \leqq n \tag{4.6}$$

for $j = 0, 1, \ldots$. It thus must happen sooner or later that

$$k_{j-1} = k_j = k_{j+1} . \tag{4.7}$$

Let $j = m$ be the smallest $j \geqq 0$ such that (4.7) is true, and let

$$k := k_m . \tag{4.8}$$

We now define Kneser's descent function at z by

$$f(z) := z + h , \tag{4.9}$$

where $h = \rho e^{i\phi}$ $(\rho > 0)$ with

$$\rho := \gamma^k \tau \tag{4.9a}$$

and with ϕ chosen such that

$$\arg b_k h^k = \arg(-b_0) . \tag{4.9b}$$

If $k > 1$ there are several possible choices for ϕ ; any choice will do for what follows.

THEOREM 4. <u>For any polynomial</u> p <u>of the form</u> (4.1), <u>for any</u> $\gamma \in (0,1/3]$ <u>and for any</u> $z \in \mathbb{C}$, <u>Kneser's function</u> f <u>defined by</u> (4.9) <u>satisfies</u>

$$|p(f(z))| \leqq \theta_n |p(z)| , \tag{4.10}$$

<u>where</u>

$$\theta_n := 1 - \frac{1}{1-\gamma} \gamma^{n^2-1} < 1 . \tag{4.11}$$

Proof. We need to consider only the case where $p(z) \neq 0$. From (4.5) and (4.9a) there follows

$$\begin{aligned} |b_k|\rho^k &= |b_k|(\gamma^k\tau)^k \\ &= \gamma^{k^2}|b_k|\tau^k \\ &= \gamma^{k^2}|b_0| \leqq |b_0| , \end{aligned} \tag{4.12}$$

hence, considering (4.9b),

$$|b_0 + b_k h^k| = |b_0| - |b_k|\rho^k .$$

We thus have

$$\begin{aligned} |p(z+h)| &\leqq |b_0 + b_k h^k| + \sum_{\substack{j=1 \\ j\neq k}}^{n} |b_j h^j| \\ &= |b_0| - |b_k|\rho^k + \sum_{\substack{j=1 \\ j\neq k}}^{n} |b_j|\rho^j . \end{aligned} \tag{4.13}$$

In the last sum, we estimate separately the terms where $j < k$ and those where $j > k$. If $j < k$, then in view of $k_{m+1} = k_m = k$

$$\begin{aligned} |b_j|\rho^j &= \gamma^{-j}|b_j|(\gamma\rho)^j \\ &\leqq \gamma^{-j}\,\mu(\gamma\rho) \\ &= \gamma^{-j}|b_k|(\gamma\rho)^k \\ &= \gamma^{k-j}|b_k|\rho^k , \end{aligned}$$

and there follows

$$\begin{aligned} \sum_{1\leqq j<k} |b_j|\rho^j &\leqq (\gamma^{k-1} + \gamma^{k-2} + \ldots + \gamma)|b_k|\rho^k \\ &= \gamma\,\frac{1-\gamma^{k-1}}{1-\gamma}\,|b_k|\rho^k . \end{aligned}$$

If $j > k$ then since $k_{m-1} = k_m = k$

$$\begin{aligned}|b_j|\rho^j &= \gamma^j|b_j|\left(\frac{\rho}{\gamma}\right)^j\\ &\leqq \gamma^j\ \mu\left(\frac{\rho}{\gamma}\right)\\ &= \gamma^j|b_k|\left(\frac{\rho}{\gamma}\right)^k\\ &= \gamma^{j-k}|b_k|\rho^k\ ,\end{aligned}$$

and there similarly follows

$$\begin{aligned}\sum_{k<j\leqq n}|b_j|\rho^j &\leqq (\gamma+\gamma^2+\ldots+\gamma^{n-k})|b_k|\rho^k\\ &= \gamma\,\frac{1-\gamma^{n-k}}{1-\gamma}\,|b_k|\rho^k\ .\end{aligned}$$

Thus from (4.13),

$$\begin{aligned}|p(z+h)| &\leqq |b_0|-|b_k|\rho^k\\ &\quad+\left(\gamma\frac{1-\gamma^{k-1}}{1-\gamma}+\gamma\frac{1-\gamma^{n-k}}{1-\gamma}\right)|b_k|\rho^k\\ &= |b_0|-\frac{1-3\gamma+\gamma^{k-1}+\gamma^{n-k}}{1-\gamma}|b_k|\rho^k\\ &\leqq |b_0|-\frac{\gamma^{k-1}}{1-\gamma}|b_k|\rho^k\ ,\end{aligned}\tag{4.14}$$

where we have used $\gamma\leqq\frac{1}{3}$.

To complete the proof, it remains to relate

$$|b_k|\rho^k=\mu(\rho)\quad\text{to}\quad|b_0|=\mu(\tau)\ .$$

For $j=0,1,2,\ldots$ we have

$$\mu(\gamma^j\tau)=|b_{k_j}|(\gamma^j\tau)^{k_j}$$

and

$$\mu(\gamma^{j+1}\tau)\geqq|b_{k_j}|(\gamma^{j+1}\tau)^{k_j}=\gamma^{k_j}\,\mu(\gamma^j\tau)\ .$$

Thus there follows

$$\mu(\rho) = \mu(\gamma^k \tau) \geqq \gamma^s \mu(\tau)$$

where

$$s := k_0 + k_1 + \ldots + k_{m-1} .$$

In view of (4.6) and of the definition of k_m we have

$$k_0 \leqq n ,$$

$$k_1 \leqq n - 1 ,$$

$$k_2 \leqq n - 1 ,$$

and generally

$$k_j \leqq n - \left[\frac{j+1}{2}\right] , \quad j = 0, 1, \ldots .$$

Since $k_{m-1} = k$ there follows

$$s \leqq n + 2(n-1) + 2(n-2) + \ldots + 2(k+1) + k$$

$$= n^2 - k^2 .$$

Hence from (4.13), using $s + k - 1 \leqq n^2 - k^2 + k - 1 \leqq n^2 - 1$,

$$|p(z+h)| \leqq |b_0|\left(1 - \frac{1}{1-\gamma}\gamma^{n^2}\right) = \theta_n |p(z)| ,$$

as was to be shown.

THEOREM 5. Kneser's descent function is continuous at the zeros of p.

Proof. Let w be a zero of multiplicity ℓ of p. Then there exist positive constants κ_1, κ_2, and ε such that if $|z - w| < \varepsilon$,

$$|b_0| \leqq \kappa_1 |z - w|^{\ell},$$

$$|b_j| \geqq \kappa_2 |z - w|^{\max(\ell - j, 0)} , \tag{4.15}$$

$j = 1, 2, \ldots, n$. It thus follows from (4.12) that

$$\rho^k \leqq \frac{|b_0|}{|b_k|} \leqq \frac{\kappa_1}{\kappa_2}|z - w|$$

and thus that

$$|f(z) - f(w)| \leq \rho + |z - w| \to 0$$

as $z \to w$, establishing Theorem 5.

The inequality (4.10) implies, of course, that Kneser's descent function is range-convergent. This, together with Theorem 3 and the result just proved, shows that any iteration sequence $\{z_m\}$ generated by Kneser's descent function converges to a zero of p. Concerning the speed of convergence, Theorem 4 merely states that

$$|p(z_m)| \leq \theta_n^m |p(z_0)| .$$

Thus, only linear convergence is established; moreover, since θ_n is close to one, this convergence is established only with a very unfavorable convergence factor. The following result may therefore be of interest.

THEOREM 6. <u>In a neighborhood of any simple zero of</u> p, <u>the descent algorithm based on Kneser's function is identical with Newton's method</u>, <u>and thus converges quadratically</u>.

<u>Proof</u>. Let w be a simple zero of p. By the estimates (4.15) we then have for $|z - w| < \varepsilon$

$$|b_0| \leq \kappa_1 |z - w| , \quad |b_1| \geq \kappa_2$$

and

$$|b_j| \leq \kappa_3 , \quad j = 2, 3, \ldots , n$$

where κ_1, κ_2, κ_3, and ε are positive. Thus for sufficiently small $\sigma \geq 0$, the maximum of the terms

$$|b_j|\sigma^j , \ j = 1, 2, \ldots , n$$

is achieved for $j = 1$, and thus

$$\mu(\sigma) = |b_1|\sigma . \tag{4.16}$$

Thus if $|z - w|$ is sufficiently small, the solution of the equation (4.5) is

$$\tau = \frac{|b_0|}{|b_1|} .$$

Moreover, the representation (4.16) still holds for $\sigma = \gamma^{-1}\tau$, and it follows that $k_{-1} = k_0 = 1$. In this situation, $k = k_0 = 1$, and by (4.9b) we see that

$$f(z) = z - \frac{b_0}{b_1} = z - \frac{p(z)}{p'(z)} ,$$

which is the iteration function for Newton's method.

It can be shown that in the neighborhood of a zero of multiplicity $\ell > 1$ Kneser's algorithm yields linear convergence, with a convergence factor $1 - 1/\ell$. Whether the method can be so modified as to yield quadratic convergence also in the presence of multiple zeros is an open question.

5. IMPLEMENTATION OF KNESER'S METHOD

The evaluation of Kneser's descent function for a given monic polynomial p at a given point z requires the following steps:

(i) Representation of $p(z + h)$ as a power series in h . This may be done by the Horner algorithm, either in its classical form requiring about $n^2/2$ additions and multiplications (see Henrici (1974), §6.1), or in the fast form suggested by Shaw and Traub (1974) which requires only $n^2/2$ additions and $3n$ multiplications. (The algorithm proposed by Aho, Steiglitz, and Ullmann (1975), which requires only $O(n\log n)$ operations, is not numerically stable.)

(ii) Computation of the function $\mu(\sigma)$. Let

$$0 = \sigma_1 < \sigma_2 < \dots < \sigma_m \quad (m \leq n)$$

denote the points of discontinuity of the function $k(\sigma)$, and let

$$k(\sigma) = j_i , \; \sigma_i \leq \sigma < \sigma_{i+1}$$

($\sigma_{m+1} := \infty$).Then the function μ is represented by

$$\mu(\sigma) = \beta_i \sigma^{j_i} , \; \sigma_i \leq \sigma < \sigma_{i+1} ,$$

where

$$\beta_i := |b_{j_i}| .$$

The required numbers σ_i and j_i are computed by the following algorithm: Set

$$\sigma_1 := 0$$

$$j_1 := \min \{j \geq 1 : b_j \neq 0\}$$

and for $i = 1, 2, \ldots$ do as follows: If $j_i = n$, stop. If not, let

$$\sigma_{i+1} := \min_{j>j_i} \left| \frac{b_{j_i}}{b_j} \right|^{\frac{1}{j-j_i}} , \tag{5.1}$$

and let j_{i+1} be the largest index for which the minimum (5.1) is assumed. Because the integers j_i form an increasing sequence, the algorithm terminates after at most n steps. Along with the σ_i and the j_i, it is convenient to compute the β_i as well as the quantities

$$\mu_i := \mu(\sigma_i) = \beta_i \, \sigma_i^{j_i} , \tag{5.2}$$

$i = 1, 2, \ldots, m$.

Normally, it will not be necessary to carry the algorithm (5.1) to its conclusion, because the function μ is required only to compute the τ defined by (4.5). We thus may stop at the first i for which

$$\mu_i > |b_0|$$

and then have

$$\tau = (\beta_0/\beta_{i-1})^{1/j_{i-1}} .$$

Then

$$k_0 = j_{i-1} .$$

The value σ_i is required in order to check whether $\gamma^{-1}\tau \leq \sigma_i$, in which case $k_{-1} = k_0 = j_{i-1}$.

(iii) Determination of $k = k_m$. By definition, m is the smallest index h for which the three numbers

$$\gamma^{m-1}\tau \ , \ \gamma^{m}\tau \ , \ \gamma^{m+1}\tau$$

all lie in the same interval $[\sigma_h, \sigma_{h+1})$. If that interval is found, then

$$k = j_h \ .$$

Having determined m and k , then the remaining part of the algorithm is trivially implemented by (4.9).

BIBLIOGRAPHY

A. V. Aho, K. Steiglitz, and J. D. Ullmann (1975): Evaluating polynomials at fixed sets of points. SIAM J. Comput. 4, 533 - 539.

P. Henrici (1974): Applied and Computational Complex Analysis, vol. I. Wiley, New York.

M. Kneser (1981): Ergänzung zu einer Arbeit von Hellmuth Kneser über den Fundamentalsatz der Algebra. Math. Z. 177, 285 - 287.

D. H. Martin (1976): On continuous descent functions for polynomial equations. Z. angew. Math. Physik 27, 863 - 866.

M. Shaw and J. F. Traub (1974): On the number of multiplications for the evaluation of a polynomial and some of its derivatives. J. Assoc. Comput. Mach. 21, 161 - 167.

TOPICS IN COMPUTATIONAL COMPLEX ANALYSIS:
II. NEW DEVELOPMENTS CONCERNING THE QUOTIENT-DIFFERENCE ALGORITHM.

Peter Henrici
ETH-Zentrum, CH-8092 Zürich

Abstract. The quotient-difference (qd) algorithm, originally introduced by Rutishauser in 1954, serves a variety of purposes, such as the determination of zeros and poles of analytic functions, the construction of Padé approximants, and the summation of asymptotic series. Here we describe, among others, the following recent results concerning the algorithm: (i) Seewald's (1981) simple convergence proof for Rutishauser's rules concerning the case of several equimodular poles; (ii) Stokes' (1980) stable construction of the qd scheme using difference tables.

1. INTRODUCTION

Let

$$F = c_0 + c_1 x + c_2 x^2 + \dots \tag{1.1}$$

be a formal power series with complex coefficients, and let

$$H_k^{(n)} = \begin{vmatrix} c_n & c_{n+1} & \cdots & c_{n+k-1} \\ c_{n+1} & c_{n+2} & \cdots & c_{n+k} \\ \cdot & \cdot & \cdot & \cdot \\ c_{n+k-1} & c_{n+k} & \cdots & c_{n+2k-2} \end{vmatrix} \tag{1.2}$$

($n = 0, 1, \dots ; k = 1, 2, \dots$) denote the Hankel determinants formed with the coefficients of F. The

H. Werner et al. (eds.), Computational Aspects of Complex Analysis, 149–168.

series F is called <u>normal</u> if

$$H_k^{(n)} \neq 0 \tag{1.3}$$

for all $n \geq 0$ and all $k \geq 1$; it is called m-<u>normal</u> if (1.3) holds for all $n \geq 0$ and for $k = 1, 2, \ldots ,$ m , and it is called <u>ultimately normal</u> or <u>ultimately</u> m-<u>normal</u> if the foregoing statements hold at least for sufficiently large n .

If F is normal, then a two-dimensional array of numbers

$$q_k^{(n)} \quad \text{and} \quad e_k^{(n)} ,$$

called the <u>quotient-difference</u> (qd) <u>scheme</u> associated with F , can be constructed according to the following rules:

$$e_0^{(n)} := 0 , \quad q_1^{(n)} := \frac{c_{n+1}}{c_n} , \; n = 0, 1, \ldots ; \tag{1.4}$$

$$e_k^{(n)} + q_k^{(n)} = q_k^{(n+1)} + e_{k-1}^{(n+1)} , \tag{1.5a}$$

$$e_k^{(n)} q_{k+1}^{(n)} = e_k^{(n+1)} q_k^{(n+1)} , \tag{1.5b}$$

$$n = 0, 1, 2, \ldots , k = 1, 2, \ldots$$

These rules should of course not be remembered in terms of their indices, but in terms of their graphical interpretation:

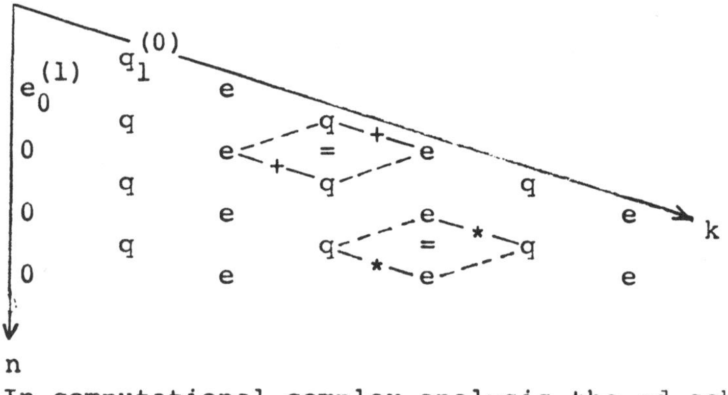

In computational complex analysis the qd scheme has the following two superficially distinct applications:

(i) The determination of zeros and poles of analytic functions;

(ii) The construction of the continued fractions "corresponding" (in a technical sense) to the remainders of the series F, and thus of the Padé table associated with F.

A detailed exposition of these applications is given in our books ((1974), chapter 7; (1977), chapter 12). Here we describe some recent results which deepen our understanding of the performance, either analytical or numerical, of the algorithm.

2. A PROOF OF RUTISHAUSER'S RULES

Let F be the Taylor series at 0 of a function f which is analytic at 0, meromorphic in the disk $|z| < \rho$ $(0 < \rho \leqq \infty)$, and whose poles in that disk are at the points $z_i =: u_i^{-1}$. It is assumed that the poles are arranged in the increasing order of their moduli, so that

$$|u_1| \geqq |u_2| \geqq \ldots , \tag{2.1}$$

and that a pole of order k occurs k times in this enumeration.

It is well known that if the series F is normal, then for each index k such that

$$|u_{k-1}| > |u_k| > |u_{k+1}|$$

$(u_0 := \infty)$, there holds

$$\lim_{n\to\infty} q_k^{(n)} = u_k . \tag{2.2}$$

More generally, if a two-dimensional array of polynomials

$$p_k^{(n)}(u)$$

is defined by the relations

$$p_0^{(n)}(u) := 1 , \ n \geqq 0 ; \tag{2.3}$$

$$p_k^{(n)}(u) := u\, p_{k-1}^{(n+1)}(u) - q_k^{(n)} p_{k-1}^{(n)}(u) , \qquad n \geqq 0 , \quad k \geqq 1 , \tag{2.4}$$

as indicated in the scheme below,

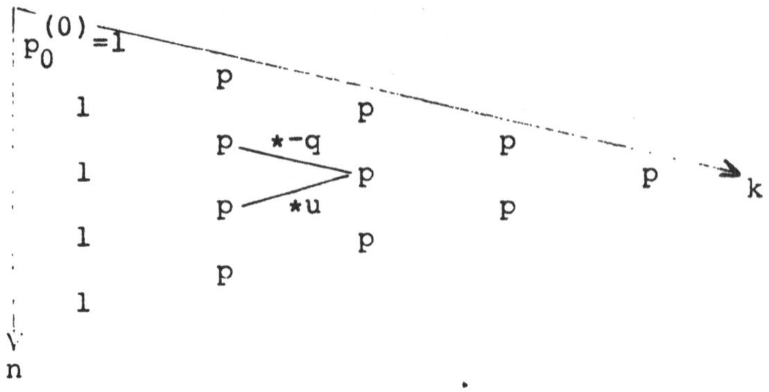

then for each k such that $|u_{k+1}| < |u_k|$,

$$\lim_{n\to\infty} p_k^{(n)}(u) = p_k(u) \qquad (2.5)$$

where

$$p_k(u) := (u - u_1)(u - u_2)\dots(u - u_k) \ . \qquad (2.6)$$

The limit (2.5) holds uniformly on compact sets of the complex plane, and therefore (by Cauchy's formula) also for the coefficient sequences of the polynomials $p_k^{(n)}$. Henrici (1974) failed to notice that the result (2.5) is closely linked to a classical result of de Montessus de Ballore (1902) (quoted after Perron (1957), p. 265).

The qd algorithm thus may be used, with due caution, to find poles of analytic functions. We do not mention here the adaptation of the algorithm to the problem of finding zeros, and how to modify it so that the zeros themselves, and not their reciprocals, emerge as limits of the q-columns. Nor do we discuss at this point the required measures with regard to numerical stability.

The result (2.2) does not indicate how to determine the poles when several poles have the same modulus, say when

$$|u_m| > |u_{m+1}| = |u_{m+2}| = \dots = |u_{m+h}| > |u_{m+h+1}| \ . \qquad (2.7)$$

(The case $h = 2$ arises frequently through the presence of complex conjugate zeros or poles.) In the situation (2.7) de Montessus de Ballore's result (2.5) provides a realistic solution if $m = 0$, for it allows us to find $u_1, u_2, \ldots, u_h$ as zeros of the polynomial

$$p_h(u) := \lim_{n\to\infty} p_h^{(n)}(u) .$$

If $m > 0$, de Montessus de Ballore's Theorem theoretically permits us to find the equimodular zeros as the zeros of the quotient polynomial

$$\frac{p_{h+m}(u)}{p_m(u)} ;$$

however, for values of m that are not very small this does not seem to be a realistic possibility.

Rutishauser (1954) gave the following much simpler rule for how to deal with the case (2.7) if $m > 0$: For $k = 0, 1, \ldots, h,$ let polynomials

$$\hat{p}_k^{(n)}(u)$$

be defined by

$$\hat{p}_0^{(n)}(u) := 1 , \; n \geqq 0 ; \tag{2.8}$$

$$\hat{p}_k^{(n)}(u) := u\, \hat{p}_{k-1}^{(n+1)}(u) - q_{m+k}^{(n)}\, \hat{p}_{k-1}^{(n)}(u) , \tag{2.9}$$

$$n \geqq 0 , \; 1 \leqq k \leqq h .$$

(Except for matters of notation, the recurrence relation (2.9) and the starting condition (2.8) are identical with (2.4) and (2.3), respectively. The only difference is that the construction is now started in the m-th column rather than in the 0-th column.) Rutishauser then asserted that

$$\lim_{n\to\infty} \hat{p}_h^{(n)}(u) = \hat{p}_h(u) , \tag{2.10}$$

where

$$\hat{p}_h(u) := (u - u_{m+1})\ldots(u - u_{m+h}) , \tag{2.11}$$

again uniformly on compact sets and therefore coefficientwise. In the practically most frequent case $h = 2$,

$$\hat{p}_2^{(n)}(u) = u^2 - (q_{m+1}^{(n+1)} + q_{m+2}^{(n)})u + q_{m+1}^{(n)} q_{m+2}^{(n)} ,$$

and Rutishauser's assertion is equivalent to

$$q_{m+1}^{(n+1)} + q_{m+2}^{(n)} \to u_{m+1} + u_{m+2} ,$$

$$q_{m+1}^{(n)} q_{m+2}^{(n)} \to u_{m+1} u_{m+2}$$

as $n \to \infty$.

Rutishauser's rule is very easy to apply, at least in the case $h = 2$, and it is easy to see by means of numerical examples that it obviously works. However, Rutishauser himself did not give a mathematical proof of (2.10). An erroneous proof was given by Henrici (1958). Later, Henrici (1974) gave a correct but rather involved proof by means of determinants; here for $h > 2$ an additional assumption was required which is difficult to verify in concrete situations. Stewart (1971) gave a proof by means of functional analysis. Here we wish to present a very simple proof due to Seewald (1981) which does not require additional assumptions.

Let

$$\hat{p}_k^{(n)}(u) = \sum_{j=0}^{k} a_{k,j}^{(n)} u^j , \quad k = 0, 1, \ldots , h$$

$$\hat{p}_h(u) = \sum_{j=0}^{h} a_{h,j} u^j .$$

The key to Seewald's proof is the observation that the original polynomials $p_k^{(n)}$ (without hat) satisfy

$$p_{m+k}^{(n)}(u) = \sum_{j=0}^{k} a_{k,j}^{(n)} u^j p_m^{(n+j)}(u) , \quad k = 0,\ldots,h. \tag{2.12}$$

We omit the very easy induction proof of (2.12). The relation (2.12) implies for $k = h$

$$p_m(u) \hat{p}_h^{(n)}(u) - p_{m+h}^{(n)}(u)$$

$$= \sum_{j=0}^{h} a_{h,j}^{(n)} u^j \{p_m(u) - p_m^{(n+j)}(u)\} . \tag{2.13}$$

For $n = 0, 1, 2, \ldots$, let

$$\beta_n := \max_{0 \leq j \leq h} |a_{h,j}^{(n)} - a_{h,j}| .$$

If

$$1 > \lambda > \max \left(\frac{|u_{m+1}|}{|u_m|} , \frac{|u_{m+h+1}|}{|u_{m+h}|} \right)$$

and if $K \subset \mathbb{C}$ is any compact set, there follows from the proof of de Montessus de Ballore's theorem (see Henrici (1974), Theorem 7.7f) the existence of a constant μ such that

$$|p_m^{(n)}(u) - p_m(u)| \leqq \mu\lambda^n ,$$
$$|p_{m+h}^{(n)}(u) - p_{m+h}(u)| \leqq \mu\lambda^n$$

for $u \in K$. Therefore from (2.13),

$$|p_m(u)\ \hat{p}_h^{(n)}(u) - p_{m+h}^{(n)}(u)|$$
$$\leqq \sum_{j=0}^{h} |a_{h,j} + (a_{h,j}^{(n)} - a_{h,j})| \mu\lambda^n |u|^j$$
$$\leqq \sum_{j=0}^{h} (|a_{h,j}| + \beta_n)\mu\lambda^n |u|^j ,$$

hence, for some appropriate $\mu_1 > 0$,

$$|p_m(u)\ \hat{p}_h^{(n)}(u) - p_{m+h}(u)|$$
$$\leqq |p_m(u)\ \hat{p}_h^{(n)}(u) - p_{m+h}^{(n)}(u)| + |p_{m+h}^{(n)}(u) - p_{m+h}(u)|$$
$$\leqq \mu\lambda^n \{ \sum_{j=0}^{h} (|a_{h,j}| + \beta_n)|u|^j + 1 \}$$
$$\leqq \mu_1\lambda^n(1 + \beta_n) , \quad u \in K .$$

Deleting from K small neighborhoods of $u_1, \ldots, u_m$ and dividing by $p_m(u)$, there follows

$$|\hat{p}_h^{(n)}(u) - \hat{p}_h(u)| \leqq \mu_2(1 + \beta_n)\lambda^n$$

for some $\mu_2 > 0$. By the principle of the maximum, this holds for all $u \in K$. By Cauchy's formula the corresponding estimate holds for the coefficients:

$$|a_{h,j}^{(n)} - a_{h,j}| \leqq \mu_3(1 + \beta_n)\lambda^n ,$$

$j = 0, 1, \ldots , h$. Thus by the definition of β_n ,

$$\beta_n \leqq \mu_3(1 + \beta_n)\lambda^n .$$

If n is sufficiently large, then $\mu_3\lambda^n < 1$, hence

$$(1 - \mu_3\lambda^n)\beta_n \leqq \mu_3\lambda^n$$

and

$$\beta_n \leqq \frac{\mu_3}{1 - \mu_3\lambda^n}\lambda^n .$$

It follows that $\beta_n \to 0$, which implies (2.10).

3. CONSTRUCTION OF THE PADE TABLE VIA THE qd ALGORITHM

Let r be a rational function,

$$r = \frac{p}{q} ,$$

where p and q are polynomials. It is convenient to define the degree of r as the pair of non-negative integers (m,n) , where

$$m := \deg p , \quad n := \deg q .$$

We call (m,n) the precise degree of r if p and q have no common factors, and if both m and n are the precise degrees of p and q .

Let now

$$P = c_0 + c_1x + c_2x^2 + \ldots$$

be a formal power series, and let m and n be arbitrary non-negative integers. Then the (m,n)-th Padé approximant $P_{m,n}$ of P is the uniquely determined rational function $r = p/q$ of degree (m,n) such that formally

$$Pq - p = O(x^{m+n+1}) . \tag{3.1}$$

Although $P_{m,n}$ as a rational function is uniquely determined, the polynomials p and q in general are not. They <u>are</u> unique (apart from constant factors), and (m,n) is the precise degree of p/q, if the power series P is <u>hypernormal</u>, i.e. if the Hankel determinants defined by (1.1) satisfy

$$H_k^{(n)} \neq 0 \tag{3.2}$$

for $k = 1, 2, \ldots$ and for $n \geq -k+1$. (To define $H_k^{(n)}$ for $n < 0$ we assume $c_n = 0$ for $n < 0$.) We assume in the following that P is hypernormal. We then surely have $q(0) \neq 0$, and (3.1) may be written

$$P - P_{m,n} = O(x^{m+n+1}) .$$

To understand the construction of Padé approximants by means of the qd scheme, it is necessary to recall some basic facts of continued fraction theory. Let the series P be merely normal. Then for every $n = 0, 1, 2, \ldots$ there exists a continued fraction

$$C^{(n)} = \frac{a_1^{(n)}|}{|1} + \frac{a_2^{(n)}x|}{|1} + \frac{a_3^{(n)}x|}{|1} + \ldots \tag{3.3}$$

which corresponds to the series

$$c_n + c_{n+1}x + c_{n+2}x^2 + \ldots$$

in the sense that for $k = 1, 2, \ldots$ the rational function

$$C_k^{(n)} = \frac{a_1^{(n)}|}{|1} + \frac{a_2^{(n)}x|}{|1} + \ldots + \frac{a_k^{(n)}x|}{|1} , \tag{3.4}$$

called the k-th <u>approximant</u> of $C^{(n)}$, satisfies

$$c_n + c_{n+1}x + \ldots - C_k^{(n)}(x) = O(x^k) . \tag{3.5}$$

Now it is well known that the approximant (3.4) is a rational function of precise degree

$$\left(\left[\frac{k-1}{2}\right], \left[\frac{k}{2}\right] \right) .$$

Thus $C_k^{(n)}$ is a certain diagonal or subdiagonal Padé approximant of $c_n + c_{n+1}x + \ldots$. By writing (3.5) as

$$P - \{c_0 + c_1 x + \ldots + c_{n-1}x^{n-1} + x^n C_k^{(n)}(x)\} = O(x^{n+k})$$

and noting that the expression in { } is a rational function of precise degree

$$(n + \left[\frac{k-1}{2}\right], \left[\frac{k}{2}\right]) ,$$

we see that it equals the Padé approximant of degree

$$(n + \left[\frac{k-1}{2}\right], \left[\frac{k}{2}\right]) \quad \text{of} \quad P .$$

Thus the continued fractions $C_k^{(n)}$ yield all Padé approximants $P_{m,n}$ with $m \geq n$, in two different ways even according to whether $k = 2n$ or $k = 2n + 1$.

The connection with the qd algorithm is now established through the well-known fact (see Henrici (1977), §12.4) that

$$C^{(n)} = \frac{c_n \rvert}{\lvert 1} - \frac{q_1^{(n)} x \rvert}{\lvert 1} - \frac{e_1^{(n)} x \rvert}{\lvert 1} - \frac{q_2^{(n)} x \rvert}{\lvert 1} - \frac{e_2^{(n)} x \rvert}{\lvert 1} - \dots \tag{3.6}$$

Thus if $m \geq n$, $P_{m,n}$ is given by the explicit formulas

$$\begin{aligned} P_{m,n}(x) &= c_0 + c_1 x + \dots + c_{m-n} x^{m-n} \\ &+ \frac{c_{m-n+1} x^{m-n+1} \rvert}{\lvert 1} - \frac{q_1^{(m-n+1)} x \rvert}{\lvert 1} - \frac{e_1^{(m-n+1)} x \rvert}{\lvert 1} - \dots \\ &- \frac{q_n^{(m-n+1)} x \rvert}{\lvert 1} - \frac{e_n^{(m-n+1)} x \rvert}{\lvert 1} \\ &= c_0 + c_1 x + \dots + c_{m-n-1} x^{m-n-1} \\ &+ \frac{c_{m-n} x^{m-n} \rvert}{\lvert 1} - \frac{q_1^{(m-n)} x \rvert}{\lvert 1} - \frac{e_1^{(m-n)} x \rvert}{\lvert 1} - \dots \\ &- \frac{e_n^{(m-n)} x \rvert}{\lvert 1} - \frac{q_{n+1}^{(m-n)} x \rvert}{\lvert 1} . \end{aligned} \tag{3.7}$$

If the series P is hypernormal, then so is the series

$$\hat{P} := \frac{1}{P} = \hat{c}_0 + \hat{c}_1 x + \hat{c}_2 x^2 + \dots .$$

All the above can be carried through for the series $\hat{P}$;

we obtain continued fractions

$$\hat{C}^{(n)} = \frac{\hat{a}_1|}{|1} + \frac{\hat{a}_2 x|}{|1} + \frac{\hat{a}_3 x|}{|1} + \ldots$$

such that their approximants $\hat{C}_k^{(n)}$ satisfy

$$\hat{P} - \{\hat{c}_0 + \hat{c}_1 x + \ldots + \hat{c}_{n-1} x^{n-1} + x^n \hat{C}_k^{(n)}(x)\} = O(x^{n+k}) .$$

Thus the expression in $\{\ \}$ is the Padé approximant of degree

$$(n + \left[\frac{k-1}{2}\right], \left[\frac{k}{2}\right]) \quad \text{of} \quad \hat{P} ,$$

and its reciprocal consequently is the Padé approximant of degree

$$(\left[\frac{k}{2}\right], n + \left[\frac{k-1}{2}\right]) \quad \text{of} \quad P .$$

Thus the fractions $\hat{C}^{(n)}$ furnish all Padé approximants $P_{m,n}$ of P where $m \leq n$, again each fraction in two ways.

The continued fractions $\hat{C}^{(n)}$ can of course be expressed in terms of the qd scheme for the series $\hat{P}$, resulting in quantities $\hat{q}_k^{(n)}$, $\hat{e}_k^{(n)}$. However, thanks to a remarkable formal result this scheme need not actually be constructed. For if the series P is hypernormal, its qd scheme may be continued to the horizontal line $n = -k + 1$, still using the "rhombus rules" (1.5). There will result a scheme of the form

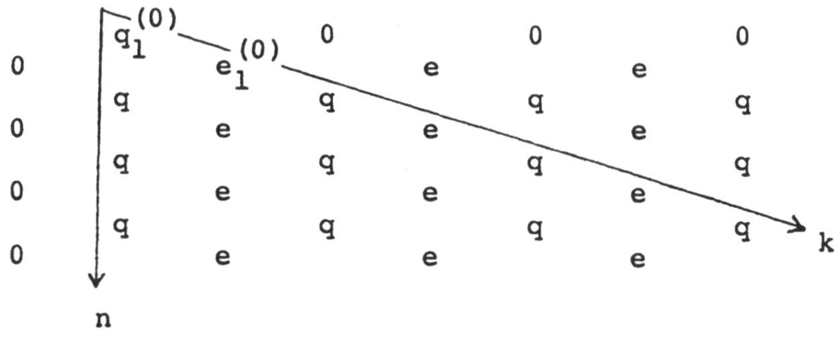

The connection with the qd scheme for $\hat{P}$ is given by the formulas (see Henrici (1974), Theorem 7.6d)

$$\hat{q}_1^{(0)} = -q_1^{(0)} ; \tag{3.8a}$$

$$\hat{q}_k^{(n)} = e_{n+k-1}^{(1-n)} , \quad \hat{e}_k^{(n)} = q_{n+k}^{(1-n)} . \tag{3.8b}$$

The following explicit representations for the Padé approximants $P_{m,n}$ result if $m < n$:

$$P_{m,n}(x)$$

$$= 1/\{\hat{c}_0 + \hat{c}_1 x + \ldots + \hat{c}_{n-m} x^{n-m} + \frac{\hat{c}_{n-m+1} x^{n-m+1}|}{|1} - \frac{e_{n-m+1}^{(m-n)} x|}{|1} - \frac{q_{n-m+2}^{(m-n)} x|}{|1} - \ldots - \frac{e_n^{(m-n)} x|}{|1} - \frac{q_{n+1}^{(m-n)} x|}{|1} \}$$

$$= 1/\{\hat{c}_0 + \hat{c}_1 x + \ldots + \hat{c}_{n-m-1} x^{n-m-1} + \frac{\hat{c}_{n-m} x^{n-m}|}{|1} - \frac{e_{n-m}^{(m-n+1)} x|}{|1} - \frac{q_{n-m+1}^{(m-n+1)} x|}{|1} - \ldots - \frac{q_n^{(m-n+1)} x|}{|1} - \frac{e_n^{(m-n+1)} x|}{|1}$$

(3.9)

If $m = n$, the quantity $e_{n-m}^{(m-n+1)}$ in the second formula according to (3.8b) must be replaced by

$$- q_1^{(0)} = c_1/c_0 .$$

Both formulas (3.9) are then valid for $m = n$, and are easily seen to be identical with the corresponding formulas (3.7).

As shown by (3.7) and (3.9), all Padé approximants $P_{m,n}$ of a hypernormal series P can be expressed in terms of the extended qd scheme of P . Only a single two-dimensional array of numbers (and not of functions) is thus necessary to compute the Padé approximants for arbitrary values of x . For a sequence of Padé approximants where $n - m = \ell$, a constant, only the elements

$$q_k^{(\ell)} , e_k^{(\ell)}$$

of the scheme are required. Once the qd scheme is known, only $m + n - 1$ multiplications and additions are required to evaluate $P_{m,n}$ if $m \geq n$, and only $2n - 1$ divisions. These numbers could even be reduced somewhat by writing the continued fractions in contracted form, which requires some additional preliminary arithmetic.

Although the qd algorithm is mentioned in Baker's treatise on Padé theory (1975), its basic connection with Padé approximants, as expressed by the formulas (3.7) and (3.9), is not pointed out there.

4. NUMERICAL INSTABILITY OF THE qd ALGORITHM

By converting a given power series, or one of its remainders, into a corresponding continued fraction, a divergent series can often be turned into a converging expression. This for instance is always so if the coefficients c_k of the given series are moments,

$$(-1)^k c_k = \int_0^\infty t^k \, d\mu(t) \quad , \; k = 1, 2, \ldots , \tag{4.1}$$

where $d\mu$ is a non-negative measure on $[0,\infty)$ whose support has infinitely many points. The series

$$P = c_0 + c_1 x + c_2 x^2 + \ldots \tag{4.2}$$

then is, in general, divergent. It is, however, normal, and if the $|c_k|$ don't grow too strongly - a condition such as

$$\sum_{k=1}^{\infty} |c_k|^{-1/2k} = \infty$$

is sufficient - the continued fraction (3.6) converges, for all $x \in \mathbb{C}$ not on the axis of negative reals, to the function

$$C^{(n)}(x) := (-1)^n \int_0^\infty \frac{t^n \, d\mu(t)}{1 + xt}$$

which is asymptotically represented by $c_n + c_{n+1}x + \ldots$ as $x \to 0$. Equivalently, there holds the identity

$$\int_0^\infty \frac{d\mu(t)}{1 + xt} = c_0 + c_1 x + \ldots + c_{n-1} x^{n-1} + x^n C^{(n)}(x)$$

for every x in the cut complex plane. This method for "summing" an asymptotic series is applied frequently, especially in computational physics, even in cases where it cannot be rigorously justified.

Rigorous examples for the above technique abound in classical analysis; see e.g. Henrici (1977), §12.12, §12.13. One of the simplest examples is given by

$$d\mu(t) = e^{-t}dt ,$$

so that

$$c_k = (-1)^k \int_0^\infty t^k e^{-t} dt = (-1)^k k!, \; k = 0, 1, 2, \ldots \quad (4.3)$$

For the power series

$$P = 1 - 1!x + 2!x^2 - 3!x^3 + \ldots$$

the qd scheme can be exhibited explicitly. We have

$$q_1^{(n)} = \frac{c_{n+1}}{c_n} = - n - 1 \; , \; n = 0, 1, 2, \ldots ,$$

and using the basic relations (1.5) it is easily seen that the complete scheme looks as suggested below:

$$\begin{array}{lllllll}
q_1^{(0)} = -1 & & & & & & \\
 & -1 & & & & & \\
-2 & & -2 & & & & \\
 & -1 & & -2 & & & \\
-3 & & -3 & & -3 & & \\
 & -1 & & -2 & & -3 & \\
-4 & & -4 & & -4 & & -4 \\
 & -1 & & -2 & & -3 & \\
-5 & & -5 & & -5 & & -5
\end{array} \quad (4.4)$$

From this, the continued fractions $C^{(n)}$ can be written down immediately, and the process yields convergent representations of the function

$$E(x) := \int_0^\infty \frac{e^{-t}}{1 + xt} dt ,$$

which is a form of the exponential integral.

What makes our example especially interesting, however, is the fact that it enables us to study the numerical performance of the qd algorithm. While it is true that in our example the scheme can be described by explicit formulas,

$$q_k^{(n)} = - n - k \; , \quad e_k^{(n)} = - k \; , \quad (4.5)$$

no such formulas exist in general, and the scheme must be constructed numerically. If the c_k are such that the scheme qualitatively behaves as in our example, then the errors that must be expected are similar to the errors incurred if the scheme (4.4) were generated numerically disregarding the fact that the elements are integers. While the product rule (1.5b) is harmless, the sum rule (1.5a) in our case yields

$$e_k^{(n)} = e_{k-1}^{(n+1)} + (q_k^{(n+1)} - q_k^{(n)})$$
$$= -(k-1) + [-(n+1+k) + (n+k)] .$$

The value in brackets arises as a difference of large numbers, and thus is subject to loss of accuracy. Since the value of the difference is of the order of 1, working in the number system with base b produces a loss of approximately $\mathrm{Log}_b(n + k + k)$ digits.

To see the consequences, assume that the elements

$$q_1^{(0)}, e_1^{(0)}, \ldots, q_m^{(0)}$$

of the scheme are to be computed. This requires the construction of a triangular segment of the qd scheme which in the k-th column extends down to the element

$$e_k^{(2m-2k-1)} .$$

The calculation of this element involves a loss of $\mathrm{Log}_b(2m - k)$ digits. The total number of lost digits therefore equals

$$\mathrm{Log}(2m-1) + \mathrm{Log}(2m-2) + \ldots + \mathrm{Log}(m+1)$$
$$= \mathrm{Log}\frac{(2m-1)!}{m!}$$
$$\sim \mathrm{Log}(2^{2m-1/2}\, m^{m-1}\, e^{-m})$$

where all logarithms are to the base b. If $b = 2$, about

$$2m - \frac{1}{2} + (m-1)\,\mathrm{Log}_2 m - m\,\mathrm{Log}_2 e$$

digits will be lost, which of course cannot be tolerated.

A situation like the one described occurs, for instance, with the hypergeometric series

$$_2F_0(\alpha,\beta;\ -x)\ . \tag{4.6}$$

All asymptotic series for confluent hypergeometric functions can be reduced to this form; the special case $\alpha = 1/2 + \nu$, $\beta = 1/2 - \nu$ corresponds to the Bessel functions of order ν (see Gargantini and Henrici (1967)). For the series (4.6),

$$c_k = (-1)^k \frac{(\alpha)_k(\beta)_k}{k!}\ ,$$

[$(\alpha)_k$ is the Pochhammer symbol], and consequently

$$\begin{aligned} q_1^{(n)} &= -\frac{(n+\alpha)(n+\beta)}{n+1} \\ &= -n-\alpha-\beta+1-\frac{(\alpha-1)(\beta-1)}{n+1}\ . \end{aligned} \tag{4.7}$$

It is easy to see that the resulting qd scheme behaves qualitatively like (4.5), and the resulting loss of accuracy predicted above has been observed experimentally.

5. STABLE WAYS OF GENERATING THE qd SCHEME

A radical way to combat numerical instability in the generation of the qd scheme consists in performing all arithmetic operations in the field of rationals, i.e., of equivalence classes of pairs of integers. This is feasible, in principle, whenever the quotients $q_1^{(n)}$ are rational numbers. Due to the usually rapid growth of the numerators and of the denominators rational arithmetic soon becomes costly in terms of computing time and storage. Because (as opposed to other ways of constructing the Padé table) the qd table has to be constructed only once, a high initial cost for setting up the scheme might be tolerated. To avoid the nuisance of rational arithmetic, especially if approximants of very high order are desired, it still is of interest to study stable methods for constructing the qd scheme in conventional arithmetic. One such method was recently proposed by Stokes (1980).

In his algorithm Stokes successively constructs ordinary difference tables of the columns of the qd table. We describe the algorithm in terms of the usual forward difference operator Δ, which here always acts on the superscript of a two-dimensional array $c_k^{(n)}$. Thus,

$$\Delta c_k^{(n)} := c_k^{(n+1)} - c_k^{(n)} ,$$

and similarly for higher powers of Δ. The following Leibniz type formula for the differences of a sequence of products is easily established by induction with respect to k :

$$\Delta^k(a^{(n)}b^{(n)}) = \sum_{j=0}^{k} \binom{k}{j} \Delta^j a^{(n+k-j)} \Delta^{k-j} b^{(n)} . \quad (5.1)$$

We also recall that the full difference table of a segment

$$a^{(n)}, a^{(n+1)}, \ldots , a^{(n+k)}$$

of a sequence may be constructed from the first diagonal

$$a^{(n)}, \Delta a^{(n)}, \ldots , \Delta^k a^{(n)}$$

of the difference table by "filling up" the table.

We now describe Stokes' algorithm formally. Later we shall find conditions under which it can be expected to behave stably. The aim is to construct the first $2m-1$ elements of the first diagonal of the qd scheme,

$$q_1^{(0)}, e_1^{(0)}, q_2^{(0)}, e_2^{(0)}, \ldots , q_m^{(0)} . \quad (5.2)$$

from the first $2m-1$ first elements

$$q_1^{(0)}, q_1^{(1)}, \ldots , q_1^{(2m-2)} \quad (5.3)$$

of the first column of the scheme. Assume that the full (ordinary) difference table of the array (5.3) is known. From (1.5a), since $e_0^{(n)} = 0$,

$$e_1^{(n)} = \Delta q_1^{(n)} ,$$

thus the difference table of the array

$$e_1^{(0)}, e_1^{(1)}, \ldots , e_1^{(2m-3)}$$

then is also known. From (1.5b),

$$q_1^{(n+1)} e_1^{(n+1)} = e_1^{(n)} q_2^{(n)} ,$$

thus

$$\Delta^k(q_1^{(1)} e_1^{(1)}) = \Delta^k(e_1^{(0)} q_2^{(0)})$$

$k = 0, 1, \ldots$, hence using (5.1),

$$\sum_{j=0}^{k} \binom{k}{j} \Delta^{k-j} q_1^{(1)} \Delta^j e_1^{(k-j+1)} = \sum_{j=0}^{k} \binom{k}{j} \Delta^{k-j} q_2^{(0)} \Delta^j e_1^{(k-j)} , \qquad (5.4)$$

$k = 0, 1, \ldots, 2m-4$. If $e_1^{(k)} \neq 0$, these relations may be solved for the differences

$$q_2^{(0)}, \Delta q_2^{(0)}, \ldots, \Delta^{2m-4} q_2^{(0)} ,$$

which permit us to construct the full difference table of the array

$$q_2^{(0)}, q_2^{(1)}, \ldots, q_2^{(2m-4)} . \qquad (5.5)$$

From (1.5a),

$$e_2^{(n)} = e_1^{(n+1)} + \Delta q_2^{(n)} ,$$

thus

$$\Delta^k e_2^{(n)} = \Delta^k e_1^{(n+1)} + \Delta^{k+1} q_2^{(n)} , \qquad (5.6)$$

and the difference table of the array

$$e_2^{(0)}, e_2^{(1)}, \ldots, e_2^{(2m-5)}$$

can be found from the difference tables that already have been constructed. Proceeding in this manner, we successively find the difference tables (of diminishing size) of the arrays

$$q_k^{(0)}, q_k^{(1)}, \ldots, q_k^{(2m-2k)} ; \qquad (5.7)$$

$$e_k^{(0)}, e_k^{(1)}, \ldots, e_k^{(2m-2k-1)} , \qquad (5.8)$$

$k = 1, 2, \ldots, m-1$, until after a last use of (1.5b) we arrive at the array consisting of the sole element $q_m^{(0)}$. Any of the difference tables thus constructed contains, in particular, the first element of the array. Thus the goal of constructing the array (5.2) has been achieved.

If this roundabout way of generating the qd scheme is to be numerically stable, the following conditions should be satisfied.

(i) The difference table of the original array (5.3) must be known exactly. Since, as is well known, the numerical generation of a difference table is an unstable process, this means that it must be possible to represent the differences of the sequence $q_1^{(n)}$ by a stable analytical formula. This assumption is not as unrealistic as it might appear at first sight, see the example given below.

(ii) In any of the difference tables generated, differences of a fixed order should have a constant sign. This is required because whole difference tables are reconstructed from a diagonal. Unless differences of fixed order have a constant sign, cancellation may occur, as it does in the formation of a difference table from a first column.

(iii) In order to avoid cancellation in sums like (5.4), either all differences should have the same sign, or the signs of the differences in consecutive columns of any difference table should alternate.

One example where the foregoing conditions can at least be partially verified is given by (4.7). It is well known that the differences of the sequence $\{c^{(n)}\}$ where

$$c^{(n)} = \frac{1}{n + a} \tag{5.9}$$

($a \neq 0, -1, \ldots$) are given by

$$\Delta^k c^{(n)} = (-1)^k \frac{k!}{(n + a)_{k+1}} . \tag{5.10}$$

Thus for the $q_1^{(n)}$ defined by (4.7) we have, letting $\gamma := (\alpha-1)(\beta-1)$,

$$\Delta q_1^{(n)} = -1 + \gamma\frac{1}{(n+1)_2},$$
$$\Delta^k q_1^{(n)} = (-1)^{k+1} \gamma\frac{k!}{(n+1)_{k+1}} \tag{5.11}$$

$k = 2, 3, \ldots$. Thus we see that condition (i) is satisfied, and that the conditions (ii) and (iii) hold at least for the two initial difference tables.

Starting from the formulas (5.11), Stokes in the case $\alpha = \beta = 1/2$ was able to generate the array (5.2) for $m = 35$ to full machine precision, working in ordinary single precision arithmetic.

Of the three conditions mentioned, (i) is not only satisfied in the special case (4.7). For if P is any hypergeometric series (convergent or not), $q_1^{(n)}$ has the form

$$q_1^{(n)} = r(n) , n = 0, 1, \ldots ,$$

where r is a rational function. If the poles of r are distinct, $q_1^{(n)}$ thus can be represented as a polynomial in n plus a finite sum of terms of the form (5.9), and the full difference table can be formed by means of (5.10). If r has repeated poles, a method of confluence would have to be used.

BIBLIOGRAPHY

G. A. Baker jr. (1975): Essentials of Padé approximants. Academic Press, New York.

R. de Montessus de Ballore (1902): Sur les fractions continues algébriques. Bull. Soc. Math. Française 30.

I. Gargantini and P. Henrici (1967): A continued fraction algorithm for the computation of higher transcendental functions. Math. Comp. 21, 18 - 29.

P. Henrici (1958): The quotient-difference algorithm. Appl. Math. Ser. 49, 26 - 46. National Bureau of Standards, Washington, D. C.

P. Henrici (1974, 1977): Applied and Computational Complex Analysis, vols. 1, 2,. Wiley, New York.

O. Perron (1957): Die Lehre von den Kettenbrüchen, Band II. Teubner, Stuttgart.

H. Rutishauser (1954): Der Quotienten-Differenzen-Algorithmus. Z. angew. Math. Physik 5, 233 - 251.

W. Seewald (1981): Quotienten-Differenzen-Algorithmus: Beweis der Regeln von Rutishauser. Report 81-08, Sem. f. angew. Math., ETH Zürich

G. W. Stewart (1971): On a companion operator for analytic functions. Numer. Math. 18, 26 - 43.

A. N. Stokes (1980): A stable quotient-difference algorithm. Math. Comp. 34, 515 - 519.

TOPICS IN COMPUTATIONAL COMPLEX ANALYSIS:
III. THE ASYMPTOTIC BEHAVIOR OF BEST APPROXIMATIONS TO ANALYTIC FUNCTIONS ON THE UNIT DISK.

Peter Henrici
ETH-Zentrum, CH-8092 Zurich

Abstract. Let $f(z) = \Sigma a_k z^k$ be analytic for $|z| \leq 1$, let $n \geq 1$ be an integer, and let p^* denote the best polynomial approximation of degree $n-1$ to f in the Chebyshev norm. We study the asymptotic behavior of p^* as the parameter $\rho := \sup_{k>0} |a_{n+k}/a_n|^{1/k}$ tends to zero. Essential use is made of results of Trefethen (1981) and of Stewart (1973).

1. INTRODUCTION

We call A_∞ the algebra of functions

$$g(z) = \sum_{k=-\infty}^{\infty} a_k z^k \tag{1.1}$$

analytic in a neighborhood of $|z| = 1$, normed by

$$\|g\| := \sup_{|z|=1} |g(z)| , \tag{1.2}$$

and we let $A_0 \subset A_\infty$ denote the subalgebra of functions g analytic throughout the disk $|z| \leq 1$. Let, for $n = 0, 1, \ldots$, P_n denote the polynomials of degree n. Our basic problem is the approximation of a given function $g \in A_0$ by polynomials $p \in P_{n-1}$. It is well known (see Walsh (1969)) that for any $g \in A_0$ and for any $n > 0$ there exists a unique $p^* \in P_{n-1}$ such that $\|g - p^*\|$ is minimized. Until recently, no very efficient algorithms were known for constructing this "best" approximation p^*. Trefethen (1981) showed how to apply

H. Werner et al. (eds.), Computational Aspects of Complex Analysis, 169–191.

a classical result due to Carathéodory, Fejer, and Schur to obtain an approximation p^{CF} that in a certain sense approximates p^* very closely, and he demonstrated the numerical effectiveness of the "Carathéodory-Fejer" approximation p^{CF}. In this report we shall describe some aspects of Trefethen's work and, in addition, explicitly compute the asymptotic behavior of p^{CF} and, hence, of p^*.

The author is indebted to M. Gutknecht, G. W. Stewart, and L. N. Trefethen for helpful discussions and suggestions.

2. NORMALIZATION OF THE APPROXIMATION PROBLEM

We may assume without loss of generality that the function $g \in A_0$ to be approximated is of the form

$$g(z) = \sum_{k=n}^{\infty} a_k z^k \tag{2.1}$$

because if $p \in P_{n-1}$ approximates (2.1), an approximation to the function

$$g(z) = \sum_{k=0}^{\infty} a_k z^k$$

with the same error is given by $p(z) + \Sigma_{k=0}^{n-1} a_k z^k$. We shall only consider approximations to functions (2.1) such that $a_n \neq 0$. We then may assume without loss of generality that g is normalized so that $a_n = 1$. The best approximation to a non-normalized function g such that $a_n \neq 0$ obviously is $a_n p^*$, where p^* is the best approximation to the normalized function $a_n^{-1} g(z)$.

We finally note that the norms of the functions $g(z)$ and $z^m g(z)$ are the same for any integer m. Instead of approximating a normalized function g of the form (2.1) by a polynomial $p \in P_{n-1}$ we thus may approximate

$$z^{-n} g(z) = 1 + \sum_{k=1}^{\infty} a_{n+k} z^k \tag{2.2}$$

by a polynomial of degree n in z^{-1} which has the special form

$$h(z) = \sum_{k=1}^{n} b_k z^{-k} . \tag{2.3}$$

In the interest of economy of notation, we write the function (2.2) to be approximated in the form

$$f_0(z) = \sum_{k=0}^{\infty} a_k z^k , \tag{2.4}$$

where it is always understood that the series is normalized, i.e., that $a_0 = 1$. Similarly, the approximating polynomial (2.3) is now written as

$$h(z) = - \sum_{k=-n}^{-1} a_k z^k \tag{2.5}$$

The error $f := f_0 - h$ of the approximation of f_0 by h is then given by

$$f(z) = \sum_{k=-n}^{\infty} a_k z^k . \tag{2.6}$$

Denoting by A_n the space of functions in A_∞ whose only singularity in $|z| < 1$ is a pole of order $\leq n$ at $z = 0$, our approximation problem now may be stated thus: <u>Given a normalized power series</u> $f_0 \in A_0$, <u>to extend it to a Laurent series</u> $f \in A_n$ <u>such that</u> $\|f\|$ <u>becomes as small as possible</u>.

The sense in which we shall study the asymptotic behavior of best and of near-best approximations can now be described as follows: We consider normalized power series

$$f_0(z) = \sum_{k=0}^{\infty} a_k z^k$$

for which

$$\rho := \sup_{k \geq 1} |a_k|^{1/k} \leq 1 , \tag{2.7}$$

so that

$$|a_k| \leq \rho^k , \; k = 1, 2, \ldots , \tag{2.8}$$

and it is our endeavor to describe the behavior of these approximations and their errors as $\rho \to 0$.

Although most of our methods carry over to the general case, we shall, for simplicity of exposition, consider only functions f_0 such that the series (2.4) has <u>real</u> coefficients.

3. MARGIN OF ERROR; STATEMENT OF RESULTS

In the above normalization, the approximation of a given function g by one of its Taylor polynomials corresponds to the trivial extension of the series f_0 so that $f = f_0$. The norm of the error in this case is

$$\begin{aligned}\|f\| &= \left\|\sum_{k=0}^{\infty} a_k z^k\right\| \\ &= 1 + |a_1| + O(\rho^2) .\end{aligned} \tag{3.1}$$

On the other hand, if

$$f(z) = \sum_{k=-\infty}^{\infty} a_k z^k$$

is any extension of f_0 whatsoever (even an extension involving infinitely many terms with negative exponents), then by Laurent's formula

$$a_k = \frac{1}{2\pi i} \int_{|z|=1} f(z) z^{-k-1} dz ,$$

hence

$$|a_k| \leq \|f\|$$

for all integers k. In view of $a_0 = 1$ there follows

$$\|f\| \geq 1 . \tag{3.2}$$

By the definition of ρ, $|a_1| \leq \rho$. Thus as $\rho \to 0$, the margin between (3.1) and (3.2) is not very great. The following results should be seen in the light of these remarks.

THEOREM 1. Let $f_0(z) = 1 + \Sigma_{k=1}^{\infty} a_k z^k$ ($a_k \in R$) be in A_0, let n be an integer ≥ 5, and let $f^* = f_0 - h^*$ be the minimum norm extension of f_0 into A_n. If the ρ defined by (2.7) satisfies $\rho < 1/36$, then

$$\left| \|f^*\| - (1+a_1^2+a_2^2-a_1^2 a_2+a_1^4) \right| \leq 0.57 \; (18\rho)^6 . \tag{3.3}$$

Similarly if $n \geq 3$,

$$\left| \|f^*\| - (1+a_1^2) \right| \leq 0.15 \; (18\rho)^4 .$$

Thus in particular, $\|f^*\| = 1 + O(\rho^2)$, which is a qualitative improvement over (3.1). Concerning the asymptotic behavior of h^* we prove

THEOREM 2. _If_ $\rho \leq 0.07\ 3^{-n}$, _then_

$$\|h^* - h\| \leq 1.19\ (36\rho)^3 , \tag{3.4}$$

where

$$h(z) := a_1 z^{-1} - (a_1^2 - a_2) z^{-2} . \tag{3.5}$$

It should be noted that the foregoing results hold for fixed n ; they describe the asymptotic behavior of the best approximation as $\rho \to 0$. It is remarkable, however, that in the normalization given in §2 this behavior does not depend on n .

Asymptotic formulas for $\|f^*\|$ and f^* as $n \to \infty$ have been given by Bernstein (1926) and Saff (1973). In those cases where both results are applicable, our results are more precise.

The proof of the above Theorems will be given in the remaining sections of this paper. In addition to proving our asymptotic formulas, we present an algorithm that permits, in principle, to obtain asymptotic formulas with error terms up to order $O(\rho^{n+1})$. However, the algebraic manipulations required to obtain the explicit form of these expansions would require MACSYMA or some similar such system.

4. THE CARATHEODORY-FEJER APPROXIMATION

If f_0 is a polynomial,

$$f_0(z) = 1 + \sum_{k=1}^{N} a_k z^k , \tag{4.1}$$

the problem of extending f_0 to a series $f \in A_\infty$,

$$f(z) = \sum_{k=-\infty}^{N} a_k z^k \tag{4.2}$$

such that $\|f\|$ becomes as small as possible was completely solved by Carathéodory and Fejer (see Goluzin (1969)). They found that there exists a unique function f minimizing $\|f\|$. It is characterized by the property that for $|z| = 1$

$$|f(z)| = \text{const} = \|f\| . \tag{4.3}$$

If the coefficients of the polynomial (4.1) are real, the solution of the Caratheodory-Fejer problem may be found as follows (Schur (1917)). Let λ be an eigenvalue of largest absolute value of the real symmetric matrix

$$\underline{A} := \begin{pmatrix} 1 & a_1 & a_2 & \dots & a_{N-1} & a_N \\ a_1 & a_2 & a_3 & \dots & a_n & 0 \\ a_2 & a_3 & a_4 & \dots & 0 & 0 \\ & \cdot & \cdot & \cdot & \cdot & \cdot \\ a_N & 0 & 0 & & 0 & 0 \end{pmatrix} , \tag{4.4}$$

and let

$$\underline{b} = (b_0, b_1, \dots , b_N)^T \tag{4.5}$$

be a corresponding eigenvector, so that

$$\underline{A}\,\underline{b} = \lambda \underline{b} . \tag{4.6}$$

Let

$$q(z) := b_0 + b_1 z + \dots + b_N z^N ; \tag{4.7}$$

all zeros z_i of q may be shown to satisfy $|z_i| \geqq 1$. The solution of the problem of Carathéodory-Fejer is then given by

$$f(z) := \lambda \frac{q(z)}{q(z^{-1})} \tag{4.8}$$

By comparing the coefficients of z^k in the identity

$$q(z^{-1}) f(z) = \lambda\, q(z)$$

we obtain for $k \geqq 0$ the system of equations (4.6); for $k < 0$ there results the recurrence relation

$$a_k b_0 + a_{k+1} b_1 + \dots + a_{k+N} b_N = 0 , \; k = -1, -2, \dots , \tag{4.9}$$

which allows to determine the coefficients of the series (4.2).

From (4.8) we see that for $|z| = 1$

$$|f(z)| = \|f\| = |\lambda| \; ; \tag{4.10}$$

the minimum of the norms of all extensions (4.2) of a given polynomial (4.1) thus equals the spectral radius of the matrix (4.4).

The connection of the Carathéodory-Fejer problem with the approximation problem stated in §1 was noted by Trefethen (1981). He showed that if the coefficients of the polynomial (4.1) satisfy $|a_k| \leqq \rho^k$ $(k = 1, 2, \ldots, N)$, then the coefficients a_k of the series (4.2) as defined by (4.8) satisfy

$$|a_{-k}| \leqq 3^N |\lambda| (6\rho)^k \, , \quad k = 1, 2, \ldots \, . \tag{4.11}$$

Consequently, if ρ is sufficiently small, the truncated series

$$f^{CF}(z) := \sum_{k=-n}^{N} a_k z^k \tag{4.12}$$

on $|z| = 1$ does not differ much from f ; in fact if $\rho < 1/6$ we have

$$\|f^{CF} - f\| \leqq 3^N |\lambda| \sum_{k=n+1}^{\infty} (6\rho)^k$$

or

$$\|f^{CF} - f\| \leqq \frac{3^N |\lambda| (6\rho)^{n+1}}{1 - 6\rho} \, . \tag{4.13}$$

By definition, f^{CF} is the error of the <u>Carathéodory-Fejer approximation</u> h^{CF} (of degree $n-1$) to f_0 , thus in our normalization

$$h^{CF}(z) = - \sum_{k=-n}^{-1} a_k z^k \, . \tag{4.14}$$

Let now

$$f^*(z) := \sum_{k=-n}^{-1} a_k^* z^k + \sum_{k=0}^{N} a_k z^k$$

be the error of the <u>best</u> approximation to f_0 by a polynomial of degree $n-1$. This likewise is an extension of the series f_0 to A_∞ . Hence, since f is the extension with smallest norm,

$$\|f^*\| \geqq \|f\| \, .$$

On the other hand, since $\|f^*\|$ is the error of at least as good an approximation as h^{CF}, we have

$$\|f^*\| \leqq \|f^{CF}\| .$$

We thus find for $\|f^*\|$ the very tight estimates

$$|\lambda| \leqq \|f^*\| \leqq |\lambda|\left(1 + \frac{3^N(6\rho)^{n+1}}{1 - 6\rho}\right) , \qquad (4.15)$$

showing that for fixed n and N, $|\lambda|$ approximates $\|f^*\|$ to within $O(\rho^{n+1})$.

This does not necessarily imply that the approximations h^* and h^{CF} are pointwise close. However, Trefethen (1981) also showed that for the corresponding <u>polynomial</u> approximations, if $N = n$ and $\rho < 1/36$,

$$|z^n\{h^*(z) - h^{CF}(z)\}| \leqq (36\rho)^{n+1}$$

for all $|z| \leqq 1$. This is the same as saying that

$$\|f^* - f^{CF}\| \leqq (36\rho)^{n+1} , \qquad (4.16)$$

and by using the relation (4.13), still assuming $N = n$ and $\rho < 1/36$, and making use of the inequality

$$|\lambda| \leqq \frac{1}{1 - \rho}$$

following trivially from Gerschgorin's theorem, there now follows

$$\|f^* - f\| \leqq 1.21\ (36\rho)^{n+1} . \qquad (4.17)$$

5. EXTENSION TO NON-POLYNOMIAL f_0

The estimates of §4 have been proved under the assumption that f_0, the series which is to be extended to A_n, is a polynomial; for (4.17) it is even assumed that the degree of the polynomial is n. This restriction will now be removed.

Thus let

$$f_0(z) = 1 + \sum_{k=1}^{\infty} a_k z^k$$

be in A_0. We denote by

$$f^*(z) = \sum_{k=-n}^{-1} a_k^* z^k + 1 + \sum_{k=1}^{\infty} a_k z^k$$

the smallest norm extension of f_0 into A_n ; thus, $z^n f^*(z)$ is the error of the best polynomial approximation

$$p^*(z) := - \sum_{k=0}^{n-1} a_{k-n}^* z^k$$

($p^* \in P_{n-1}$) to $z^n f_0(z)$, and $\|f^*\|$ is the quantity which we wish to study. Let, furthermore,

$$\tilde{f}_0(z) := 1 + \sum_{k=1}^{n} a_k z^k$$

be the series f_0 truncated to degree n , let

$$\tilde{f}(z) = \sum_{k=-\infty}^{n} a_k z^k$$

be its Carathéodory-Fejer extension to A_∞ , let

$$\tilde{f}^{CF}(z) = \sum_{k=-n}^{n} a_k z^k$$

be the error of the Carathéodory-Fejer approximation to $\tilde{f}_0$, and let

$$\tilde{f}^*(z) = \sum_{k=-n}^{-1} \tilde{a}_k^* z^k + 1 + \sum_{k=1}^{n} a_k z^k$$

be the smallest norm extension of $\tilde{f}_0$ to A_n . If λ denotes the eigenvalue of largest modulus of the matrix $\underline{A}$ defined by (4.4), then if $\rho < 1/18$ we have by (4.15)

$$|\lambda| \leqq \|\tilde{f}^*\| \leqq |\lambda| + 0.50(18\rho)^{n+1} . \tag{5.1}$$

Obviously,

$$\|f_0 - \tilde{f}_0\| \leqq \sum_{k=n+1}^{\infty} |a_k| \leqq \frac{\rho^{n+1}}{1-\rho} \tag{5.2}$$

Since f^* is the extension of f_0 of smallest norm,

$$\|f^*\| \leqq \|f_0 + (\tilde{f}^* - \tilde{f}_0)\|$$
$$\leqq \|f_0 - \tilde{f}_0\| + \|\tilde{f}^*\|$$

and thus, using (5.1) and supposing $\rho < 1/36$,

$$\|f^*\| \leqq \frac{\rho^{n+1}}{1-\rho} + |\lambda| + 0.50(18\rho)^{n+1}$$
$$\leqq |\lambda| + 0.56(18\rho)^{n+1} . \tag{5.3}$$

On the other hand, $\tilde{f}^*$ is the smallest norm extension of $\tilde{f}_0$, thus

$$\begin{aligned}\|f^*\| &= \|f_0 + (f^* - f_0)\| \\ &= \|f_0 - \tilde{f}_0 + \{\tilde{f}_0 + (f^* - f_0)\}\| \\ &\geq \|\tilde{f}_0 + (f^* - f_0)\| - \|f_0 - \tilde{f}_0\| \\ &\geq \|\tilde{f}^*\| - \|f_0 - \tilde{f}_0\| ,\end{aligned}$$

hence

$$\|f^*\| \geq |\lambda| - \frac{\rho^{n+1}}{1 - \rho} .$$

Together with (5.3) this shows that

$$\big|\|f^*\| - |\lambda|\big| \leq 0.56(18\rho)^{n+1} , \tag{5.4}$$

and hence that $|\lambda|$ asymptotically up to $O(\rho^{n+1})$ approximates $\|f^*\|$ as well as $\|\tilde{f}^*\|$.

We similarly wish to show that

$$\|f^* - f\| = O(\rho^{n+1}) . \tag{5.5}$$

To this end we require a result of Trefethen ((1981), Theorem 6) which in our notation and adapted to our normalization of the approximation problem reads as follows: Let $f_0 \in A_0$, and let f^* be the minimum norm extension of f_0 to A_n . If g is any extension of f_0 into A_0 such that

$$\alpha := \|g - 1\| < 1 \tag{5.6}$$

then, letting

$$\delta := \|g\| - \min_{|z|=1} |g(z)| , \tag{5.7}$$

there holds

$$\|g - f^*\| \leq \frac{2^n \delta}{\sqrt{1 - \alpha^2}} . \tag{5.8}$$

We apply this result to

$$g := \tilde{f}^{CF} + (f_0 - \tilde{f}_0) , \tag{5.9}$$

that is, to

$$g(z) = \sum_{k=-n}^{-1} a_k z^k + 1 + \sum_{k=1}^{\infty} a_k z^k .$$

By (4.11),

$$\alpha = \|g - 1\| \leq \sum_{k=-1}^{-n} |a_k| + \sum_{k=1}^{\infty} |a_k|$$

$$\leq 3^n |\lambda| \frac{6\rho}{1 - 6\rho} + \frac{\rho}{1 - \rho}$$

$$\leq 8.41 \cdot 3^n \rho ,$$

and condition (5.6) is seen to be satisfied for sufficiently small values of ρ . To estimate δ , we write

$$g(z) = \sum_{k=-\infty}^{n} a_k z^k + \sum_{k=n+1}^{\infty} a_k z^k - \sum_{k=-\infty}^{-n-1} a_k z^k$$

and use the fact that the first term on the right has constant modulus $|\lambda|$ on $|z| = 1$. Again using (4.11), this yields

$$\delta \leq \frac{2\rho^{n+1}}{1 - \rho} + 2|\lambda|\frac{1}{3} \frac{1}{1 - 6\rho} (18\rho)^{n+1}$$
$$\leq 0.94 \ (18\rho)^{n+1} .$$

Restricting ρ to values such that $\sqrt{1 - \alpha^2} \geq \frac{1}{2}$, that is, to the range

$$\rho \leq 0.10 \cdot 3^{-n} , \qquad (5.10)$$

(5.8) now yields

$$\|g - f^*\| \leq 0.94 \ (36\rho)^{n+1} . \qquad (5.11)$$

To obtain (5.4), we write

$$f^* - \tilde{f} = f^* - g + g - \tilde{f}$$

$$= (f^* - g) + (\tilde{f}^{CF} - \tilde{f}) + (f_0 - \tilde{f}_0) .$$

Using (4.11) once again, we get by (5.2) and (5.11), if ρ satisfies (5.10),

$$\|f^* - \tilde{f}\| \leq 0.94(36\rho)^{n+1} + 0.42(18\rho)^{n+1} + 1.03\rho^{n+1}$$

or

$$\|f^* - \tilde{f}\| \leq 1.18(36\rho)^{n+1} , \qquad (5.12)$$

showing that $\tilde{f}$ asymptotically approximates f^* as well as $\tilde{f}^*$.

6. ASYMPTOTIC SOLUTION OF THE EIGENVALUE PROBLEM (4.6)

The problem of describing the asymptotic behavior of f^* and of $\|f^*\|$ is solved, in principle, by the formulas (5.12) and (5.4), which show that these quantities are asymptotic, with errors $O(\rho^{n+1})$, to f and $|\lambda| = \|\tilde{f}\|$, respectively. However, in order to take advantage of this fact for the purpose of computing near-best approximations, it is necessary to describe explicitly the asymptotic behavior of f^* or, what amounts to the same, of the solution of the eigenvalue problem (4.6). To this task we turn next.

A method due to Stewart (1973) is ideally suited to our purpose. We partition the given matrix $\underline{A}$ (where $N = n$) in the form

$$\underline{A} = \begin{pmatrix} 1 & \underline{a}^T \\ \underline{a} & \underline{C} \end{pmatrix}$$

where

$$\underline{a} := \begin{pmatrix} a_1 \\ a_2 \\ \vdots \\ a_n \end{pmatrix} , \qquad \underline{C} := \begin{pmatrix} a_2 & a_3 & \cdots & a_n & 0 \\ a_3 & a_4 & \cdots & 0 & 0 \\ & \cdot & \cdot & \cdot & \cdot \\ a_n & 0 & \cdots & 0 & 0 \\ 0 & 0 & \cdots & 0 & 0 \end{pmatrix} .$$

We seek an eigenvector $\underline{b}$ of $\underline{A}$ in the form

$$\underline{b} = \begin{pmatrix} 1 \\ \underline{x} \end{pmatrix} .$$

To get an equation for $\underline{x}$, we use the fact that a vector $\underline{c}$ is proportional to $\underline{b}$ if and only if

$$(- \underline{x} \quad \underline{I})\, \underline{c} = \underline{0} .$$

(Here and in the following $\underline{I}$ denotes the n-dimensional unit matrix.) Thus $\underline{x}$ must satisfy

$$(-\underline{x} \quad \underline{I}) \begin{pmatrix} 1 & \underline{a}^T \\ \underline{a} & \underline{C} \end{pmatrix} \begin{pmatrix} 1 \\ \underline{x} \end{pmatrix} = \underline{0}$$

or

$$(\underline{I} - \underline{C})\underline{x} = \underline{a} - \underline{x}\,\underline{a}^T\underline{x} \quad . \tag{6.1}$$

This equation for the non-trivial components of the eigenvector $\underline{b}$, although nonlinear, has the advantage of not containing the unknown eigenvalue λ. Moreover, if the a_k satisfy (2.7) where ρ is small, the nonlinear term on the right is small whereas the coefficient $\underline{I} - \underline{C}$ is of order unity. The equation thus may be solved by iteration which we now proceed to do rigorously.

We denote by $\|\underline{x}\|$ the Euclidean vector norm and by $\|\underline{C}\|$ the spectral norm of a matrix, i.e., the matrix norm induced by the Euclidean vector norm. It follows from more general work of Stewart (1973) that if

$$\sigma := \|(\underline{I} - \underline{C})^{-1}\|\|\underline{a}\| < \frac{1}{2} , \tag{6.2}$$

equation (6.1) has a unique solution $\underline{x}$ satisfying

$$\|\underline{x}\| \leqq 2\sigma . \tag{6.3}$$

The quantity σ is easily estimated in terms of ρ : We have

$$\|\underline{a}\| \leqq \sum_{k=1}^{n} |a_k| \leqq \frac{\rho}{1-\rho} , \tag{6.4}$$

furthermore,

$$\|(\underline{I} - \underline{C})^{-1}\| \leqq \frac{1}{1 - \|\underline{C}\|} ,$$

and since the spectral norm is dominated by the Frobenius norm,

$$\begin{aligned} \|\underline{C}\| &\leqq \{a_2^2 + 2a_3^2 + 3a_4^2 + \dots\}^{1/2} \\ &\leqq \{\rho^4 + 2\rho^6 + 3\rho^8 + \dots\}^{1/2} \\ &= \left[\frac{\rho^4}{(1-\rho^2)^2}\right]^{1/2} = \frac{\rho^2}{1-\rho^2} , \end{aligned} \tag{6.5}$$

$$\sigma \leq \frac{\rho}{1-\rho} \cdot \frac{1}{1 - \frac{\rho^2}{1-\rho^2}} = \frac{\rho(1+\rho)}{1-2\rho^2}$$

We see that (6.2) is satisfied if $\rho \leq \frac{1}{4}(\sqrt{5}-1) = 0.31$. Supposing $\rho < 1/36$ as we always shall in the following because this is required for the estimates of §5, we have

$$\sigma \leq 1.03\rho \quad . \tag{6.6}$$

We now give an algorithm for the construction of the solution $\underline{x}$ because the successive approximations obtained will be required in the following. We obtain $\underline{x}$ as $\lim \underline{x}^{(j)}$, where

$$\begin{aligned} \underline{x}^{(0)} &:= \underline{0} \ ; \\ (\underline{I} - \underline{C})\underline{x}^{(j+1)} &:= \underline{a} - \underline{x}^{(j)}\underline{a}^T\underline{x}^{(j)} \ ; \end{aligned} \tag{6.7}$$

here the recurrence relation is meaningful since $\underline{I} - \underline{C}$ is nonsingular by virtue of (6.2). We assert that for $j = 0, 1, 2, \ldots$,

$$\|\underline{x}^{(j)}\| \leq 2\sigma < 1 \ , \tag{6.8}$$

$$\|\underline{x}^{(j)} - \underline{x}^{(j-1)}\| \leq (2\sigma)^j \ . \tag{6.9}$$

Here (6.8) is certainly true for $j = 0$. If true for some $j \geq 0$, then it follows from

$$\underline{x}^{(j+1)} = (\underline{I} - \underline{C})^{-1}\{\underline{a} - \underline{x}^{(j)}\underline{a}^T\underline{x}^{(j)}\}$$

that

$$\begin{aligned} \|\underline{x}^{(j+1)}\| &\leq \|(\underline{I} - \underline{C})^{-1}\|\|\underline{a} - \underline{x}^{(j)}\underline{a}^T\underline{x}^{(j)}\| \\ &\leq \|(\underline{I} - \underline{C})^{-1}\|\|\underline{a}\|(1 + \|\underline{x}^{(j)}\|^2) \\ &\leq 2\sigma \ . \end{aligned}$$

Similarly, (6.9) holds for $j = 1$ and, if true for some $j \geq 1$, it follows from

$$\begin{aligned} (\underline{I} - \underline{C})(\underline{x}^{(j+1)} - \underline{x}^{(j)}) &= - \underline{x}^{(j)}\underline{a}^T\underline{x}^{(j)} + \underline{x}^{(j-1)}\underline{a}^T\underline{x}^{(j-1)} \\ &= - (\underline{x}^{(j)} - \underline{x}^{(j-1)})\underline{a}^T\underline{x}^{(j)} - \underline{x}^{(j-1)}\underline{a}^T(\underline{x}^{(j)} - \underline{x}^{(j-1)}) \end{aligned}$$

using (6.8), that

$$\|\underline{x}^{(j+1)} - \underline{x}^{(j)}\|$$
$$\leqq \|(\underline{I} - \underline{C})^{-1}\|\|\underline{a}\|\|\underline{x}^{(j)} - x^{(j-1)}\|(\|\underline{x}^{(j)}\|+\|\underline{x}^{(j-1)}\|)$$
$$\leqq 2\sigma \cdot (2\sigma)^{j} ,$$

establishing (6.9) for the next higher j .

It immediately follows from (6.8) and (6.9) that $\underline{x} := \lim \underline{x}^{(j)}$ exists and satisfies (6.3), and by letting $j \to \infty$ in (6.7) we see that $\underline{x}$ is a solution of (6.1). It easily follows from (6.2) that $\underline{x}$ is the only solution of norm $\leqq 2\sigma$. We shall require estimates for the components of $\underline{x}$. To this end we multiply (6.1) from the left by $\underline{e}_k^T$, where $\underline{e}_k$ is the k-th unit coordinate vector. Writing

$$\underline{x} = \begin{pmatrix} x_1 \\ x_2 \\ \vdots \\ x_n \end{pmatrix}, \qquad \underline{a}_k := \begin{pmatrix} a_{k+1} \\ a_{k+2} \\ \vdots \\ 0 \end{pmatrix},$$

the result is

$$x_k - \underline{a}_k^T\underline{x} = a_k - x_k\underline{a}^T\underline{x} .$$

This yields

$$x_k = \frac{a_k + \underline{a}_k^T\underline{x}}{1 + \underline{a}^T\underline{x}} .$$

We note that using (6.6)

$$|\underline{a}_k^T\underline{x}| \leqq \|\underline{a}_k\|\|\underline{x}\| \leqq \frac{2\sigma\rho^{k+1}}{1 - \rho} = 2.12\ \rho^{k+2} \tag{6.10}$$

$(k = 0\ [\underline{a}_0 := \underline{a}],\ 1,\ 2,\ \ldots,\ n)$, and we thus find

$$|x_k| \leqq 1.01\ \rho^k ,\ k = 1, 2, \ldots , n, \tag{6.11}$$

and also

$$|x_k - a_k| \leqq 4.25\ \rho^{k+2} . \tag{6.12}$$

Thus the components of $\underline{x}$ decrease at a geometric rate, and they closely resemble the corresponding components of $\underline{a}$.

We also need to estimate the components of the error vector

$$\underline{d}^{(j)} := \underline{x}^{(j)} - \underline{x} = \begin{pmatrix} d_1^{(j)} \\ \vdots \\ d_n^{(j)} \end{pmatrix} \tag{6.13}$$

For the norm of the whole vector we find from (6.9) by summing a geometric series and using (6.6),

$$\|\underline{d}^{(j)}\| \leqq \frac{(2\sigma)^{j+1}}{1 - 2\sigma} \leqq 1.08 \ (2\sigma)^{j+1} . \tag{6.14}$$

Subtracting (6.1) from (6.7) yields

$$\begin{aligned}&(\underline{I} - \underline{C})\underline{d}^{(j+1)} \\ &= \underline{x}\ \underline{a}^T\underline{x} - (\underline{x} + \underline{d}^{(j)})\underline{a}^T(\underline{x} + \underline{d}^{(j)}) \\ &= \underline{x}\underline{a}^T\underline{d}^{(j)} - \underline{d}^{(j)}\underline{a}^T\underline{x} - \underline{d}^{(j)}\underline{a}^T\underline{d}^{(j)} .\end{aligned}$$

Multiplying from the left by $\underline{e}_k^T$ we obtain

$$d_k^{(j+1)} - \underline{a}_k^T\underline{d}^{(j+1)} = - x_k\underline{a}^T\underline{d}^{(j)} - d_k^{(j)}\underline{a}^T\underline{x} - d_k^{(j)}\underline{a}^T\underline{d}^{(j)} ,$$

$k = 1, 2, \ldots , n$, and there follows

$$\begin{aligned}|d_k^{(j+1)}| \leqq (|\underline{a}^T\underline{x}| + |\underline{a}^T\underline{d}^{(j)}|)|d_k^{(j)}| \\ + |\underline{a}_k^T\underline{d}^{(j+1)}| + |x_k||\underline{a}^T\underline{d}^{(j)}| .\end{aligned}$$

Using (6.10) and (6.14) this implies

$$|d_k^{(j+1)}| \leqq 4.41\ \rho^2|d_k^{(j)}| + 2.24\ \rho^{k+1}(2\sigma)^{j+1} .$$

By elementary estimates, using the fact that

$$|d_k^{(0)}| = |x_k| \leqq 1.01\ \rho^k ,$$

one now obtains

$$|d_k^{(j)}| \leqq 1.01\ \rho^k(2.10\rho)^{2j} + 2.39\ \rho^{k+1}(2.06\rho)^j . \tag{6.15}$$

In particular, if $j \geqq 1$,

$$d_k^{(j)} = O(\rho^{k+j+1}) , \ \rho \to 0 .$$

7. ASYMPTOTIC BEHAVIOR OF $\|f^*\|$

By (5.4), the asymptotic behavior of $\|f^*\|$ is up to $O(\rho^{n+1})$ identical with the asymptotic behavior of the leading eigenvalue λ. To study the latter, we form Rayleigh quotients with the approximate eigenvectors constructed in §6. For any vector $\underline{v} \neq \underline{0}$, let

$$R(\underline{v}) := \frac{\underline{v}^T \underline{A}\, \underline{v}}{\underline{v}^T \underline{v}} \tag{7.1}$$

denote its Rayleigh quotient. If $\underline{b}$ is the eigenvector constructed in §6, then

$$R(\underline{b}) = \lambda . \tag{7.2}$$

If $\underline{c} = \underline{b} + \underline{e}$ is any other vector $\neq \underline{0}$, we find for the difference

$$R(\underline{c}) - \lambda = \frac{(\underline{b} + \underline{e})^T \underline{A}(\underline{b} + \underline{e})}{(\underline{b} + \underline{e})^T (\underline{b} + \underline{e})} - \frac{\underline{b}^T \underline{A}\, \underline{b}}{\underline{b}^T \underline{b}}$$

$$= \|\underline{b} + \underline{e}\|^{-2} \|\underline{b}\|^{-2} \{2\underline{e}^T \underline{A}\, \underline{b} \cdot \underline{b}^T \underline{b}$$

$$+ \underline{e}^T \underline{a}\, \underline{e} \cdot \underline{b}^T \underline{b} - 2\underline{e}^T \underline{b} \cdot \underline{b}^T \underline{A}\, \underline{b} - \underline{e}^T \underline{e} \cdot \underline{b}^T \underline{A}\, \underline{b}\}$$

which on account of $\underline{A}\, \underline{b} = \lambda\, \underline{b}$ simplifies to

$$R(\underline{c}) - \lambda = \|\underline{b} + \underline{e}\|^{-2}\, \underline{e}^T (\underline{A} - \lambda \underline{I})\underline{e} .$$

If the discrepancy $\underline{e}$ has the form

$$\underline{e} = \begin{pmatrix} 0 \\ \underline{d} \end{pmatrix} \tag{7.3}$$

and if $\underline{A}$ is partitioned as in §6, then by virtue of

$$\|\underline{b} + \underline{e}\| \geqq 1$$

this yields

$$|R(\underline{c}) - \lambda| \leqq (\lambda + \|\underline{C}\|)\|\underline{d}\|^2 .$$

By Gerschgorin's theorem there holds

$$\lambda \leqq \frac{1}{1 - \rho} .$$

Using (6.5), we thus have for vectors $\underline{e}$ of the form (7.3)

$$|R(\underline{b} + \underline{e}) - \lambda| \leq \frac{1+\rho+\rho^2}{1-\rho^2}\|\underline{d}\|^2 \leq 1.03\|\underline{d}\|^2 . \qquad (7.4)$$

The result (7.4) could be applied, for instance, to any of the vectors $\underline{x}^{(j)}$ constructed in §6, the required estimates for

$$\underline{d}^{(j)} = \underline{x}^{(j)} - \underline{x}$$

being given by (6.14). Instead, we choose to apply (7.4) to the vector

$$\underline{c} = \begin{pmatrix} 1 \\ \underline{a} \end{pmatrix} .$$

By (6.12),

$$\|\underline{d}\| = \|\underline{a} - \underline{x}\| \leq 4.25 \frac{\rho^3}{(1-\rho^2)^{1/2}} \leq 4.26 \rho^3 .$$

There follows

$$|R(\underline{c}) - \lambda| \leq 18.70 \rho^6 . \qquad (7.5)$$

In order to obtain an asymptotic formula for λ, it is necessary to expand $R(\underline{c})$ up to terms $O(\rho^6)$. Writing

$$R(\underline{c}) = \frac{1 + 2\underline{a}^T\underline{a} + \underline{a}^T\underline{C}\,\underline{a}}{1 + \underline{a}^T\underline{a}}$$

$$= 1 + \underline{a}^T\underline{a} - (\underline{a}^T\underline{a})^2 + \underline{a}^T\underline{C}\,\underline{a} + \frac{\underline{a}^T\underline{a}\{(\underline{a}^T\underline{a})^2 - \underline{a}^T\underline{C}\,\underline{a}\}}{1 + \underline{a}^T\underline{a}}$$

it is easy to see that

$$|R(\underline{c}) - (1 + a_1^2 + a_2^2 - a_2 a_1^2 + a_1^4)| \leq 7.02 \rho^6 .$$

We thus have

$$|\lambda - (1 + a_1^2 + a_2^2 - a_2 a_1^2 + a_1^4)| \leq 25.72 \rho^6 . \qquad (7.6)$$

Combining this with (5.4) there follows for $n \geq 5$ and $\rho < 1/36$ the asymptotic result of Theorem 1,

$$\|f^*\| = 1 + a_1^2 + a_2^2 - a_2 a_1^2 + a_1^4 + 0.57\theta(18\rho)^6 \qquad (7.7)$$

where $|\theta| \leq 1$. Its corollary,

$$\|f^*\| = 1 + a_1^2 + 0.15\theta(18\rho)^4 , \qquad (7.8)$$

is a cruder form of the same result.

8. THE ASYMPTOTIC BEHAVIOR OF f^*

By (5.12),

$$\|f^* - \tilde{f}\| \leq 1.18\ (36\rho)^{n+1}, \tag{8.1}$$

and it suffices to describe the asymptotic behavior of

$$\tilde{f}(z) = \lambda \frac{q(z)}{q(z^{-1})}$$

where

$$q(z) = 1 + b_1 z + b_2 z^2 + \dots + b_n z^n,$$

λ being the leading eigenvalue and

$$\underline{b} = \begin{pmatrix} 1 \\ \underline{x} \end{pmatrix} = \begin{pmatrix} 1 \\ b_1 \\ \vdots \\ b_n \end{pmatrix}$$

the corresponding eigenvector of the matrix $\underline{A}$ defined by (4.1) where $N = n$.

Let

$$\hat{f}(z) = \lambda \frac{\hat{q}(z)}{\hat{q}(z^{-1})}$$

be an approximation to f in the sense that $\|\hat{q} - q\|$ can be estimated,

$$\hat{q}(z) = 1 + c_1 z + \dots + c_n z^n.$$

Plain algebra yields

$$\hat{f}(z) - \tilde{f}(z) = \lambda \frac{\hat{q}(z^{-1})\{\hat{q}(z) - q(z)\} - \hat{q}(z)\{\hat{q}(z^{-1}) - q(z^{-1})\}}{q(z^{-1})\hat{q}(z^{-1})}$$

so that

$$\|\hat{f} - \tilde{f}\| \leq \frac{2\lambda}{\gamma\hat{\gamma}} \|\hat{q} - q\|, \tag{8.2}$$

where

$$\gamma := \min_{|z|=1} |q(z)|, \quad \hat{\gamma} := \min_{|z|=1} |\hat{q}(z)|$$

Assuming that the zeros of q and of $\hat{q}$ all lie outside $|z| = 1$ and letting

$$\tilde{f}(z) = \sum_{k=-\infty}^{n} a_k z^k ,$$

$$\hat{f}(z) = \sum_{k=-\infty}^{n} \hat{a}_k z^k ,$$

it follows by Cauchy's coefficient estimate that likewise

$$|\hat{a}_k - a_k| \leq \frac{2\lambda}{\gamma\hat{\gamma}}\|\hat{q} - q\| \tag{8.3}$$

for all k .

Given any vector that approximates the eigenvector $\underline{b}$ of $\underline{A}$ with an error $O(\rho^m)$ $(1 \leq m \leq n)$, these results could be used to obtain an asymptotic representation for f , and hence for f^* , that is accurate to within $O(\rho^m)$. To illustrate the procedure, we choose to work with

$$\hat{q}(z) := 1 + a_1 z + a_2 z^2 . \tag{8.4}$$

By (6.11) and (6.12),

$$\|\hat{q} - q\| \leq \sum_{k=1}^{2} |a_k - x_k| + \sum_{k=3}^{n} |x_k| \leq 5.42\ \rho^3 ;$$

furthermore,

$$\gamma \geq 1 - \sum_{k=1}^{n} |x_k| \geq 0.96 ,$$

$$\hat{\gamma} \geq 1 - \sum_{k=1}^{2} |a_k| \geq 0.97 .$$

In view of $\lambda \leq 1.03$, (8.2) now yields

$$\|\hat{f} - \tilde{f}\| \leq 12.00\ \rho^3 \tag{8.5}$$

and consequently,

$$|\hat{a}_k - a_k| \leq 12.00\ \rho^3$$

for all k .

To obtain the desired asymptotic representation, we have to compute and estimate the coefficients $\hat{a}_k$. We write

$$\hat{f}(z) = \sum_{k=-\infty}^{2} \hat{a}_k z^k = \lambda \frac{\hat{q}(z)}{\hat{q}(z^{-1})}$$

$$= \lambda(1 + a_1 z + a_2 z^2)\{1 - (a_1 z^{-1} + a_2 z^{-2}) + \dots\}$$

and use the fact, following from (7.6), that

$$\lambda = 1 + a_1^2 + 3.02\theta\rho^4 ,$$

where here and in the following θ is used as a generic symbol for a number of modulus not exceeding 1 that may differ from one equation to the next. Comparing coefficients then yields

$$\begin{aligned} \hat{a}_2 &= a_2 + 1.01\theta\rho^4 \\ \hat{a}_1 &= a_1 + 2.01\theta\rho^3 \\ \hat{a}_0 &= 1 + 6.02\theta\rho^4 \\ \hat{a}_{-1} &= -a_1 + 4.01\theta\rho^3 \\ \hat{a}_{-2} &= -a_2 + a_1^2 + 7.01\theta\rho^4 \end{aligned} \tag{8.6}$$

The remaining coefficients are estimated by Cauchy's formula. If $|z| \geqq 2\rho$, then

$$|\hat{q}(z^{-1})| \geqq 1 - \frac{\rho}{2\rho} - \frac{\rho^2}{4\rho^2} > 0 ,$$

hence the zeros of $\hat{q}(z^{-1})$ lie in $|z| < 2\rho$, and if $|z| = 3\rho$,

$$\left|\frac{\hat{q}(z)}{\hat{q}(z^{-1})}\right| \leqq |3(1 + 6\rho)|^2 .$$

Therefore by Cauchy's estimate, if $\rho < 1/36$,

$$|\hat{a}_{-k}| \leqq 10.80\ (3\rho)^k , \quad k = 1, 2, \dots .$$

From this we conclude

$$\left\|\sum_{k=-\infty}^{-3} \hat{a}_k z^k\right\| \leqq 318.11\ \rho^3 \tag{8.7}$$

It emerges from (8.6) that

$$h(z) := a_1 z^{-1} + (a_2 - a_1^2) z^{-2} \tag{8.8}$$

is the desired asymptotic approximation to

$$h^*(z) = f_0(z) - f^*(z) \ .$$

Indeed we have

$$\left\| \sum_{k=-2}^{2} \hat{a}_k z^k + h(z) - f_0(z) \right\| \leqq 7.44 \, \rho^3 \tag{8.9}$$

and thus, using (8.1), (8.5), (8.7), (8.9),

$$\begin{aligned} \|h - h^*\| &= \|f^* - f_0 + h\| \\ &\leqq \|f^* - \tilde{f}\| + \|\tilde{f} - \hat{f}\| + \left\|\hat{f} - \sum_{k=-2}^{2} \hat{a}_k z^k\right\| \\ &\quad + \left\| \sum_{k=-2}^{2} \hat{a}_k z^k + h - f_0 \right\| \\ &\leqq 1.18 \, (36\rho)^{n+1} + 12 \, \rho^3 + 318.11 \, \rho^3 + 7.44\rho^3 \\ &\leqq 1.19 \, (36\rho)^3 \ , \end{aligned}$$

which is our Theorem 2 .

BIBLIOGRAPHY

S. Bernstein (1926): Leçons sur les propriétés extremales et la meilleure approximation des fonctions analytiques d'une variable réelle. Gauthier-Villars, Paris.

G. M. Goluzin (1969): Geometric theory of functions of a complex variable. Translations of Mathematical Monographs vol. 26. Amer. Math. Soc., Providence.

E. B. Saff (1973): On the zeros of the error function for Tchebycheff approximation on a disk. J. Approx. Theory 9, 112 - 117.

I. Schur (1917): Über Potenzreihen, die im Innern des Einheitskreises beschränkt sind. J. reine angew. Math. 147, 205 - 232.

G. W. Stewart (1973): Error and perturbation bounds for subspaces associated with certain eigenvalue problems. SIAM Rev. 15, 727 - 764.

L. Trefethen (1981): Near-circularity of the error curve in complex Chebyshev approximation. J. Approx. Theory 31, 344 - 367.

J. L. Walsh (1969): Interpolation and approximation by rational functions in the complex domain. American Math. Soc. Colloquium Publications vol. 20, Amer. Math. Soc., Providence, Rhode Island.

TOPICS IN COMPUTATIONAL COMPLEX ANALYSIS:
IV. THE LAGRANGE-BÜRMANN FORMULA FOR SYSTEMS OF FORMAL POWER SERIES

Peter Henrici
ETH - Zentrum, CH-8092 Zürich

Abstract. Let g and h be analytic at 0, $g(0) = 0$, $g'(0) \neq 0$. The classical Lagrange-Bürmann formula expresses the Taylor coefficients of $h \circ g$ in terms of the Taylor coefficients of h and of f, where f is the inverse function of g. P. Henrici (1964, 1974) gave a version of the formula that is valid for formal power series, and M.-L. Henrici (1976) extended the formal result to systems of power series. Here we present a more general formal result where a condition of "normality" is dropped.

1. INTRODUCTION

Let the function f of a single complex variable be analytic at 0, $f(0) = 0$, $f'(0) \neq 0$, and let f near 0 be represented by the power series F:

$$f(z) = F(z) = \sum_{k=1}^{\infty} f_k z^k .$$

The inverse function $g = f^{[-1]}$ is likewise analytic at 0, and hence represented by a power series:

$$g(z) = G(z) = \sum_{k=1}^{\infty} g_k z^k .$$

A classical result due to Lagrange (1770) and Bürmann (1799) expresses the g_k in terms of the f_k. More generally, if h is any function analytic at 0,

H. Werner et al. (eds.), Computational Aspects of Complex Analysis, 193–215.

$$h(z) = \sum_{k=0}^{\infty} h_k z^k ,$$

the formula expresses the coefficients of the series representing $h \circ g$,

$$h(g(z)) = \sum_{k=0}^{\infty} a_k z^k ,$$

in terms of the coefficients h_k and f_k .

Here we consider the Lagrange-Bürmann formula, and its generalization to systems of n power series in n variables, in the context of formal power series.

2. ONE VARIABLE

Let $\mathcal{F}$ be a field. A formal power series (fps) over $\mathcal{F}$ is a sequence $\{p_k\}_0^\infty$ of elements of $\mathcal{F}$, written suggestively as

$$P = \sum_{k=0}^{\infty} p_k x^k \tag{2.1}$$

No meaning is attached to the powers x^k except that of a placekeeper. No topology is defined, we thus cannot talk about convergence, and no value is attached to P . If sums and products are defined in the obvious manner (see Henrici (1974), chapter 1), the totality of all fps over $\mathcal{F}$ forms an integral domain which we call $\mathcal{P}$. The neutral element in $\mathcal{P}$ is the fps $I := 1$ corresponding to the sequence $\{1, 0, 0, \dots\}$. The <u>units</u> in $\mathcal{P}$, i.e. the series P for which $Q = P^{-1} \in \mathcal{P}$ exists such that $PQ = I$, are precisely the fps (2.1) such that $p_0 \neq 0$. We call the series $Q = P^{-1}$ the <u>reciprocal</u> of P .

If $P, Q \in \mathcal{P}$ and if Q is a non-unit in $\mathcal{P}$, the <u>composition</u> $P \circ Q$ may be defined formally by substituting Q for x in (2.1), i.e. by forming

$$p_0 + p_1 Q + p_2 Q^2 + \dots ,$$

and collecting coefficients of like powers of x . Since the series for Q^k begins, at the earliest, with the term in x^k , for each fixed power of x only a finite number of terms needs to be collected, and the composition with a non-unit is thus always defined.

We call <u>almost unit</u> (au) a non-unit

$$P = \sum_{k=1}^{\infty} p_k x^k$$

such that $p_1 \neq 0$. It is easily verified that the composition of two au's is again an au, and that in fact the totality of all au's forms a group whose neutral element is the au $X = x$ corresponding to the sequence $\{0, 1, 0, 0, \ldots\}$. The inverse of an au P with respect to composition, i.e. the uniquely determined au Q such that

$$P \circ Q = Q \circ P = X ,$$

is called the inverse of P and is denoted by $P^{[-1]}$.

In addition to $\mathcal{P}$, we have to consider formal Laurent series (fLs). These are bilaterally infinite sequences $\{r_k\}_{-\infty}^{\infty}$, subject to the condition that only finitely many r_k with $k < 0$ are $\neq 0$. We write a fLs suggestively as

$$R = \sum_{k=-\infty}^{\infty} r_k x^k . \tag{2.2}$$

Sums and differences of fLs may be defined in the obvious manner, and so may products in view of the above finiteness condition. It is clear that under these operations the totality of fLs again forms an integral domain. However, since every nonzero fLs may be written in the form

$$R = X^m Q$$

where m is a suitable integer and Q is a unit in $\mathcal{P}$, it possesses the reciprocal

$$R^{-1} = X^{-m} Q^{-1}$$

and thus is a unit. We conclude that the totality of fLs over $\mathcal{F}$ forms a field, which we denote by $\mathcal{L}$.

It is clear that if R is a fLs and Q is a non-zero non-unit in $\mathcal{P}$, the composition $R \circ Q$ can be formed, and is again a fLs.

The derivative of a fLs (2.2) is defined by

$$R' := \sum_{k=-\infty}^{\infty} k r_k x^{k-1}$$

where, for $k > 0$, kr_k is an abbreviation for $r_k + r_k + \ldots + r_k$ (k terms) and, for $k < 0$, $kr_k :=$ $-|k|r_k$. Obviously R' is again a fLs. It is easily checked that the usual differentiation rules of calculus, such as $(RS)' = R'S + RS'$, hold.

If $R = \Sigma\, r_k x^k$ is any fLs, we call r_{-1} the <u>residue</u> of P and write

$$r_{-1} = \text{res } R \;.$$

At this point we assume that the coefficient field $\mathcal{F}$ has <u>characteristic zero</u>. Then $kr_k = 0$ only if either $k = 0$ or $r_k = 0$. The special role of the residue is illustrated by

LEMMA 1. <u>The residue of a fLs R over a field of characteristic 0 is zero if and only if R is a derivative, i.e., if there exists $S \in \mathcal{L}$ such that $R = S'$</u> .

The "if" part is a consequence of the definition, even without the assumption that $\mathcal{F}$ has characteristic zero. To prove the "only if" part, let R have the form (2.2) where $r_{-1} = 0$. Then clearly $R = S'$ where

$$S = \sum_{k \neq 0} \frac{r_{k-1}}{k} x^k \;.$$

We are now prepared to give a very simple proof of the Lagrange-Bürmann theorem for formal Laurent series. Let $R \in \mathcal{L}$, let P be an au in $\mathcal{P}$, and let $Q := P^{[-1]}$. The Lagrange-Bürmann theorem expresses the coefficients c_k in the expansion

$$R \circ Q = \sum_k c_k x^k \qquad (2.4)$$

in terms of R and of P . - Forming the composition of (2.4) with P we get

$$R = \sum_k c_k P^k \;.$$

This we multiply with the fLs $P^{-m-1}P'$, where m is any integer:

$$R\, P^{-m-1}P' = \sum_k c_k\, P^{k-m-1}P' \;.$$

Here we take the residue. Since for $k \neq m$

$$P^{k-m-1}P' = \frac{1}{k-m}\,(P^{k-m})'$$

is a derivative, the residues in all terms on the right are zero except for $k = m$. It is easily checked that

$$\text{res } P^{-1}P' = 1 .$$

There follows

$$c_k = \text{res}(R\ P^{-k-1}P') \qquad (2.5)$$

or

THEOREM 2 (Lagrange-Bürmann theorem). <u>Let $\mathcal{F}$ be a field with characteristic zero, let R be a fLs over $\mathcal{F}$, let P be an almost unit in the integral domain of fps over $\mathcal{F}$, and let Q be the inverse of P . Then</u>

$$R \circ Q = \sum_k \text{res}(R\ P^{-k-1}P')x^k . \qquad (2.6)$$

We mention some easy consequences. In view of Lemma 1,

$$\text{res}((R\ P^{-k})') = 0 ,$$

hence

$$\text{res}(R'P^{-k} - k\ R\ P^{-k-1}P') = 0 .$$

Thus if $k \neq 0$, the coefficients (2.5) are also given by

$$c_k = \frac{1}{k}\ \text{res}(R'P^{-k}) . \qquad (2.7)$$

Consider the special case $R = X^m$, where m is any integer. To evaluate (2.7) we let

$$P = \sum_{h=1}^{\infty} a_h x^h$$

and write, for any integer k ,

$$P^k = \sum_{h=k}^{\infty} a_h^{(k)} x^h . \qquad (2.8)$$

We then have

$$\begin{aligned} \text{res}(R'P^{-k}) &= m\ \text{res}(x^{m-1}P^{-k}) \\ &= m * \text{coefficient of } x^{-m} \text{ in } P^{-k} \\ &= m\ a_{-m}^{(-k)} . \end{aligned}$$

If $k \neq 0$, there follows

$$c_k = \frac{m}{k} a_{-m}^{(-k)} .$$

Since $X^m \circ Q = Q^m$, we thus obtain for $m > 0$:

THEOREM 3 (Schur (1947)). <u>In the notation</u> (2.8), <u>if</u> P <u>is an almost unit in</u> $\mathcal{P}$ <u>and</u> $Q := P^{[-1]}$, $m > 0$,

$$Q^m = \sum_{k=m}^{\infty} \frac{m}{k} a_{-m}^{(-k)} x^k . \qquad (2.9)$$

Henrici (1964, 1974) used a matrix isomorphism of the group of almost units to first establish Theorem 3 and then derived Theorem 2 (for $R \in \mathcal{P}$) as a corollary. The approach given above not only yields a more general result but is also believed to be more direct.

A number of interesting illustrations for Theorem 2 can be found, for instance, in Polya and Szegö (1924), chapter III. Here we just give a very simple example which later will be generalized. Let

$$P = u = x - x^2 . \qquad (2.10)$$

We wish to expand x^k, where k is any integer, in powers of u. By Theorem 2 the coefficient of u^m in this expansion is

$$c_m = \operatorname{res}(R\, P^{-m-1} P') ,$$

where $R = x^k$. Thus

$$c_m = \text{coefficient of } x^{-1} \text{ in } x^k (x - x^2)^{-m-1} (1-2x)$$

$$= \text{coefficient of } x^{m-k} \text{ in } (1-x)^{-m-1} (1-2x) .$$

Letting $m = k + p$ where $p \geq 0$ we obtain

$$c_k = 1 ,$$

$$c_{k+p} = k \frac{(p+k+1)_{p-1}}{p!} , \quad p > 0 ,$$

and the desired expansion thus is

$$x^k = u^k + \sum_{p=1}^{\infty} k \frac{(p+k+1)_{p-1}}{p!} u^{k+p} . \qquad (2.11)$$

It is readily verified that for $k = 1$ this agrees with

the expansion of the solution of (2.10) for x,

$$x = \frac{1}{2}(1 - \sqrt{1-4u}) ,$$

in powers of u. Of course, (2.11) has the advantage of being valid for any power x^k.

3. FORMAL POWER SERIES IN n INDETERMINATES

Let Z_+^n be the set of vectors

$$\underline{k} = (k_1, k_2, \ldots, k_n)$$

whose components k_i are non-negative integers. Such vectors are called index vectors. If $\mathcal{F}$ is any field, a formal power series in n indeterminates over $\mathcal{F}$ is a function P defined on Z_+^n with values in $\mathcal{F}$, written suggestively as

$$P = \sum_{\underline{k} \in Z_+^n} p_{\underline{k}} \underline{x}^{\underline{k}} . \qquad (3.1)$$

Here again the symbol $\underline{x}^{\underline{k}}$ should be interpreted as a placekeeper, to be read as

$$\underline{x}^{\underline{k}} = x_1^{k_1} x_2^{k_2} \ldots x_n^{k_n} ,$$

where $x_1, x_2, \ldots, x_n$ are indeterminates. The totality of fps (3.1) will be denoted by $\mathcal{P}_n$. It is obvious how to form sums and differences of series in $\mathcal{P}_n$. Products are formed according to the rule

$$P\,Q = \sum_{\underline{k} \in Z_+^n} \Big(\sum_{\underline{h}+\underline{j}=\underline{k}} p_{\underline{h}} q_{\underline{j}} \Big) \underline{x}^{\underline{k}} . \qquad (3.2)$$

Since for any given $\underline{k} \in Z_+^n$ there are only finitely many pairs $(\underline{h}, \underline{j})$ such that $\underline{h} \in Z_+^n$, $\underline{j} \in Z_+^n$ and $\underline{h} + \underline{j} = \underline{k}$, forming products again is a purely algebraic operation. It is clear that under the operations thus defined, $\mathcal{P}_n$ is an integral domain. The units in $\mathcal{P}_n$ are the series P such that $p_{\underline{0}} \neq 0$, where $\underline{0}$ is the index vector with all components zero.

We also consider systems of fps

$$\underline{Q} = \begin{pmatrix} Q_1 \\ Q_2 \\ \vdots \\ Q_n \end{pmatrix}$$

where each $Q_i \in \mathcal{P}_n$. The totality of all such systems is denoted by $\mathcal{P}_n^n$. If $\underline{Q} \in \mathcal{P}_n^n$ and $\underline{m} \in Z_+^n$, we define

$$\underline{Q}^{\underline{m}} := Q_1^{m_1} Q_2^{m_2} \cdots Q_n^{m_n} ; \tag{3.3}$$

it will be noted that this is a single fps $\in \mathcal{P}_n$.

Let now

$$P = \sum_{\underline{k}\in Z_+^n} p_{\underline{k}} \underline{x}^{\underline{k}}$$

be in $\mathcal{P}_n$, and let

$$\underline{Q} \in \mathcal{P}_n^n$$

be a system of non-units in $\mathcal{P}_n$, i.e., let each Q_i be a non-unit. We then define $P \circ \underline{Q}$, the composition of P with the system $\underline{Q}$, by substituting $\underline{Q}$ for $\underline{x}$ in P , observing the rule (3.3), and collecting coefficients of like powers $\underline{x}^{\underline{k}}$ in the resulting sum

$$\sum_{\underline{k}\in Z_+^n} p_{\underline{k}} \underline{Q}^{\underline{k}} . \tag{3.4}$$

To see that this operation is algebraically meaningful, we call the non-negative integer

$$|\underline{k}| := k_1 + k_2 + \cdots + k_n$$

the weight of the index vector $\underline{k} = (k_1,\ldots,k_n)$, or of the power $\underline{x}^{\underline{k}}$; also, for any fps

$$R = \sum r_{\underline{k}} \underline{x}^{\underline{k}}$$

we define

$$\text{ord } R := \min\{|\underline{k}| : r_{\underline{k}} \neq 0\} ,$$

the order of R . In this terminology, a fps $R \in \mathcal{P}_n$ is a unit precisely if $\text{ord } R = 0$. If $\underline{Q}$ is a system of non-units, then for each component Q_i

$$\text{ord } Q_i \geq 1 ,$$

thus for any $\underline{k} \in Z_+^n$,

$$\text{ord } \underline{Q}^{\underline{k}} \geq |\underline{k}| .$$

Thus if $\underline{m}$ is any index vector, only the powers $\underline{Q}^{\underline{k}}$ where $|\underline{k}| \leq |\underline{m}|$ in the sum (3.4) can contribute to $\underline{x}^{\underline{m}}$. Since there are only finitely many such index vectors, the sum of these contributions is always defined, and so is therefore the composition $\underline{P} \circ \underline{Q}$.

Finally, if $\underline{P}$ is in $\mathcal{P}_n^n$,

$$\underline{P} = \begin{pmatrix} P_1 \\ P_2 \\ \vdots \\ P_n \end{pmatrix} , \tag{3.5}$$

and if $\underline{Q}$ is a system of non-units in $\mathcal{P}_n^n$, we define the composition of $\underline{P}$ with $\underline{Q}$ by

$$\underline{P} \circ \underline{Q} = \begin{pmatrix} P_1 \circ \underline{Q} \\ P_2 \circ \underline{Q} \\ \vdots \\ P_n \circ \underline{Q} \end{pmatrix} .$$

Let

$$\underline{X} := \begin{pmatrix} x_1 \\ x_2 \\ \vdots \\ x_n \end{pmatrix} .$$

The system $\underline{X}$ evidently acts as a neutral element under composition. That is, there holds

$$\underline{P} \circ \underline{X} = \underline{P}$$

for any system $\underline{P} \in \mathcal{P}_n^n$, and

$$\underline{X} \circ \underline{Q} = \underline{Q}$$

for any system of non-units. It is therefore natural to ask: Under what conditions on the system (3.5) does there exist a system of non-units $\underline{Q} \in \mathcal{P}_n^n$ such that

$$\underline{P} \circ \underline{Q} = \underline{X} \text{ ?} \tag{3.6}$$

Systems $\underline{P}$ with this property will be called <u>invertible</u>, and the solution $\underline{Q}$ of (3.6) will be denoted by

$$\underline{Q} = \underline{P}^{[-1]}.$$

Let

$$P_h = \sum p_{h,\underline{k}}\, \underline{x}^{\underline{k}} , \; h = 1, 2, \ldots , n,$$

and put tentatively

$$Q_i = \sum q_{i,\underline{k}}\, \underline{x}^{\underline{k}} , \; i = 1, 2, \ldots , n.$$

Condition (3.6) requires that the terms of zero weight in all components of the composition $\underline{P} \circ \underline{Q}$ should be zero. Since $p_{h,0}$ is the only term of zero weight in the h-th component, (3.6) requires that these terms are all zero. It follows that $\underline{P}$ itself must be a system of non-units.

We next consider the terms of weight 1 . These can arise only from the terms

$$p_{h,\underline{k}}\, \underline{Q}^{\underline{k}} \tag{3.7}$$

where $|\underline{k}| = 1$, i.e. where $\underline{k}$ equals one of the unit vectors

$$\underline{e}_i = (e_{i,1}, e_{i,2}, \ldots, e_{i,n}) , \; i = 1,\ldots,n$$

where $e_{i,j} = 1$ if $j = i$ and $e_{i,j} = 0$ otherwise. We put, for simplicity,

$$\begin{aligned} p_{h,i} &:= p_{h,\underline{e}_i} \\ q_{i,j} &:= q_{i,\underline{e}_j} \end{aligned} \tag{3.8}$$

The terms of weight 1 in (3.7) then are $p_{h,i}\, q_{i,j}\, x_j$, and the contribution of all these terms to x_j clearly is

$$\left(\sum_{i=1}^{n} p_{h,i}\, q_{i,j}\right) x_j .$$

By (3.6) this contribution should be x_j for $j = h$ and 0 for $j \neq h$. This means that the $n \times n$ matrices

$$p := (p_{h,i}) \quad \text{and} \quad q := (q_{i,j}) \tag{3.10}$$

should be inverses of each other, and hence that the matrix p must be non-singular.

We thus have found: A necessary condition for the relation (3.6) to hold is that $\underline{P}$ is a system of non-units such that the matrix p defined by (3.10) is non-singular.

It turns out that these conditions are also sufficient. Suppose that we have shown the unique existence of the terms up to weight $w - 1 \geq 1$ in the system of non-units

$$Q_i = \sum q_{i,\underline{m}}\, \underline{x}^{\underline{m}} , \quad i = 1, \ldots , n$$

such that (3.6) holds. Then if $|\underline{k}| > 1$, all terms of weight $\leq w$ in the series $\underline{Q}^{\underline{k}}$ are fully determined. Let now $\underline{m} \in Z_+^n$ such that $|\underline{m}| = w$. In the sum

$$\sum p_{h,\underline{k}}\, \underline{Q}^{\underline{k}}$$

the contributions of all terms with $|\underline{k}| > 1$ to the coefficient of $\underline{x}^{\underline{m}}$ thus are fixed, and the only terms where unknown coefficients of weight w occur are those where $|\underline{k}| = 1$. For these we find the system of equations

$$\sum_{i=1}^{n} p_{h,i}\, q_{i,\underline{m}} =$$

$$= - \text{coefficient of } \underline{x}^{\underline{m}} \text{ in } \sum_{|\underline{k}| \geq 2} p_{h,\underline{k}}\, \underline{Q}^{\underline{k}} ,$$

$h = 1, \ldots , n$. The coefficient matrix of the system being p, which by assumption is non-singular, this can be solved uniquely for the vector

$$(q_{1,\underline{m}}, \ldots , q_{n,\underline{m}}) .$$

Since this holds for any $\underline{m}$ such that $|\underline{m}| = w$, the induction step is complete.

In the case of systems of convergent power series, the matrix p would be the Jacobian matrix at $\underline{0}$ of the system $\underline{P}$. It is convenient to retain this terminology also for systems of fps. We call a system of fps non-singular if its Jacobian at $\underline{0}$ is non-singular. We then have proved:

THEOREM 4. A system $\underline{P} \in \mathcal{P}_n^n$ of non-units is invertible if and only if it is non-singular.

It is clear that for systems of non-units the operation of composition is associative. It thus follows that for any fixed n the totality of non-singular systems $\underline{P}$ of non-units under the operation of composition forms a group. Our generalization of the Lagrange-Bürmann formula in this context will provide a representation for

$$\underline{Q} := \underline{P}^{[-1]}$$

in terms of $\underline{P}$.

4. FORMAL LAURENT SERIES IN SEVERAL INDETERMINATES

Let Z^n denote the set of n-tuples (again called index vectors)

$$\underline{k} = (k_1, k_2, \ldots, k_n)$$

the components of which are arbitrary integers. We again call

$$|\underline{k}| := k_1 + k_2 + \ldots + k_n$$

the weight of the index vector $\underline{k}$; since the k_i now are no longer restricted to the non-negative integers, $|\underline{k}|$ may be negative.

A formal Laurent series in n indeterminates over $\mathcal{F}$ is a function R from Z^n to $\mathcal{F}$,

$$R : \underline{k} \in Z^n \longrightarrow r_{\underline{k}} \in \mathcal{F},$$

subject to the condition that for any integer m only finitely many $r_{\underline{k}}$ with $|\underline{k}| < m$ are different from zero. We again use the notation

$$R = \sum_{\underline{k}} r_{\underline{k}} \underline{x}^{\underline{k}} \tag{4.1}$$

to suggest the appropriate algebraic operations.

EXAMPLES. We choose $n = 2$ and write $x := x_1$, $y := x_2$ for simplicity. The series

$$1 + xy^{-1} + x^2y^{-2} + x^3y^{-3} + \dots$$

is not a fLs in two indeterminates, because it has infinitely many terms of weight 0. The series

$$1 + x^2y^{-1} + x^4y^{-2} + x^6y^{-3} + \dots \tag{4.2}$$

is a fLs, because for each integer m it has at most $|m|$ terms of weight $< m$. -

We see from the second example that in the case $n > 1$ there is no lower bound on the exponents k_i of the individual indeterminates x_i. For every fLs (4.1), however, there exists a largest integer m such that $|\underline{k}| \geq m$ for all $\underline{k}$ such that $r_{\underline{k}} \neq 0$. This number m (which may be negative) is again called the order of R; we write

$$m = \operatorname{ord} R .$$

EXAMPLE. For the fLs (4.2), $m = 0$.

The totality of fLs in n indeterminates over a given field $\mathcal{F}$ will be denoted by $\mathcal{L}_n$. It is clear that $\mathcal{L}_n$ is closed under addition. The product of two fLs $R, S \in \mathcal{L}_n$ is defined by

$$RS = \sum_{\underline{m}} \Big(\sum_{\underline{k}+\underline{\ell}=\underline{m}} r_{\underline{k}} s_{\underline{\ell}} \Big) \underline{x}^{\underline{m}} .$$

Here the inner sum exists for every $\underline{m}$, because $\underline{k} + \underline{\ell} = \underline{m}$ implies $|\underline{k}| + |\underline{\ell}| = |\underline{m}|$. Thus either $|\underline{k}| \leq |\underline{m}|$ or $|\underline{\ell}| \leq |\underline{m}|$, and since there are only finitely many nonzero $r_{\underline{k}}$ such that $|\underline{k}| \leq |\underline{m}|$ and only finitely many nonzero $s_{\underline{\ell}}$ such that $|\underline{\ell}| \leq |\underline{m}|$, the sum is finite. A similar argument shows that $RS \in \mathcal{L}_n$. Thus $\mathcal{L}_n$ is closed under addition and multiplication, and it is easy to verify that in fact $\mathcal{L}_n$ is an integral domain.

In the one-dimensional case, we saw that every nonzero element of $\mathcal{L} = \mathcal{L}_1$ is a unit, and hence that is a field. Does the same hold when $n > 1$? The example

$$R = x^{-1} - y^{-1} \in \mathcal{L}_2$$

shows that this is not so. The reciprocal of R, computed formally, is either

$$x + x^2y^{-1} + x^3y^{-2} + x^4y^{-3} + \dots$$

or

$$-y - y^2x^{-1} - y^3x^{-2} - y^4x^{-3} - \dots$$

neither of which is in $\mathcal{L}_2$, since both series have infinitely many terms of weight 1.

THEOREM 5. <u>The units of $\mathcal{L}_n$ are precisely the nonzero fLs which have precisely one term of lowest weight</u>.

<u>Proof</u>. A construction similar to the foregoing examples shows that fLs with several terms of lowest weight cannot have a reciprocal in $\mathcal{L}_n$. Now let the series

$$R = \sum_{\underline{k}} r_{\underline{k}}\, \underline{x}^{\underline{k}}$$

have precisely one nonzero term of lowest weight, and let $r_{\underline{k}_0}$ be its coefficient. Then

$$R = r_{\underline{k}_0}\, \underline{x}^{\underline{k}_0}(1 - P),$$

where

$$-P = \sum_{|\underline{k}| > |\underline{k}_0|} r_{\underline{k}_0}^{-1}\, r_{\underline{k}}\, \underline{x}^{\underline{k}-\underline{k}_0}$$

is a fLs whose order is positive. We assert that the reciprocal of R is

$$R^{-1} = r_{\underline{k}_0}^{-1}\, \underline{x}^{-\underline{k}_0}(1 + P + P^2 + \dots\,). \qquad (4.3)$$

Here the geometric series has a meaning, because in view of

$$\operatorname{ord} P^k \geqq k$$

only finitely many terms can possibly contribute to any fixed power $\underline{x}^{\underline{m}}$. For the same reason, the product RR^{-1} may be computed formally and yields, of course, the value 1. ∎

The composition of an $R \in \mathcal{L}_n$,

$$R = \sum_{\underline{k}} r_{\underline{k}} \underline{x}^{\underline{k}},$$

with a system $\underline{P}$ of non-units P_i in $\mathcal{P}_n$ can be defined formally by

$$R \circ \underline{P} = \sum_{\underline{k}} r_{\underline{k}} \underline{P}^{\underline{k}}, \tag{4.4}$$

provided that each P_i is also a unit in $\mathcal{L}_n$. Since

$$\operatorname{ord} \underline{P}^{\underline{k}} \geqq |\underline{k}|,$$

only a finite number of terms can contribute to any fixed power $\underline{x}^{\underline{m}}$, and the sum (4.4) thus is defined.

5. DIFFERENTIATION IN $\mathcal{L}_n$

Let

$$R = \sum_{\underline{k}} r_{\underline{k}} \underline{x}^{\underline{k}} \tag{5.1}$$

be in $\mathcal{L}_n$. The partial derivative of R with respect to x_i is defined by

$$D_i R := \sum_{\underline{k}} k_i r_{\underline{k}} \underline{x}^{\underline{k}-\underline{e}_i}, \tag{5.2}$$

where $\underline{e}_i$ is the i-th unit coordinate vector, and where $\underline{k} = (k_1, \ldots, k_n)$. Obviously, $D_i R \in \mathcal{L}_n$, and the usual formal rules of differentiation hold.

We seek an analog of Lemma 1 stating that the residue of the derivative of a fLs in $\mathcal{L}_1$ is zero. What, first of all, should be the residue of a fLs in $\mathcal{L}_n$?

DEFINITION. The <u>residue</u> of the series (5.1) is the coefficient $r_{-\underline{e}}$ where $\underline{e} := (1, 1, \ldots, 1)$.

Let

$$\underline{R} = \begin{pmatrix} R_1 \\ \vdots \\ R_n \end{pmatrix} \tag{5.3}$$

be a system of fLs in $\mathcal{L}_n$. We call <u>Jacobian determinant</u> of the system $\underline{R}$ the determinant

$$\underline{R}' := \det(D_i R_j) ,$$

where $i, j = 1, \ldots, n$. Evidently $\underline{R}'$, being a finite sum of products of elements of $\mathcal{L}_n$, is in $\mathcal{L}_n$.

LEMMA 6. <u>The residue of the Jacobian determinant of any system of fLs in $\mathcal{L}_n$ is zero</u>.

<u>Proof</u>. Our proof is computational; it follows the proof of a somewhat more special result by M. L. Henrici (1976). Let

$$R_j = \sum_{\underline{k}} r_{j,\underline{k}} \underline{x}^{\underline{k}} .$$

Then

$$D_i R_j = \sum_{\underline{k}} k_i r_{j,\underline{k}} \underline{x}^{\underline{k}-\underline{e}_i} .$$

By the addition theorem for determinants, $\underline{R}'$ is the sum of all determinants

$$\det(D_i r_{j,\underline{k}(j)} \underline{x}^{\underline{k}(j)})$$

taken with respect to all possible n-tuples of index vectors

$$(\underline{k}(1), \underline{k}(2), \ldots , \underline{k}(n)) .$$

Carrying out the differentiation and pulling out common factors the above determinant equals

$$\underline{x}^{\sum_{j=1}^{n} \underline{k}(j)-\underline{e}} \prod_{j=1}^{n} r_{j,\underline{k}(j)} \det(k_{ij}) , \qquad (5.4)$$

where

$$\underline{k}(j) = (k_{1j}, k_{2j}, \ldots , k_{nj}) .$$

A contribution to $\operatorname{res} \underline{R}'$ arises from those determinants where

$$\sum_{j=1}^{n} \underline{k}(j) = \underline{0} . \qquad (5.5)$$

These determinants are obviously zero, because (5.5) shows that their columns are linearly dependent. ■

6. THE LAGRANGE-BÜRMANN THEOREM FOR SYSTEMS OF FORMAL POWER SERIES

Let $R \in \mathcal{L}_n$, and let $\underline{P}$ be a system of non-units in $\mathcal{P}_n$ such that the Jacobian matrix at 0 of $\underline{P}$ is a non-singular diagonal matrix. The system $\underline{P}$ then is invertible; moreover, since each component satisfies

$$P_i = p_i x_i + \text{terms of weight} \geq 2 \tag{6.1}$$

where all $p_i \neq 0$, each P_i is a unit in $\mathcal{L}_n$.

Let $\underline{Q} := \underline{P}^{[-1]}$. We wish to obtain a formula, analogous to the classical Lagrange-Bürmann formula of Theorem 2, for the coefficients $c_{\underline{k}}$ in the expansion

$$R \circ \underline{Q} = \sum_{\underline{k}} c_{\underline{k}} \underline{x}^{\underline{k}} \tag{6.2}$$

where $R \in \mathcal{L}_n$ is arbitrary. For coefficient fields of characteristic zero, the answer is

THEOREM 7 (Lagrange-Bürmann formula for systems). <u>Let $\mathcal{F}$ have characteristic zero, let $R \in \mathcal{L}_n$, and let $\underline{P}$ be an invertible system in $\mathcal{P}_n^n$ whose Jacobian matrix at 0 is diagonal. If $\underline{Q}$ is the inverse system of $\underline{P}$, then</u>

$$R \circ \underline{Q} = \sum_{\underline{k}} \operatorname{res}(R\, \underline{P}^{-\underline{k}-\underline{e}} \underline{P}')\, \underline{x}^{\underline{k}} \tag{6.3}$$

<u>Proof</u>. Substituting $\underline{P}$ for $\underline{x}$ in (6.2) yields

$$R = \sum_{\underline{k}} c_{\underline{k}} \underline{P}^{\underline{k}}, \tag{6.4}$$

where on the right only a finite number of terms contribute to any given power $\underline{x}^{\underline{m}}$. Let $\underline{m} \in Z^n$ be arbitrary, and multiply (6.4) by

$$\underline{P}^{-\underline{m}-\underline{e}}\, \underline{P}' .$$

This yields

$$R\, \underline{P}^{-\underline{m}-\underline{e}}\, \underline{P}' = \sum_{\underline{k}} c_{\underline{k}}\, \underline{P}^{\underline{k}-\underline{m}-\underline{e}}\, \underline{P}' .$$

On either side of this identity we take the residue. On the right the residue may be taken termwise because only finitely many terms can contribute to it. Theorem 7 will then evidently be proved if the following Lemma is established:

LEMMA 8. For any index vectors $\underline{k}$, $\underline{m} \in Z^n$,

$$\operatorname{res}(\underline{P}^{\underline{k}-\underline{m}-\underline{e}}\,\underline{P}') = \begin{cases} 1, & \underline{m} = \underline{k}, \\ 0, & \underline{m} \neq \underline{k}. \end{cases} \tag{6.5}$$

Proof. Writing $\underline{k}$ in place of $\underline{k} - \underline{m}$, it suffices to prove the lemma for $\underline{m} = \underline{0}$. Let

$$\underline{k} = (k_1, k_2, \ldots, k_n).$$

Then

$$\underline{P}^{\underline{k}-\underline{e}}\,\underline{P}' = \prod_{j=1}^{n} P_j^{k_j - 1} \det(D_i P_j)$$

or, distributing the factors

$$P_j^{k_j-1}$$

among the columns of the determinant,

$$\underline{P}^{\underline{k}-\underline{e}}\,\underline{P}' = \det(P_j^{k_j-1} D_i P_j). \tag{6.6}$$

With a view towards Lemma 6, which states that the residue of a Jacobian determinant is zero, we try to find, for each j, a fLs F_j such that, for all i,

$$P_j^{k_j-1} D_i P_j = D_i F_j. \tag{6.7}$$

This is easy if $k_j \neq 0$, for then (6.7) evidently holds if

$$F_j = \frac{1}{k_j} P_j^{k_j}; \tag{6.8}$$

here we have used the assumption that $\mathfrak{F}$ has characteristic zero.

The construction is a little more complicated if $k_j = 0$. We let ourselves be guided by the convergent case and thus have to define the analog of the logarithm of a unit in $\mathcal{L}_n$. By (6.1) we may write

$$P_j = p_j x_j (1 + Q_j), \tag{6.9}$$

where $\operatorname{ord} Q_j \geq 1$, $Q_j \in \mathcal{L}_n$. We now let

$$F_j := Q_j - \frac{1}{2}Q_j^2 + \frac{1}{3}Q_j^3 - \frac{1}{4}Q_j^4 + \ldots .$$

The series makes sense formally since only the first $|\underline{m}|$ terms can contribute to any term $\underline{x}^{\underline{m}}$. Again the assumption that $\mathcal{F}$ has characteristic zero has been used. We have

$$\begin{aligned} D_iF_j &= (1 - Q_j + Q_j^2 - \ldots)D_iQ_j \\ &= (1 + Q_j)^{-1}\ D_iQ_j . \end{aligned}$$

From (6.9), denoting Kronecker's δ by δ_{ij},

$$\begin{aligned} D_iP_j &= \delta_{ij}p_j(1 + Q_j) + p_jx_j\ D_iQ_j \\ &= \delta_{ij}p_j(1 + Q_j) + p_jx_j(1 + Q_j)D_iF_j \\ &= P_j(\delta_{ij}x_j^{-1} + D_iF_j) , \end{aligned}$$

thus

$$P_j^{-1}D_iP_j = \delta_{ij}x_j^{-1} + D_iF_j ,$$

and we have almost, but not quite, succeeded in satisfying (6.7). Letting

$$\varepsilon_j := \begin{cases} 1 , & k_j = 0 \\ 0 , & k_j \neq 0 \end{cases} \tag{6.10}$$

we have

$$\det(P_j^{k_j-1}\ D_iP_j) =$$

$$= \det \begin{pmatrix} \varepsilon_1x_1^{-1} + D_1F_1 & D_1F_2 & \ldots & D_1F_n \\ D_2F_1 & \varepsilon_2x_2^{-1} + D_2F_2 & \ldots & D_2F_n \\ \cdot\ \cdot\ \cdot & \cdot\ \cdot\ \cdot & \ldots & \cdot\ \cdot\ \cdot \\ D_nF_1 & D_nF_2 & \ldots & \varepsilon_nx_n^{-1} + D_nF_n \end{pmatrix} \tag{6.11}$$

Let $W := \{1, 2, \ldots , n\}$, and let

$$V := \{j \in W : \varepsilon_j = 1\} .$$

Then by expanding the determinant it is found to equal

$$\sum_{U \subset V} \prod_{i \in U} x_i \cdot \det(D_i F_j)_{i,j \in W \setminus U} , \tag{6.12}$$

where the sum is taken with respect to all subsets U of V. Our task is to compute the residue of (6.12), and this we do by taking the residue of each term. For each term where $U \neq W$, the coefficient of $\underline{x}^{-\underline{e}}$ in view of the factor

$$\Pi_{i \in U} x_i^{-1}$$

equals the residue of

$$\det(D_i F_j)_{i,j \in W \; U}$$

with respect to the remaining variables x_i where $i \in W \setminus U$. This residue is zero by Lemma 6 ; here the series F_j must be considered as fLs in the variables x_i $(i \in W \setminus U)$ with coefficients that are fLs in x_i $(i \in U)$. A nonzero residue can arise only when $U = W$. This is possible only when $V = W$, i.e., when all $\varepsilon_j = 1$, which means that $\underline{k} = \underline{0}$. In this one case

$$\operatorname{res}(\underline{P}^{-\underline{e}} \underline{P}') = \operatorname{res} \prod_{j=1}^{n} x_j^{-1} = 1 . \blacksquare$$

Some historical remarks concerning Theorem 7 follow. (For more details see M.-L. Henrici (1976).) For systems of convergent series, the first version of a Lagrange-Bürmann type formula was probably given, without proof, by Laplace (1780). Jacobi (1830) gave a formal proof according to the standards of his time. Darboux (1869), Stieltjes (1885), and Poincaré commented on Laplace's result. A modern proof was given by I. J. Good (1960). A Lagrange-Bürmann theorem for formal series was given by M.-L. Henrici (1976).

Even apart from the fact that it is valid for formal series, our Theorem 7 is more general than all these results in the following two aspects: (i) All the papers mentioned deal only with systems $\underline{P}$ that are "normal" in the sense that

$$P_i = x_i Q_i ,$$

where Q_i is a unit in $\mathcal{P}_n$. (P_i "contains the factor

x_i"; in order to express the residue as a derivative, Q_i^1 is then frequently expressed in terms of its reciprocal.) No such assumption is made here. (ii) In all the papers mentioned, R is either special (Laplace, Jacobi: $R = x_i$; Darboux: $R = x_1x_2$), or they deal with $R \in \mathcal{F}_n^1$. In Theorem 7, R is an arbitrary series in $\mathcal{L}_n$.

Our proof differs from the technique of M.-L. Henrici (1976) by not using a matrix isomorphism which was crucial there.

7. AN EXAMPLE

We let $n = 2$ and write

$$\underline{x} = \begin{pmatrix} x \\ y \end{pmatrix}, \quad \underline{P} = \underline{u} = \begin{pmatrix} u \\ v \end{pmatrix}, \quad \underline{k} = \begin{pmatrix} k \\ \ell \end{pmatrix}, \quad \underline{m} = \begin{pmatrix} m \\ n \end{pmatrix}.$$

Consider the problem of expanding $\underline{x}^{\underline{k}} = x^k y^\ell$ in terms of u and v , where

$$\begin{aligned} u &= x - y^2 \\ v &= y - x^2 \end{aligned} \tag{7.1}$$

(This system is not normal.) By Theorem 7, the coefficient of $\underline{u}^{\underline{m}} = u^m v^n$ in that expansion is

$$c_{m,n} = \operatorname{res}(x^k y^\ell u^{-m-1} v^{-n-1} \underline{P}'),$$

where

$$\underline{P}' = \det \begin{pmatrix} u_x & v_x \\ u_y & v_y \end{pmatrix} = \det \begin{pmatrix} 1 & -2x \\ -2y & 1 \end{pmatrix}$$
$$= 1 - 4xy .$$

By the binomial theorem,

$$u^{-m-1} = x^{-m-1}\left(1 - \frac{y^2}{x}\right)^{-m-1} = x^{-m-1} \sum_{p=0}^{\infty} \frac{(m+1)_p}{p!} y^{2p} x^{-p},$$

$$v^{-n-1} = y^{-n-1} \sum_{q=0}^{\infty} \frac{(n+1)_q}{q!} x^{2q} y^{-q}.$$

Thus $c_{m,n}$ equals the coefficient of $x^{m-k}y^{n-\ell}$ in

$$(1-4xy) \sum_{p,q=0}^{\infty} \frac{(m+1)_p}{p!} \frac{(n+1)_q}{q!} x^{2q-p} y^{2p-q} .$$

A nonzero contribution arises if

$$2q - p = m - k \ , \ 2p - q = n - \ell \tag{7.2}$$

has a solution in non-negative integers (p,q). We use (7.2) to parametrize these solutions and thus let

$$m = 2q - p + k \ , \ n = 2p - q + \ell \ ,$$

where $p, q, = 0, 1, 2, \ldots$. For such values of m and n,

$$\begin{aligned} c_{m,n} &= \frac{(m+1)_p (n+1)_q}{p!q!} - 4 \frac{(m+1)_{p-1} (n+1)_{q-1}}{(p-1)!(q-1)!} \\ &= \frac{(2q-p+k+1)_p (2p-q+\ell+1)_q}{p!q!} - 4 \frac{(2q-p+k+1)_{p-1} (2p-q+\ell+1)_q}{(p-1)!(q-1)!} \end{aligned}$$

or, using a conventional notation in terms of binomial coefficients,

$$c_{m,n} = \binom{2q+k}{p}\binom{2p+\ell}{q} - 4\binom{2q+k-1}{p-1}\binom{2p+\ell-1}{q-1} ,$$

where binomial coefficients with a negative lower parameter are defined to be zero. Thus finally, if (7.1) holds,

$$x^k y^\ell = \sum_{p,q=0}^{\infty} \left\{ \binom{2q+k}{p}\binom{2p+\ell}{q} - 4\binom{2q+k-1}{p-1}\binom{2p+\ell-1}{q-1} \right\} \cdot u^{k+2q-p} v^{\ell+2p-q} .$$

BIBLIOGRAPHY

(Bürmann) .(1799): Rapport sur deux mémoires d'analyse du professeur Burmann, signé Lagrange, Legendre. Aus: Mémoires de l'Institut national des sciences et arts; sciences mathématiques et physiques; tome second. Paris, an VII.

G. Darboux (1869): Sur la série de Laplace. Comptes Rendus Acad. Sci. Paris 68, 324 - 327.

I. J. Good (1965): The generalization of Lagrange's expansion and the enumeration of trees. Proc. Camb. Phil. Soc. 61, 499 - 517.

M.-L. Henrici (1976): Die Lagrange-Bürmann-Formel bei Systemen von formalen Potenzreihen. Diss. ETH 5773. Juris-Verlag, Zürich.

P. Henrici (1964): An algebraic proof of the Lagrange-Bürmann formula. J. MAth. Anal. Appl. 8, 218 - 224.

P. Henrici (1974): Applied and Computational Complex Analysis, vol I. Wiley, New York.

C. G. J. Jacobi (1830): De resolutione aequationum per series infinitas. Crelle J. reine angew. Math. 6, 257 - 286 = Gesammelte Werke 6, Berlin 1891, S. 26 - 61.

J. L. Lagrange (1770): Nouvelle méthode pour résoudre les équations littérales par la moyen des séries. Mémoire de l'Académie royale des Sciences et Belles-Lettres de Berlin, t. XXIV = Oeuvres, Deuxième Section (suite), tome troisième, Paris MDCCCLXIX, p. 5 - 73.

P. S. Laplace (1780): Mémoire sur l'usage du calcul aux différences partielles dans la théorie des suites. Mémoire de l'Académie royale des Sciences de Paris = Oeuvres, tome 9ième, 1893, p. 313 - 335.

H. Poincaré (1887): Sur les résidues des intégrales doubles. Acta Math. 9. 321 - 380.

G. Polya and G. Szegö (1924): Aufgaben und Lehrsätze der Analysis, Bd. I. Springer, Berlin.

I. Schur (1947): Identities in the theory of power series. Amer. J. Math. 69, 14 - 26.

T.-J. Stieltjes (1885): Sur une généralisation de la série de Lagrange. Ann. scient. éc. normale sup., 3ième série, 2, 93 - 98.

A SIMPLIFIED PRESENTATION OF THE ADAMJAN-AROV-KREIN APPROXIMATION THEORY

Jean Meinguet

Université Catholique de Louvain
B-1348 Louvain-la-Neuve (Belgium)

ABSTRACT

The relatively recent approximation-theoretic work we consider here deals with infinite bounded Hankel operators and approximation problems associated with them. In addition to its intrinsic mathematical interest, this material also proves very useful for solving fundamental problems of linear systems theory. Our purpose here is to contribute to a more thorough understanding of the fine mathematics involved in the original and difficult papers [1], [5], [6], by relying on basic principles and avoiding technicalities.

1. INTRODUCTION

By "Adamjan-Arov-Kreĭn approximation theory", we mean here the

H. Werner et al. (eds.), Computational Aspects of Complex Analysis, 217–248.

impressive collection of remarkable results obtained by these authors (see specially [1], § 0 and § 1) in their systematic investigations dealing with *infinite bounded Hankel operators and approximation problems associated with them.* In actual fact, some of those results were obtained earlier by D.N. Clark [5], who gave further a most satisfactory characterization of the eigenvalues of *real* Hankel matrices, which indicates links with approximation theory in a meromorphic function-theoretic setting [6].

It turns out that, in recent years, this mathematically sophisticated material has received much attention in Linear Systems Theory (specially as regards finite causal linear discrete-time systems). This is due largely to the efficiency and elegance of the resulting model reduction method; inasmuch as this method is based essentially upon the well known singular value decomposition of the (possibly complex) Hankel matrix (associated with the impulse response of the system), it is clear that its algorithmic implementation must prove reasonably fast and exceptionally robust (remember Weyl's inequalities!). The interested reader is referred specially to [4] and [9], both for an excellent introduction to this important field of applications and for complementary references (see also [11]).

Our main purpose here is to contribute (at least in part!) to a more thorough understanding of the fine mathematics involved in the rather difficult papers [1], [5], [6] (related contributions to numerical linear algebra and algorithmics are being prepared for publication elsewhere). It is obvious that one of the first fundamental topics to be investigated in the above spirit is the famous Nehari criterion for boundedness (in the ℓ^2 sense) of Hankel matrices. Section 3 is essentially devoted to a detailed study of the somewhat mysterious relationship that exists between a Hankel operator and its defining symbol function. Unlike other authors, we have tried throughout to avoid technicalities and to rely systematically on *basic principles* (borrowed mainly from functional analysis and approximation theory). In particular, the so-called duality relations in normed linear spaces (whose background is recalled in Section 2) shall play a prominent role. In Section 4, we turn to some of the most significant results of [1], [5], [6], to wit : the far-reaching equivalence between the generalized Takagi (function-theoretic) problem and the Hankel operator-theoretic best approximation problem, and the characterization of singular values of Hankel matrices in relation with the number of zeros (inside the unit circle) of the associated singular functions (or elements of Schmidt pairs).

Most of the functions to be used in this paper can be regarded as defined on the unit circle and belong to the classical Banach spaces L^p or H^p $(1 \leqslant p \leqslant \infty)$ provided with their usual norm :

$$\|f\|_p := \{(2\pi)^{-1} \int_0^{2\pi} |f(e^{it})|^p dt\}^{1/p}, \text{ if } 1 \leqslant p < \infty, \tag{1a}$$

$$\|f\|_\infty := \operatorname{ess.sup}_t |f(e^{it})|, \text{ if } p = \infty. \tag{1b}$$

The *Hardy space* H^p, whose theory is excellently covered by the books [7] and [15] (among other references!) can be defined as the space of L^p functions f on the unit circle whose Fourier coefficients

$$\hat{f}(n) := (2\pi)^{-1} \int_0^{2\pi} f(e^{it})e^{-int}dt \text{ for all } n \in \mathbb{Z}, \tag{2a}$$

vanish on the negative integers; it should be noted that the recipe

$$\bar{f}(z) := \sum_{n\geqslant 0} \hat{f}(n)z^n \text{ for } |z| < 1, \tag{2b}$$

associates with each $f \in H^p$ a self-evident analytic extension into the unit disc (whose "radial limit" or "boundary function" is precisely f). For $p = 2$, the splitting of the Fourier series representing any L^2 function f into a nonnegative and a negative "half-series" amounts to the orthogonal direct sum decomposition

$$L^2 = H^2 \oplus (H^2)^{\perp}, \tag{3}$$

each of these Hilbert spaces being isometrically isomorphic to the space ℓ^2 of square-summable sequences (Riesz-Fischer theorem). On the other hand, this splitting into half-series does not define a continuous projector of L^p onto H^p if $p = 1$ or if $p = \infty$; in which cases, it can even be proved that there does not exist any bounded projector at all of L^p onto H^p (see [15], pp. 154-155).

As for notations, * stands for complex conjugation (or for "dual"), ^{T} for transposition and $^\sim$ for conjugate transposition (of matrices).

2. AN APPROXIMATION-THEORETIC BACKGROUND

By expressing the equality of a certain "sup" and a certain "inf", the so-called duality relations in normed linear spaces pair up an extremal problem in a space with a corresponding extremal problem in the dual space. Credit for observing that extremal problems arise naturally in "resonant" pairs, and that functional analysis can assist in their mutual solution, probably goes back to Kreĭn (1938); though this was rediscovered subsequently, and exploited in a variety of ways, by a number of other mathematicians working independently (in particular, by H.S. Shapiro, whose excellent Chapter 5 in [23] is specially recommended here).

As regards the specific needs of the present paper, we can look upon the matter as follows. Place the elements of L^∞, L^1 in duality via the bilinear form (or pairing)

$$<f,x> := (2\pi)^{-1} \int_0^{2\pi} f(e^{it})x(e^{it})e^{it}dt, \quad f \in L^\infty, \; x \in L^1. \qquad (4a)$$

Then, L^∞ is known to be the dual space of L^1, and its subspace H^∞ is clearly the annihilator (or "polar") of H^1 (i.e., the set of all functions $f \in L^\infty$ such that $<f,x> = 0$ for all $x \in H^1$); needless to say, the lack of symmetry inherent in the classical fact that the dual space of L^∞ is strictly larger than L^1 (which implies that, relative to the pairing (4a), the topology of L^∞ is "not admissible") may cause serious difficulties and anyhow requires special care. Most important for the following is the *basic duality relation*

$$\sup_{\substack{x\in H^1 \\ \|x\|_1 \leq 1}} |<f,x>| = \min_{g\in H^\infty} \|f-g\|_\infty \quad \text{for each } f \in L^\infty, \qquad (4b)$$

where "min" indicates that the infimum is attained. As a matter of fact, (4a, b) can be regarded as a concrete illustration of the fundamental result recalled hereafter (the property that L^∞, unlike L^1, is a *dual* Banach space forces its being interpreted as X^*, so that the roles of X, Y and $Y^\perp$ are necessarily to be played by L^1, H^1 and H^∞, respectively).

An abstract duality principle. *Let* X *be a normed linear space,* Y *a linear subspace of* X, X^* *the dual space of* X *and* $Y^\perp$ *the annihilator of* Y *in* X^*. *Then, for each fixed* $f \in X^*$,

$$\sup_{\substack{x\in Y \\ \|x\|_X \leq 1}} |f(x)| = \min_{g\in Y^\perp} \|f-g\|_{X^*}, \qquad (5)$$

where "min" indicates that the infimum is attained.

Though quite classical (see e.g. [3], p. 28), the proof of this far-reaching connection of an original *extremal problem* (stated on the left-hand side of (5)) with what is called appropriately (see [7], p. 130) the *dual extremal problem* (stated on the right) is so beautifully simple that it is worth reproducing here for the reader's convenience. Let $x \in Y$ and $g \in Y^{\perp}$. Then

$$|f(x)| = |(f-g)(x)| \leq \|f-g\|_{X^*}\|x\|_X$$

and consequently

$$\|f\|_{Y^*} := \sup_{\substack{x \in Y \\ \|x\|_X \leq 1}} |f(x)| \leq \inf_{g \in Y^{\perp}} \|f-g\|_{X^*}.$$

To show that equality holds, and that "inf" can be replaced by "min", just appeal to the *Hahn-Banach extension theorem*, according to which the restriction of f to Y, whose norm appears on the left, can be extended to X as a linear functional f^o without increasing the norm; since $f-f^o \in Y^{\perp}$, it can be substituted for g on the right-hand side of the last inequality, which establishes (5), and proves accordingly that *the quotient space* $X^*/Y^{\perp}$ (provided with its usual norm, i.e., the norm of cosets of X^* modulo $Y^{\perp}$) *is isometrically isomorphic to the dual space* Y^* *of* Y.

The *existence*, asserted by the above duality principle, of a closest element $g \in Y^{\perp}$ to each fixed $f \in X^*$ or, equivalently, the fact that $Y^{\perp}$ is *proximinal*, is such an important property that we feel justified in outlining here a (more) direct (and possibly original) proof. As emphasized by Shapiro in [23] (see p. 72), what is really relevant here is that $Y^{\perp}$ is a weak*-closed linear subspace of X^* (as intersection of the kernels of the weak*-continuous functionals on X^* : $f \mapsto |f(x)|$ for each fixed $x \in Y$). This is due to the following useful (and non-trivial!) principle regarding the attainment of extrema : *every element* f *in a dual Banach space* X^* *has a closest element in each weak***-closed linear subspace of* X^*. A remarkably simple proof (which compares advantageously with those given or mentioned in [24], see pp. 101-102) goes as follows. Observe first that, in view of the triangle inequality, the search for a best approximant to f may be limited to the closed ball of all approximants of norm $\leqslant 2\|f\|_{X^*}$, which is known to be $\sigma(X^*,X)$-compact by virtue of the *Banach-Alaoglu theorem.* Secondly, *the norm on* X^* *is a weakly** *lower semicontinuous function* : indeed, whenever (g_α) is a *net* in X^* that converges weakly* to g, the "generic" inequality

$$|g_\alpha(x)|/\|x\|_X \leqslant \|g_\alpha\|_{X^*} \text{ for all } 0 \neq x \in X,$$

implies

$$\|g\|_{X^*} \leqslant \liminf_\alpha \|g_\alpha\|_{X^*},$$

simply by taking lim inf of both sides. But it is well known that any lower semicontinuous (real-valued) function defined on a compact (non-empty) set must achieve its infimum; as a matter of fact, this is a useful generalization of a most classical result due to Weierstrass, which further can be established quite similarly (see e.g. [22], p. 108). For our proof of the existence principle stated above to be complete, it remains only to provide X^* with its weak* topology. Hence finally the important "meta-principle" : wherever possible, choose the ambient linear space to be a dual space; this should normally facilitate the solution of existence questions for best approximants.

3. THE NEHARI THEOREM : A CONSTRUCTIVE APPROACH

3.1 A criterion for the boundedness of Hankel matrices

For a long time, the rather mysterious relationship that exists between a Hankel operator and its defining symbol function was not fully understood. As a matter of fact, it is only in 1957 (see [19]) that a necessary and sufficient condition has been given for determining whether or not an infinite Hankel matrix is bounded in ℓ^2, and if so, finding its spectral norm (denoted $\mathrm{lub}_2(.)$ hereafter). This most interesting result can be stated as follows.

Theorem (Z. Nehari). *Let* $(h_j)_{j=1}^{\infty}$ *be a square-summable sequence of complex numbers or, equivalently, let*

$$h(e^{it}) := \sum_{j=1}^{\infty} h_j\, e^{-ijt},\ 0 \leqslant t < 2\pi, \tag{6a}$$

be the $(H^2)^{\perp}$ *function whose (non-trivial) Fourier coefficients are*

$$h_j = \hat{h}(-j) := (2\pi)^{-1} \int_0^{2\pi} h(e^{it})e^{ijt}dt \text{ for } j=1,2,\ldots \tag{6b}$$

Let

$$F := h + H^2 \tag{7}$$

denote the linear variety of all equivalent L^2 *symbols* f *of the infinite Hankel matrix* $H = H(f) := (H_{jk})_{j,k=0}^{\infty}$ *with* $H_{j,k} := h_{j+k+1}$. *Then*

$$\mathrm{lub}_2(H) \leqslant \|f\|_{\infty} (\leqslant \infty) \text{ for all } f \in F, \tag{8}$$

and H *is bounded in* ℓ^2 *if and only if there exists an essentially bounded symbol in the equivalence class* F; *in which case, there exists a* L^∞ *function* $f^0 \in F$ *with the property that*

$$\text{lub}_2(H) = \|f^0\|_\infty = \text{dist}_\infty(h, H^2), \qquad (9)$$

that is a symbol of minimal L^∞ *norm.*

In other words, a sequence $(h_j)_{j=1}^{\infty}$ defines a ℓ^2 bounded Hankel matrix if and only if it is the sequence of *negative* Fourier coefficients of an essentially bounded measurable function f on the unit circle, in which case,

$$\text{lub}_2(H) = \min_{g \in H^\infty} \|f-g\|_\infty \qquad (9\text{bis})$$

for every symbol $f \in L^\infty$.

It turns out that Nehari's original proof is rather complicated; it is based on a progressive reduction of the problem, by standard density and compactness arguments, to considerations of a finite Hankel matrix (determined by $h_1, \ldots, h_n$), which in turn amount to solving the associated *Carathéodory-Fejér approximation problem* : given the polynomial $h_n + h_{n-1} z + \ldots + h_1 z^{n-1}$, how can it be extended to a power series representing a H^∞ function of minimal norm. According to S.C. Power (see his recent survey paper [21]), there appear to be now two lines of proof, which may be referred to as *analytic* and *geometric*. The proof of Nehari's theorem we present in Section 3.2, and the proof of the ensuing characterization result in Section 3.3, seem to be particularly transparent; they are also reasonably self-contained : apart from the material collected in Section 2, they require only some familiarity with basic results from the theory of Hardy spaces (essentially, the F. Riesz factorization theorem and the F. and M. Riesz theorem).

A non-trivial illustration : the Hilbert matrix. A famous inequality of Hilbert, dating back to 1908 (see e.g. [13], p. 212) asserts that for $h_j := 1/j$ ($j=1,2,\ldots$) the associated Hankel matrix, commonly known as Hilbert's matrix, is bounded in ℓ^2. Unlike the "natural" symbol (6a) :

$$h(e^{it}) = -\ln(1-e^{-it}),\ 0 \leqslant t < 2\pi,$$

the suitably modified symbol

$$f(e^{it}) = i(t-\pi),\ 0 \leqslant t < 2\pi, \tag{10}$$

is a L^∞ function; it then follows from (8) that the spectral norm of Hilbert's matrix is bounded from above by $\pi = \|f\|_\infty$, which amounts to the famous *Hilbert inequality*

$$\left|\sum_{j,k=0}^{\infty} \frac{u_j v_k}{j+k+1}\right| \leqslant \pi \|u\|_2 \|v\|_2 \text{ for all } u,v \in \ell^2; \tag{11}$$

incidentally, it was Schur who showed in 1911 that π is the best possible constant in (11). As investigated recently by Power (see [20]), the fact that a Hankel matrix has a *piecewise continuous symbol* may lead to more precise conclusions; as regards Hilbert's matrix, whose symbol defined by (10) is of that class, these reduce to W. Magnus's result (1950) that the spectrum is the interval $[0,\pi]$. This is in very sharp contrast with that most classical and important result known as *P. Hartman's theorem* [14], which asserts that *a Hankel matrix is compact if and only if there exists a continuous symbol.*

3.2 A transparent analytic proof

Consider the *bilinear form*

$$v^T Ru := \lim_{n\to\infty} \sum_{j=0}^{n} \sum_{k=0}^{n} R_{jk}\, u_k\, v_j , \tag{12}$$

where $R = (R_{jk})$ is an infinite matrix and $u := (u_0, u_1, \ldots)$ (resp. v) denotes the square-summable sequence of the (non-trivial) Fourier coefficients

$$u_k = \hat{u}(k) = (2\pi)^{-1} \int_0^{2\pi} u(e^{it})e^{-ikt}dt \text{ for } k=0,1,\ldots \tag{13}$$

of an arbitrary H^2 function $u(e^{it})$ (resp. $v(e^{it})$). By definition, such a bilinear form is *bounded in the sense of Hilbert* (or, equivalently, R is bounded in ℓ^2) if

$$\mathrm{lub}_2(R) := \sup_{\substack{\|u\|_2\leqslant 1 \\ \|v\|_2\leqslant 1}} |v^T Ru| < \infty, \tag{14a}$$

which actually provides a *bilinear characterization* of the spectral norm of an arbitrary matrix. Whenever R is *symmetric* (which is quite different, in the complex case, from being Hermitian!), it turns out that (14a) strictly amounts to a *quadratic characterization* of the spectral norm, viz.,

$$\mathrm{lub}_2(R) = \sup_{\|u\|_2\leqslant 1} |u^T Ru| ; \tag{14b}$$

this is readily verified by combining a polarization identity with the parallelogram theorem (see e.g. [13], p. 208). As for the "modern", general definition

$$\mathrm{lub}_2(R) := \sup_{\|u\|_2\leqslant 1} \|Ru\|_2, \tag{14c}$$

it directly follows from (14a) for $v := (Ru)^* / \|Ru\|_2$; indeed, by Schwarz's inequality, choosing such a v simply amounts to maximizing $|v^T Ru|$, for u arbitrarily fixed, over the unit v ball. This last definition leads, in particular, to the most interesting conclusion that *the spectral norm of any (possibly infinite) matrix* R *that is bounded in* ℓ^2 *is the supremum of the spectrum of the positive semidefinite Hermitian matrix*

$$S := (\tilde{R}\, R)^{1/2}; \tag{15}$$

indeed, the convex hull of the spectrum of any normal bounded matrix (which is simply a compact interval in the Hermitian case) is known to coincide with the closure of its numerical range (see e.g. [12], p. 112).

Now let us take for R the infinite Hankel matrix introduced above in association with the function $h \in (H^2)^\perp$ defined by (6a, b). An explicit calculation, taking the definition (7) into account, shows that

$$\sum_{j=0}^{n} \sum_{k=0}^{n} H_{jk}\, u_k\, v_j = (2\pi)^{-1} \int_0^{2\pi} f(e^{it}) x_n(e^{it}) e^{it} dt, \tag{16a}$$

with

$$x_n(e^{it}) := \Big(\sum_{k=0}^{n} u_k\, e^{ikt}\Big) \Big(\sum_{j=0}^{n} v_j\, e^{ijt}\Big) \tag{16b}$$

and for every $f \in F$ and every $n \in \mathbb{N}$. But it is well known (see e.g. [26], Vol. 1, p. 275) that the set of products $u(e^{it})v(e^{it}) =: x(e^{it})$ where u and v belong to the unit H^2 sphere exactly covers the unit H^1 sphere; this can be regarded as a corollary of the *F. Riesz factorization theorem* (1923) according to which the zeros of any nonnull H^p function can be divided out (simply by dividing by the corresponding Blaschke product!) without modifying the norm. It thus follows from (16a, b), in view of the definition (14a), that H is bounded in ℓ^2 if and only if there exists a function $f \in F$ such that the linear functional on H^1 :

$$x \mapsto (2\pi)^{-1} \int_0^{2\pi} f(e^{it})x(e^{it})e^{it}dt \tag{17}$$

is bounded or, equivalently, if and only if there exists for H an essentially bounded symbol f (the most general bounded linear functional on H^1 can indeed be expressed in the form (17), where $f \in L^\infty$; see e.g. [7], p. 129). In which case, the so-called (bilinear) *Toeplitz relation*

$$v^T Hu = (2\pi)^{-1} \int_0^{2\pi} f(e^{it})u(e^{it})v(e^{it})e^{it}dt \tag{18}$$

holds for all $u,v \in \ell^2$ and all $f \in F$, which implies

$$\sup_{\substack{\|u\|_2 \leq 1 \\ \|v\|_2 \leq 1}} |v^T Hu| = \sup_{\substack{x \in H^1 \\ \|x\|_1 \leq 1}} (2\pi)^{-1} \left|\int_0^{2\pi} f(e^{it})x(e^{it})e^{it}dt\right|$$

and thereby establishes (9, 9bis) by appealing again to the basic duality relation (4a, b). This completes our proof of Nehari's theorem.

3.3 Characterization of symbols of minimal norm

It turns out that a most satisfactory characterization of such symbols can be given whenever the Hankel matrix H has the property that

$$\mathrm{lub}_2(H) = |(u^o)^T Hu^o| \text{ for some } u^o \in \ell^2 \text{ of unit norm.} \tag{19}$$

This amounts to assuming that the supremum is attained in the definition (14b) of the spectral norm of H (regarded as a *symmetric* matrix) at some u^o, and therefore also at $e^{i\alpha} u^o$ for every real α (so that $|.|$ may be dropped in (19), by a suitable normalization). It should be realized that this assumption is trivially satisfied whenever H is compact or, equivalently (by

Hartman's theorem), whenever there exists for H a continuous symbol.

Now, under the assumption (19), the basic result (9) (where f^o denotes any symbol of minimal L^∞ norm of H) and the quadratic variant of the Toeplitz relation (18) (rewritten for $f = f^o$) yield the following remarkable identity :

$$\| f^o \|_\infty = (2\pi)^{-1} \int_0^{2\pi} f^o(e^{it})[u^o(e^{it})]^2 \, e^{it} dt \ (= \mathrm{lub}_2(H)), \qquad (20)$$

where, in P. Duren's terminology (see [7], p. 132), $x := (u^o)^2 \in H^1$ with $\| x \|_1 = 1$ (since $u^o \in H^2$ with $\| u^o \|_2 = 1$) is called a *(normalized) extremal function* (or solution of the "original" extremal problem stated in (5)) and $f^o \in L^\infty$ is called an *extremal kernel* (or solution of the "dual" extremal problem stated in (5)). By combining the condition for equality to hold in (the degenerate case of) Hölder's inequality applied to the right-hand side of (20), with that important variant of the *F. and M. Riesz theorem* (1916) which states that a function $u \in H^2$ cannot vanish on a set of positive measure without vanishing identically, we get the following nice result.

<u>Theorem</u> <u>(Characterization of extremal kernels and functions)</u>. *Let the assumption* (19) *be verified for a given Hankel matrix* H. *Then in order that a function* $f^o \in L^\infty$ *be a symbol of minimal norm for* H, *it is necessary and sufficient that*

$$f^o(e^{it})[u^o(e^{it})]^2 \, e^{it} \geqslant 0 \quad \text{a.e.}, \qquad (21a)$$

$$|f^o(e^{it})| \stackrel{\text{a.e.}}{=\!=} \| f^o \|_\infty \ (= \mathrm{lub}_2(H)), \qquad (21b)$$

or, equivalently, that

$$f^o(e^{it})e^{it} \overset{\text{a.e.}}{=} \text{lub}_2(H)[u^o(e^{it})]^*/u^o(e^{it}) \text{ for some } 0 \neq u^o \in H^2. \tag{22}$$

Moreover, always under assumption (19), *the minimal symbol* f^o *of* H *is unique.*

Only the *uniqueness* assertion remains to be proved. The argument is based upon the elementary fact that the set of best approximants to $f \in L^\infty$ out of H^∞ is *convex* : if there exist two distinct extremal kernels, say $f^{o,k} := h-g^{o,k}$ (k=1,2) with $g^{o,k} \in H^\infty$, then every $f^o := \lambda f^{o,1} + (1-\lambda)f^{o,2} = f^{o,2} + \lambda(g^{o,2}-g^{o,1})$ for $0 \leqslant \lambda \leqslant 1$, being an extremal kernel too, must satisfy (21b); but an explicit verification shows this to be impossible, the difference $|f^o|^2 - |f^{o,2}|^2$ depending necessarily on λ, unless $g^{o,1}$ and $g^{o,2}$ coincide.

On the other hand, if condition (19) is not fulfilled, then the extremal kernel f^o, which anyhow exists in view of the basic duality relation (4a, b), is not necessarily unique and cannot be characterized elegantly (at least to the best of our knowledge). In particular, *an extremal kernel* f^o *need not have constant modulus a.e. on the unit circle if no extremal function exists.* This negative result can be established as follows. Suppose there is a subset E of positive measure (of the unit circle) over which a.e. $|f^o(e^{it})| \leqslant \|f^o\|_\infty - \alpha$ for some $\alpha > 0$. Then, as readily verified, the fact that, according to (4a, b), the supremum of the right-hand side of (20) for $f = f^o$ and for u^o ranging over the unit H^2 sphere is equal to $\|f^o\|_\infty$ implies

$$\inf_{\substack{x \in H^1 \\ x(0)=1}} (2\pi)^{-1} \int_0^{2\pi} |x(e^{it})| c_E(t)dt = 0, \tag{23a}$$

where $c_E(t)$ denotes the characteristic function of E; by a most remarkable theorem due to Szegö (1920), (23a) can be rewritten in the form

$$G(c_E) := \exp\{(2\pi)^{-1} \int_0^{2\pi} \ln c_E(t)dt\} = 0, \tag{23b}$$

where $G(c_E)$ denotes the so-called *geometric mean* of $c_E(t)$; since the condition (23b) is nothing else than the *Kolmogoroff-Kreĭn criterion* for the sequence of functions $(e^{ikt})_{k=0}^{\infty}$ to be closed in $L^1(c_E(t)dt;-\pi,\pi)$ (see e.g. [2], p. 261), there is clearly no contradiction to expect from the existence of the set E defined above.

4. CHARACTERIZATION OF SINGULAR VALUES OF HANKEL MATRICES

4.1 The Szegö kernel for the unit disc

The Hardy space H^2, regarded as the space of analytic functions in the open unit disc $D : |z| < 1$ with square-summable Taylor series or, equivalently, with bounded concentric L^2 norms, is known to be a *functional Hilbert space*, i.e., a Hilbert space of complex-valued functions u in D which is such that, for each $a \in D$, the *evaluation functional* (on H^2) $\delta_a : u \mapsto u(a)$ is bounded. It turns out that the "reproducing" formula for H^2 is essentially the Cauchy integral formula for the unit disc :

$$u(a) = \frac{1}{2\pi i} \int_{|z|=1} \frac{u(z)dz}{z-a} \text{ for } |a| < 1, \tag{24}$$

which is known to hold for every $u \in H^2$ if, on the right-hand side, u is interpreted as *boundary function* (or radial limit a.e.). As a matter of fact, (24) is readily rewritten in the required form, viz.,

$$u(a) = (u,K_a)_2 := (2\pi)^{-1} \int_0^{2\pi} u(e^{it})[K_a(e^{it})]^* dt, \tag{25a}$$

where $(.,.)_2$ denotes the usual Hermitian pairing in H^2 (not to be confused with the general duality bracket $<.,.>$ in (4a, b), which is only bilinear) and

$$K_a(z) = K(z,a) := (1-za^*)^{-1} \text{ for } |a| < 1 \tag{25b}$$

denotes the *Fréchet-Riesz representer of the Dirac measure* δ_a, the corresponding bivariate function $K(z,a)$ on $D \times D$ being known as the *reproducing kernel* of H^2, also called the *Szegö kernel function* (for the unit disc or circle).

Consider now the *finite Blaschke product*

$$B_n(z) := \prod_{j=1}^{n} \frac{z-a_j}{1-za_j^*} \text{ where } |a_j| < 1 \text{ for } j=1,\dots,n; \tag{26}$$

for convenience, we shall assume throughout that the points a_j are all distinct (otherwise, the following is to be modified in a classical, but tedious, way). In view of the natural identifications :

$$B_nH^2 = \{u \in H^2 : u(a_j) = 0 \text{ for } j=1,\dots,n\}, \tag{27a}$$

$$(B_nH^2)^\perp = \text{span}\{K_{a_1},\dots,K_{a_n}\}, \tag{27b}$$

we must have in H^2 a *representation formula* of the form

$$u = \sum_{j=1}^{n} c_j K_{a_j} + B_n v, \tag{28}$$

where v is some (uniquely defined) H^2 function and the coefficients c_j are determined by a Cramer system of linear equations, viz.,

$$\sum_{j=1}^{n} K(a_i,a_j)c_j = u(a_i) \text{ for } i=1,\dots,n, \tag{29}$$

whose matrix is Hermitian and positive definite (as Gram matrix of the linearly independent K_{a_j}'s). We have thus rediscovered, in a quite surprisingly simple way, the intimate connection that is known to exist in H^2 (see e.g. [25], Chap. IX and X) between *best approximation in the* L^2 *sense by rational functions with preassigned simple poles outside the disc* (denoted here $1/a_j^*$ with $|a_j| < 1$) and *interpolation in the related points* a_j.

Let us mention further, in passing, that for the class of functions $K_a(.)$ defined by (25b) with $a \in D-\{a_1,\ldots,a_n\}$, the representation formula (28) can be rewritten in closed form as follows :

$$K(z,a) := K_a(z) - \sum_{j=1}^{n} \left\{\frac{B_{n,j}(a)}{B_{n,j}(a_j)} \cdot \frac{1-|a_j|^2}{1-aa_j^*}\right\}^* K_{a_j}(z) \tag{30}$$

$$= B_n(z) \frac{[B_n(a)]^*}{1-za^*},$$

where $B_{n,j}(.)$ is simply (26) where the factor $(z-a_j)/(1-za_j^*)$ is skipped. As a matter of fact, $K(z,a)$ is exactly the reproducing kernel of the functional Hilbert space B_nH^2, so that the orthogonal projection of any $u \in H^2$ onto B_nH^2 is automatically given by

$$(B_nv)(a) = (u(.),K(.,a))_2 = \frac{B_n(a)}{2\pi i} \int_{|z|=1} \frac{u(z)dz}{(z-a)B_n(z)} \tag{31a}$$

and the complementary projection by

$$(u-B_nv)(a) = \sum_{j=1}^{n} \frac{B_{n,j}(a)}{B_{n,j}(a_j)} \cdot \frac{1-|a_j|^2}{1-aa_j^*} u(a_j); \tag{31b}$$

it should be realized that (31b) yields an *interpolant* to $u \in H^2$ which is automatically *optimal in the sense of Sard*, while (31a) gives an integral expression of the associated *remainder*. By resorting to a rather standard procedure (see e.g. [18], pp. 300-

303), every *biorthonormal expansion of Lagrange type* can be "algorithmically" converted into a *biorthonormal expansion of Newton type* (whose *permanence property* is often regarded as a significant advantage); in the present situation, this reformulation of the Lagrange interpolant (31b) to $u \in H^2$ leads to the so-called *Takenaka series* (see e.g. [25], p. 305).

4.2 Two equivalent problems

Let there be given a function $f \in L^\infty$ or, equivalently, a ℓ^2 bounded Hankel matrix $H = H(f)$ of symbol f (as defined in Section 3.1). For any finite Blaschke product B_n of the form (26), the Hankel matrix $H(B_n f)$ is clearly bounded in view of Nehari's theorem, which yields the basic result

$$\text{lub}_2(H(B_n f)) = \min_{u \in H^\infty} \|B_n f - u\|_\infty . \tag{32}$$

Again for convenience, let us *temporarily* assume that $B_n(z)$ has only simple zeros. Since $H^\infty \subset H^2$, the representation formula (28) is applicable here, with the additional precision (implied by the F. Riesz factorization theorem recalled above) that v is now some H^∞ function; hence, in view of the generic definition (25b), it follows that (32) can be rewritten in the form

$$\text{lub}_2(H(B_n f)) = \min_{\substack{v \in H^\infty \\ \{c_1,\ldots,c_n\}}} \left\| B_n(e^{it}).(f-v)(e^{it}) - \sum_{j=1}^{n} \frac{c_j}{1-e^{it} a_j^*} \right\|_\infty$$

or, by the definition (26), in the equivalent form

$$\text{lub}_2(H(B_n f)) = \min_{\substack{v \in H^\infty \\ \{d_1,\ldots,d_n\}}} \left\| f(e^{it}) - v(e^{it}) - \sum_{k=1}^{n} \frac{d_k}{e^{it} - a_k} \right\|_\infty , \tag{33}$$

where the c_j and d_k's denote complex numbers.

Now the sum on the right-hand side of (33) can be interpreted as the partial fraction representation of a rational function $r(z)$ (for $z=e^{it}$) which is *proper and of (formal) degree* n (i.e., such that the degree of the numerator is less than the degree n of the denominator, which is simply the numerator of $B_n(z)$) and *stable* (i.e., such that its poles are all inside the unit circle). By simple classical arguments, it follows that (33) must hold for every Blaschke product B_n (whether its zeros are distinct or not) provided only that the multiplicity of each distinct pole is correctly reflected in the partial fraction representation of r. We are thus led, quite naturally, to consider the following *meromorphic function-theoretic problem.*

<u>Problem</u> (I_n) : *Let* $R^{(n)}$ *denote the set of all proper and stable rational functions of degree* $\leq n$. *Given a function* $f \in L^\infty$, *it is required to find the best approximation to it in the* L^∞ *norm by functions of the form* $v+r$, *with* $v \in H^\infty$ *and* $r \in R^{(n)}$.

It may be interesting to note that, whenever f has only a *finite* number of (nonzero) *negative* Fourier coefficients (denoted $h_j = \hat{h}(-j)$, for $j=1,2,\ldots$, in Section 3.1), then Problem (I_n) reduces to the classical *T. Takagi approximation problem* (1924) (see also [2], pp. 270-280) whose specialization for n=0 is simply the *Carathéodory-Fejér approximation problem.* This remark justifies regarding Problem (I_n) as a *generalized Takagi problem.*

Most important for the following is the fact that (33) can be rewritten in the form

$$\mathrm{lub}_2(H(B_n f)) = \min_{\{d_1,\ldots,d_n\}} \mathrm{lub}_2[H(f(z) - \sum_{k=1}^{n} \frac{d_k}{z-a_k})], \qquad (34)$$

as it directly follows from Nehari's theorem. By the linearity of the assignment $f \mapsto H(f)$, it is clear that

$$H(f(z) - \sum_{k=1}^{n} \frac{d_k}{z-a_k}) = H(f) - \sum_{k=1}^{n} d_k \, H(\frac{1}{z-a_k}), \tag{35}$$

where the sum of Hankel matrices on the right-hand side must be ℓ^2 bounded and of rank $\leqslant n$, the elementary Hankel matrix

$$H(\frac{1}{z-a}) = \begin{bmatrix} 1 & a & a^2 & . & . & . \\ a & a^2 & a^3 & . & . & . \\ a^2 & a^3 & a^4 & . & . & . \\ . & . & . & . & . & . \\ . & . & . & . & . & . \\ . & . & . & . & . & . \end{bmatrix}$$

being indeed ℓ^2 bounded (if and only if $|a| < 1$) and of rank one. Let us remark in passing that this type of argument can lead to a simple proof (different from that given in [8], Vol. II, p. 182) of the important *Kronecker theorem* (1881), which essentially asserts that a (ℓ^2 bounded) Hankel matrix $H = (h_{j+k+1})_{j,k=0}^{\infty}$ is of rank n if and only if the series

$$h(z) = \sum_{j=1}^{\infty} h_j / z^j$$

determines a (stable) rational function of exact degree n. Needless to say, the same type of argument can be used to extend (34) so that, like (33), it holds for every Blaschke product B_n. We are thus finally led to the alternative, *operator-theoretic problem.*

<u>Problem</u> (II_n) : *Let* $H^{(n)}$ *denote the set of all* ℓ^2 *bounded Hankel matrices of rank* $\leqslant n$. *Given a* ℓ^2 *bounded Hankel matrix* H, *it is required to find the best approximation to it in the spectral norm by matrices* $K \in H^{(n)}$.

From their very derivation, it is clear that, *for each* n, *Problem*

(I_n) *and Problem* (II_n) *are equivalent to the problem of minimizing the spectral norm of* H(Bf) *over the set* $\mathcal{B}^{(n)}$ *of all Blaschke products* B *with at most* n *factors.*

Moreover, *these minimization problems are always solvable.* Indeed, in view of (32), $\mathrm{lub}_2(H(B_n f))$ can be interpreted as the L^∞ distance of $B_n f$ to H^∞; therefore, as recalled above (at the end of Section 2), $\mathrm{lub}_2(H(Bf))$ for $f \in L^\infty$ fixed is a weakly* lower semicontinuous function of $B \in \mathcal{B}^{(n)} \subset H^\infty$. But the set $\mathcal{B}^{(n)}$ is trivially compact for the topology of pointwise convergence (over the closed unit disc); since the elements of $\mathcal{B}^{(n)}$ are Blaschke products, it directly follows that $\mathcal{B}^{(n)}$ is weak*-compact, which virtually completes the proof. As regards more particularly Problem (II_n), a proof is outlined in [1] (see p. 32); though presented otherwise, it actually exploits the weakly lower semicontinuity of the rank (on the set of ℓ^2 bounded matrices of finite rank).

4.3 The main results : a direct proof

Let us now turn to the central problem of characterizing singular values of Hankel matrices in relation with the number of zeros (inside the unit circle) of the associated singular functions. Unlike the authors of [1], whose approach is substantially based on an operator-theoretic analysis of the properties of Schmidt pairs for Hankel operators, and unlike Clark in [5], whose process by progressive extension (from *finite*, through *compact*, to *bounded* Hankel matrices) rests upon the specific (resp. general) fact (see [14], Lemmas 1 and 2) that *any* ℓ^2 *bounded (resp. compact) Hankel matrix can be regarded as the limit of a strongly (resp. uniformly) convergent sequence of finite Hankel matrices*, we are essentially interested here in elaborating a *direct (elementary) proof of the essential results.*

Mainly for simplicity, we shall assume throughout this section that the Hankel matrix

$$H = H(f), \text{ of symbol } f \in L^{\infty}, \text{ is } \ell^2 \textit{ compact.} \tag{36}$$

However, it must be emphasized that significant extensions to the case of ℓ^2 *bounded* Hankel matrices are quite feasible (and even easy!), at least as regards the (possibly empty!) part of the spectrum of the positive semidefinite Hermitian matrix $(\tilde{H}\,H)^{1/2}$ consisting of the *isolated eigenvalues of finite multiplicity* which are greater than the supremum of the so-called *essential spectrum* (by definition, the essential spectrum, called spectrum of condensation in [1], of an operator consists of the *accumulation points of its spectrum* and its *eigenvalues of infinite multiplicity*). Such interesting extensions are considered at great length in [1] and [5].

Now, in view of the well known *Riesz-Schauder theory of compact operators*, the assumption (36) implies that H has an at most countable number of positive *singular values* (called s-numbers in the Russian literature)

$$s_j = s_j(H) := \lambda_j^{1/2}(\tilde{H}\,H) \tag{37}$$

we can enumerate in decreasing order

$$\mathrm{lub}_2(H) = s_0 \geqslant s_1 \geqslant \dots \geqslant s_j \geqslant \dots \ (> 0), \tag{38}$$

their (necessarily finite) multiplicities being taken into account; indeed, under the assumption (36), the positive semidefinite Hermitian matrix $\tilde{H}\,H$ is ℓ^2 compact, so that its spectrum, with the possible exception of the point 0, can only consist of isolated eigenvalues $\lambda_j \geqslant 0$ of finite multiplicity. As emphasized in [10], whose excellent Chapter II contains a rather complete account of the basic properties of s-numbers of linear operators, the following *best approximation property*

$$s_j(H) = \min_{C \in C^{(j)}} \mathrm{lub}_2(H-C) \text{ for } j=0,1,\ldots, \tag{39}$$

where $C^{(j)}$ denotes the set of all ℓ^2 bounded matrices of rank $\leq j$, can be taken as a new, equivalent definition of the s-numbers of *arbitrary* ℓ^2 compact matrices, which often proves to be more convenient than the original one.

Though quite classical (see e.g. [10], pp. 28-29), the proof of this most remarkable property is so illuminating that it is worth outlining here, the more so as it can be suitably extended to the case when H is only ℓ^2 bounded (see [10], pp. 61-62). From the *Courant-Fischer* (minimax) *characterization* of the (nonnegative) eigenvalues (indexed in decreasing order) of the ℓ^2 compact Hermitian matrix $\widetilde{H}\, H$, to wit :

$$\lambda_j(\widetilde{H}\, H) = \min_{C \in C^{(j)}} \max_{\substack{\|u\|_2 \leq 1 \\ Cu=0}} \|Hu\|_2^2 \text{ for } j=0,1,\ldots,$$

it follows by (37) that

$$s_j(H) \leq \max_{\substack{\|u\|_2 \leq 1 \\ Cu=0}} \|Hu\|_2 = \max_{\substack{\|u\|_2 \leq 1 \\ Cu=0}} \|(H-C)u\|_2$$

for every ℓ^2 bounded matrix C of rank $\leq j$, and consequently

$$s_j(H) \leq \mathrm{lub}_2(H-C) \text{ for all } C \in C^{(j)} \text{ and } j=0,1,\ldots$$

To complete the proof, it remains to note that

$$s_j(H) = \mathrm{lub}_2(H-C_j) \text{ for } j=0,1,\ldots, \tag{40a}$$

where

$$C_j := \sum_{k=0}^{j-1} s_k(H)V^{(k)}[U^{(k)}]^{\sim} \tag{40b}$$

can be interpreted as the j-th partial sum of the so-called *Schmidt expansion* or *singular value decomposition*

$$H = V\,\Sigma\,\tilde{U} \text{ with } \tilde{V}\,V = \tilde{U}\,U = I \tag{41}$$

of the ℓ^2 compact matrix H. It should be noted that Σ is the diagonal matrix of positive singular values s_k ($k=0,1,\ldots$) of H (ordered here according to (38)), while the associated column vectors $U^{(k)}$ (resp. $V^{(k)}$) of U (resp. V) form a ℓ^2 orthonormal system of eigenvectors of $\tilde{H}\,H$ (resp. $H\,\tilde{H}$) spanning topologically the range of $\tilde{H}$ (resp. H). As a matter of fact, it is easily seen that (41) is equivalent to the Hermitian eigenvalue problem

$$\begin{bmatrix} 0 & \tilde{H} \\ H & 0 \end{bmatrix} \begin{bmatrix} U \\ V \end{bmatrix} = \begin{bmatrix} U \\ V \end{bmatrix} \Sigma, \text{ with } \Sigma \text{ p.d. and } \tilde{V}\,V = \tilde{U}\,U = I, \tag{41bis}$$

where $(U^{(k)}, V^{(k)})$ is a *Schmidt pair* associated with the (positive) singular value $s_k(H)$, which implies further

$$\begin{aligned} (\tilde{H}\,H)U &= U\,\Sigma^2 \text{ with } \tilde{U}\,U = I, \\ (H\,\tilde{H})V &= V\,\Sigma^2 \text{ with } \tilde{V}\,V = I. \end{aligned} \tag{41ter}$$

For the applications we have in mind (which essentially involve finite rank Hankel matrices), the most profitable point of view is usually *sequential* rather than *functional*; that is why we prefer here, unlike the authors of [1], to speak of *singular value decomposition of matrices* (whose exceptionally fruitful idea apparently dates back to E. Beltrami and C. Jordan, 1873-1874, see [16], p. 78), rather than of *Schmidt expansion of operators* (which was first introduced by E. Schmidt in the quite different setting of integral equations with non-Hermitian kernels). Now, unlike the best approximation property (39) and the reminder complementing its proof, the following is conditioned by the property of H to be a Hankel matrix.

<u>Theorem</u> (<u>Main results of</u> [1] <u>and</u> [5]). *Let a given Hankel matrix* H *verify the assumption* (36), *its positive singular values* s_j *being indexed according to* (38). *Then, for each* j,

$$\begin{aligned} s_j &= \min_{B\in\mathcal{B}^{(j)}} \operatorname{lub}_2(H(Bf)) \\ &= \min_{\substack{v\in H^\infty \\ r\in R^{(j)}}} \|f-v-r\|_\infty = \min_{K\in\mathcal{H}^{(j)}} \operatorname{lub}_2(H(f)-K), \end{aligned} \tag{42}$$

where $\mathcal{B}^{(j)}$, $R^{(j)}$ *and* $\mathcal{H}^{(j)}$ *are defined in Section* 4.2. *Moreover, if* n *and* p *are such that*

$$\operatorname{lub}_2(H) = s_0 \geq s_1 \geq \ldots > s_n = s_{n+1} = \ldots = s_{n+p} > \ldots (>0) \tag{43}$$

and if $u(e^{it}) \in H^2$ *is an eigenvector of* H^*H *for* $\lambda_n = s_n^2$, *then* u(z) *has at least* n *and at most* n+p *zeros in* $|z| < 1$, *but the solution of the minimization problems* (42) *for* $n \leq j \leq n+p$ *is uniquely defined and corresponds with* n.

For simplicity, we will concentrate here on the case p=0, where λ_n is a *simple* eigenvalue of H^*H; this is clearly quite natural, and even fully justified whenever the entries of H derive from physical measurements. Then, up to a scalar factor, the eigenvector $u(e^{it})$ of H^*H corresponding to $\lambda_n = s_n^2$ is uniquely determined by the *algebraic* equation

$$H(f)u = s_n u^* \tag{44a}$$

or, equivalently, by the *functional* equation

$$f(e^{it})u(e^{it}) = s_n\ e^{-it}[u(e^{it})]^* + w(e^{it}), \tag{44b}$$

where w is simply the orthogonal projection onto H^2 of the L^2 function fu (remember that here f denotes a fixed L^∞ symbol of H). Consider the *F. Riesz factorization* of $u(e^{it}) \in H^2$, viz.,

$$u(z) = B(z)q(z), \tag{45}$$

where B(z) is a Blaschke product and q(z) is a H^2 function which does not vanish in $|z| < 1$. We essentially want to establish the nice result : k = n, where k denotes the number of factors of B. The direct proof we present now is surprisingly simple, owing to the following two classical lemmas (which are interesting in themselves).

<u>Lemma 1</u>. *Let* $f \in L^\infty$ *and* $g \in H^\infty$. *Then*

$$H(fg) = H(f)T(g) = T^T(g)H(f), \tag{46}$$

where T(g) *is the Toeplitz matrix associated with* g.

This can be readily verified by an explicit computation exploiting the property that T(g) for $g \in H^\infty$ is the *lower triangular matrix* whose (j,k)-th entry is the (j-k)-th Fourier coefficient of g (see also [5], pp. 631-632). It is interesting to note (see [1], pp. 36-37) that (46) can be interpreted as a natural generalization of the characteristic property of infinite Hankel matrices to satisfy the "commutation relation"

$$H_{j+1,k} = H_{j,k+1} \text{ for } j,k=0,1,\ldots$$

As for the following alternative result, it also has an easy proof (see e.g. [12], Problem 195).

Lemma 2. *The product of two Toeplitz matrices (associated with L^∞ functions g_1, g_2) is a Toeplitz matrix if and only if either the left factor is upper triangular (or co-analytic) or the right factor is lower triangular (or analytic). In which case*

$$T(g_1)T(g_2) = T(g_1g_2). \tag{47}$$

In view of Lemma 1, and since the H^2 function $u=Bq$ (where $q(e^{it}) \in H^2$) has the algebraic representation $T(B)q$ (where of course $q \in \ell^2$), (44a) can be rewritten in the form

$$H(Bf)q = s_n[T(B)]^* q^* ; \tag{48}$$

hence it follows that

$$\operatorname{lub}_2(H(Bf)) \geqslant s_n\|T(B)q\|_2/\|q\|_2 = s_n$$

since, by Lemma 2, $[T(B)]^\sim T(B) = T(B^*B) = T(|B(e^{it})|^2) = I$. On the other hand, the product on the left side and the first term on the right side of (44b) are trivially (qL^∞) functions, so that the H^2 function w must belong to qH^∞ (remember indeed that the H^2 function $q(z)$ does not vanish in $|z| < 1$); by appealing again to Nehari's theorem, it is clear that

$$\operatorname{lub}_2(H(Bf)) \leqslant s_n\|u^*/q\|_\infty = s_n.$$

Hence, by (34), the intermediate, but *fundamental result* :

$$s_n(H(f)) = \mathrm{lub}_2(H(Bf)) \geq \min_{K \in H^{(k)}} \mathrm{lub}_2(H(f)-K), \tag{49}$$

whose comparison with (39) implies $k \geq n$.

Now it turns out that *the multiplicity of* s_n *as singular value of* $H(Bf)$ *is* $\geq k+1$. Indeed, for every factorization of B of the form

$$B(e^{it}) = C(e^{it})D(e^{it}),$$

where C and D are Blaschke products, the premultiplication of (48) by the co-analytic Toeplitz matrix $[T(C)]^T$ and the use of Lemmas 1 and 2 finally yield the algebraic relation

$$H(Bf)[T(C)q] = s_n[T(D)q]^*.$$

This implies that the spectral norm of H(Bf) must be attained (at least) on the (k+1)-dimensional vector space of all possible H^2 functions of the above form Cq, so that

$$\mathrm{lub}_2(H(Bf)) = s_0(H(Bf)) = \ldots = s_k(H(Bf));$$

but

$$s_k(H(Bf)) \leq s_k(H(f))$$

by a corollary of the Courant-Fischer theorem, and consequently, in view of (49), we get

$$s_n(H(f)) \leq s_k(H(f)),$$

so that $k \leq n$, which completes our proof.

It should be noted that whenever H is a *finite* Hankel matrix, the well known *Schur-Cohn criterion* (see e.g. [17], p. 152) can lead to an alternative direct proof of the above basic result k=n. As for the general case when s_n is a *multiple* singular value of H and for the associated uniqueness assertion, they can be studied by adapting the above type of argument to the appropriate restriction of (41bis) substituted for the relation (44), in combination with some comparatively elementary properties of Schmidt pairs (essentially those given in [1], p. 39).

ACKNOWLEDGEMENTS

We are very grateful to Dr. A. Bultheel (Katholieke Universiteit Leuven) for providing us with a most relevant material and to Dr. A. Magnus (Université Catholique de Louvain) for stimulating and constructive discussions.

REFERENCES

1. Adamjan, V.M., Arov, D.Z. and Kreĭn, M.G. : *Analytic properties of Schmidt pairs for a Hankel operator and the generalized Schur-Takagi problem*, Math. USSR Sbornik 15 (1971), pp. 31-73.
2. Akhiezer, N.I. : *Theory of Approximation*. New York : Frederick Ungar Publishing Co. 1956.
3. Buck, R.C. : *Applications of duality in approximation theory.* Approximation of Functions (H.L. Garabedian ed.), pp. 27-42. Amsterdam : Elsevier Publishing Company 1965.
4. Bultheel, A. and Dewilde, P. : *On the Adamjan-Arov-Kreĭn approximation, identification and balanced realization of a system.* Report TW 48, Katholieke Universiteit Leuven, 1980.

5. Clark, D.N. : *On the spectra of bounded, Hermitian, Hankel matrices*, Amer. J. Math. 90 (1968), pp. 627-656.

6. Clark, D.N. : *Hankel forms, Toeplitz forms and meromorphic functions*, Trans. Amer. Math. Soc. 134 (1968), pp. 109-116.

7. Duren, P.L. : *Theory of* H^p *Spaces*. New York : Academic Press 1970.

8. Gantmacher, F.R. : *Matrizenrechnung* (Teil II). Berlin : VEB Deutscher Verlag der Wissenschaften 1959.

9. Genin, Y. : *An introduction to the model reduction problem with Hankel norm criterion.* Paper presented at the Leuven Workshop on Rational Approximation for Systems, September 1981, Leuven.

10. Gohberg, I.C. and Kreǐn, M.G. : *Introduction to the Theory of Linear Nonselfadjoint Operators.* Providence : American Mathematical Society 1969.

11. Gutknecht, M.H. and Trefethen, L.N. : *Recursive digital filter design by the Carathéodory-Fejér method,* IEEE Trans. Acoust. Speech Signal Processing (1981).

12. Halmos, P.R. : *A Hilbert Space Problem Book.* Princeton : D. Van Nostrand Company 1967.

13. Hardy, G.H., Littlewood, J.E. and Pólya, G. : *Inequalities.* London and New York : Cambridge University Press 1934.

14. Hartman, P. : *On completely continuous Hankel matrices*, Proc. Amer. Math. Soc. 9 (1958), pp. 862-866.

15. Hoffman, K. : *Banach Spaces of Analytic Functions.* Englewood Cliffs : Prentice-Hall 1962.

16. MacDuffee, C.C. : *The Theory of Matrices.* New York : Chelsea Publishing Company 1946.

17. Marden, M. : *The Geometry of the Zeros of a Polynomial in a Complex Variable.* New York : American Math. Soc. 1949.

18. Meinguet, J. : *Multivariate interpolation at arbitrary points made simple*, Journal of Applied Mathematics and Physics (ZAMP) 30 (1979), pp. 292-304.

19. Nehari, Z. : *On bounded bilinear forms*, Annals of Mathematics 65 (1957), pp. 153-162.

20. Power, S.C. : *The essential spectrum of a Hankel operator with piecewise continuous symbol*, Michigan Math. J. 25 (1978), pp. 117-121.

21. Power, S.C. : *Hankel operators on Hilbert space*, Bull. London Math. Soc. 12 (1980), pp. 422-442.

22. Schwartz, L. : *Analyse : Topologie générale et analyse fonctionnelle*. Paris : Hermann 1970.

23. Shapiro, H.S. : *Topics in Approximation Theory*. Berlin and Heidelberg : Springer-Verlag 1971.

24. Singer, I. : *Best Approximation in Normed Linear Spaces by Elements of Linear Subspaces*. Berlin and Heidelberg : Springer-Verlag 1970.

25. Walsh, J.L. : *Interpolation and Approximation by Rational Functions in the Complex Domain*. Providence : American Mathematical Society 1960.

26. Zygmund, A. : *Trigonometric Series*. London and New York : Cambridge University Press 1959.

To create is divine;
to multiply is human.
Man Ray

FAST FOURIER TRANSFORM ALGORITHMS WITH APPLICATIONS

Gerhard Merz

University of Kassel
Federal Republic of Germany

Abstract: After introduction of the discrete Fourier transform and a short description of its main properties we concentrate on a discussion of some of the various methods which have been used for the derivation of fast Fourier transform (FFT) algorithms. The paper closes with applications of FFT procedures in the numerical calculation of Fourier coefficients, fast multiplication of large integers and computations which involve circulant matrices.

0.

The discrete Fourier transform (DFT) plays a major role in the - sometimes approximate - solution of many problems arising in practice, among others in time series analysis, digital signal and picture processing, numerical conformal mapping, crystallography and geophysics. In pre-computer age and in the first years of the computer era the large scale application of DFT-methods was greatly complicated and sometimes even prohibited by the fact that the number of multiplications required for the direct calculation of a N-point DFT is proportional to N^2.

It suggests itself that in order to reduce this computational work one should take into account the symmetry and the periodicity properties of the N-th roots of unity. Whereas consideration of the symmetries only amounts to a smaller proportionality factor with N^2, it has been shown already in the first years of our century by Runge [21] , [22] that if N equals a power of two the periodicity of the sine-cosine functions may be used to split a N-point Fourier series into two N/2-point series and - by iterative application of this procedure - to arrive at a successive doubling al-

H. Werner et al. (eds.), Computational Aspects of Complex Analysis, 249–278.

gorithm with a number of multiplications proportional to $N \log_2 N$ (cf. Runge and König [23] , p. 215). This distinction was not too important for the values of N which were customary at Runge's time and it is therefore not surprising that his ideas got lost for some decades and were even overlooked in the first years of the advent of computing machines.

A new epoch in the application of DFT methods, characterized by the general availability of so-called fast Fourier transform (FFT) algorithms, initiated with the famous 1965 paper by Cooley and Tukey [3] , who indicated that any DFT of composite order $N = pq$ can be performed with a number of multiplications which is proportional to $N(p+q)$. (For $N = 2^k$ the Cooley-Tukey method happens to be essentially identical with Runge's successive doubling algorithm).

Since that date a still continuing flood of papers about time-saving algorithms for DFT methods has come to light.

After a short introduction of the DFT in Part 1 of this paper, we describe in Part 2 - without claiming any kind of completeness - some ideas which have been used for the construction of FFT algorithms of various kinds. The final Part 3 is devoted to a rather incomplete discussion of applications.

1.

Let $\{x_\nu\}_{-\infty}^{+\infty}$ be a doubly infinite N-periodic sequence of complex numbers, i.e. $x_{\nu+N}=x_\nu$, $\forall \nu \in \mathbb{Z}$. If we put $\omega_N := \exp(2\pi i/N)$ and define a new sequence $\{X_\mu\}_{-\infty}^{+\infty}$ by

$$X_\mu = \frac{1}{\sqrt{N}} \sum_{\nu=0}^{N-1} x_\nu \omega_N^{\mu\nu} , \quad -\infty<\mu<\infty , \tag{1}$$

then $X_{\mu+N} = X_\mu$, $\forall \mu \in \mathbb{Z}$, i.e. the sequence $\{X_\mu\}$ is also N-periodic.

By noting that

$$\sum_{\kappa=0}^{N-1} \omega_N^{\kappa\lambda} = \begin{cases} N \text{ for } \lambda \equiv 0 \pmod N \\ 0 \text{ for } \lambda \not\equiv 0 \pmod N \end{cases} \tag{2}$$

we obtain from (1)

$$\sum_{\mu=0}^{N-1} |X_\mu|^2 = \frac{1}{N} \sum_{\mu=0}^{N-1} \sum_{\nu=0}^{N-1} x_\nu \omega_N^{\mu\nu} \sum_{\lambda=0}^{N-1} \bar{x}_\lambda \omega_N^{-\mu\lambda} = \sum_{\nu=0}^{N-1} |x_\nu|^2$$

and this is the discrete analogue of Parseval's identity.

Moreover, multiplication of (1) by $\omega_N^{-\mu\lambda}$ and subsequent addition yields

$$\sum_{\mu=0}^{N-1} X_\mu \omega_N^{-\mu\lambda} = \frac{1}{\sqrt{N}} \sum_{\mu=0}^{N-1} \sum_{\nu=0}^{N-1} x_\nu \omega_N^{\mu(\nu-\lambda)} = \sqrt{N}\; x_\lambda \quad ,$$

i.e. the inverse of (1) can be written as

$$x_\nu = \frac{1}{\sqrt{N}} \sum_{\mu=0}^{N-1} X_\mu\, \omega_N^{-\mu\nu} \; , \quad -\infty<\nu<\infty \quad . \tag{3}$$

As noted above we may restrict μ,ν in (1) and (3) to the values $\mu,\nu = 0(1)N-1$, respectively.
As a consequence, we consider the transformation

$$\underline{X} = \underline{W}_N\, \underline{x} \tag{4}$$

where

$$\underline{x} = \begin{pmatrix} x_0 \\ \vdots \\ x_{N-1} \end{pmatrix}, \quad \underline{X} = \begin{pmatrix} X_0 \\ \vdots \\ X_{N-1} \end{pmatrix}, \quad \underline{W}_N = \frac{1}{\sqrt{N}} \left(\left(\omega_N^{\mu\nu}\right)\right)_{\mu,\nu=0}^{N-1} \; , \tag{5}$$

and denote the N-dimensional vector $\underline{X}$ defined by (4), (5) as the discrete Fourier transform (DFT) of the N-dimensional vector $\underline{x}$.

The matrix $\underline{W}_N$ in (5) has the following properties:

(i) $\underline{W}_N = \underline{W}_N^T$

(ii) $\underline{W}_N\, \underline{W}_N^{*} = \underline{I}_N$ where $\underline{I}_N$ denotes the N-dimensional unit matrix

(iii) $\underline{W}_N^2 = \underline{P}_N$ where $\underline{P}_N$ is the symmetric (N,N)-permutation matrix

$$\underline{P}_N = \begin{pmatrix} 1 & 0 & \cdots & 0 \\ & & & 1 \\ & & \cdot^{\cdot^{\cdot}} & \\ 0 & 1 & & 0 \end{pmatrix} \tag{6}$$

(iv) $\underline{W}_N^4 = \underline{I}_N$

(v) Let $\underline{T}_N$ denote the (N,N)-permutation matrix

$$\underline{T}_N = \begin{pmatrix} 0 & \cdots & 0 & 1 \\ 1 & \ddots & & \\ & & \ddots & \\ 0 & & 1 & 0 \end{pmatrix} \qquad (7)$$

and put $\underline{D}_N = \operatorname{diag}\,((\omega_N^{\mu}))_{\mu=0}^{N-1}$. Then $\underline{D}_N\,\underline{W}_N = \underline{W}_N\,\underline{T}_N$.

(vi) $\underline{W}_N^{-1} = ((\omega_N^{-\mu\nu}))_{\mu,\nu=0}^{N-1} = \underline{W}_N^3 = \underline{P}_N\,\underline{W}_N = \underline{W}_N\,\underline{P}_N$.

Remark. From $\operatorname{tr}\underline{P}_N = \begin{cases} 2 \text{ if } N \equiv 0 (\operatorname{mod} 2) \\ 1 \text{ if } N \not\equiv 0 (\operatorname{mod} 2) \end{cases}$ and

$$\operatorname{tr}\underline{W}_N = \frac{1}{\sqrt{N}} \sum_{\nu=0}^{N-1} \omega_N^{\nu^2} = \begin{cases} 1+i & \text{for } N \equiv 0 (\operatorname{mod} 4) \\ 1 & \text{for } N \equiv 1 (\operatorname{mod} 4) \\ 0 & \text{for } N \equiv 2 (\operatorname{mod} 4) \\ i & \text{for } N \equiv 3 (\operatorname{mod} 4) \end{cases}$$

one may easily calculate the multiplicities of the eigenvalues 1, -1, i, -i of $\underline{W}_N$ (cf. [17], p.324 - 326; see also [14]).

For later use we define the circular convolution $\underline{c} = \underline{a} * \underline{b}$ of two vectors

$$\underline{a} = \begin{pmatrix} a_0 \\ \vdots \\ a_{N-1} \end{pmatrix}, \quad \underline{b} = \begin{pmatrix} b_0 \\ \vdots \\ b_{N-1} \end{pmatrix} \text{ as the vector } \underline{c} = \begin{pmatrix} c_0 \\ \vdots \\ c_{N-1} \end{pmatrix}$$

with components

$$c_\kappa = \sum_{\mu+\nu \equiv \kappa (\operatorname{mod} N)} a_\mu b_\nu \quad , \qquad \kappa = 0(1)N-1 \quad . \qquad (8)$$

Obviously, $\underline{c}$ may equivalently be defined as

$$\underline{c} = \underline{C}_N(\underline{a})\underline{b} \qquad (9)$$

where $\underline{C}_N\,(\underline{a})$ is the circulant matrix

$$\underline{C}_N(\underline{a}) = \begin{pmatrix} a_0 & a_{N-1} & a_{N-2} & \cdots & a_2 & a_1 \\ a_1 & a_0 & a_{N-1} & \cdots & a_3 & a_2 \\ a_2 & a_1 & a_0 & \cdots & a_4 & a_3 \\ \cdots & \cdots & \cdots & \cdots & \cdots & \cdots \\ a_{N-2} & a_{N-3} & a_{N-4} & \cdots & a_0 & a_{N-1} \\ a_{N-1} & a_{N-2} & a_{N-3} & \cdots & a_1 & a_0 \end{pmatrix}. \quad (10)$$

(A (N,N)-matrix $\underline{C}_N$ is called circulant, if it commutes with the permutation matrix $\underline{T}_N$ defined by (7)).
The connection between circular convolution and DFT is established via the <u>convolution theorem</u>

$$\underline{a} * \underline{b} = \sqrt{N}\ \underline{W}_N^{-1}\ (\underline{W}_N\underline{a} \circ \underline{W}_N\underline{b}) \quad . \qquad (11)$$

Here $\underline{W}_N\underline{a} \circ \underline{W}_N\underline{b}$ denotes the vector which is obtained by <u>component-wise</u> multiplication of the vectors $\underline{W}_N\underline{a}$ and $\underline{W}_N\underline{b}$.

<u>Proof of the convolution theorem:</u> The κ-th components of $\underline{W}_N\underline{a}$ and $\underline{W}_N\underline{b}$ are given by

$$A_\kappa = \frac{1}{\sqrt{N}} \sum_{\mu=0}^{N-1} a_\mu \omega_N^{\kappa\mu} \text{ and } B_\kappa = \frac{1}{\sqrt{N}} \sum_{\nu=0}^{N-1} b_\nu \omega_N^{\kappa\nu} \quad ,$$

respectively. Thus the κ-th component of $\underline{W}_N\underline{a} \circ \underline{W}_N\underline{b}$ equals

$$A_\kappa B_\kappa = \frac{1}{N} \sum_{\mu=0}^{N-1} \sum_{\nu=0}^{N-1} a_\mu b_\nu \omega_N^{\kappa(\mu+\nu)} = \frac{1}{N} \sum_{\lambda=0}^{N-1} \sum_{\mu+\nu\equiv\lambda \pmod N} a_\mu b_\nu\ \omega_N^{\lambda\kappa} \quad .$$

On the other hand, the λ-th component of the circular convolution $\underline{a} * \underline{b}$ is

$$c_\lambda = \sum_{\mu+\nu\equiv\lambda \pmod N} a_\mu b_\nu \quad ;$$

therefore we obtain for the κ-th component of the DFT of $\underline{c} = \underline{a} * \underline{b}$

$$\frac{1}{\sqrt{N}} \sum_{\lambda=0}^{N-1} c_\lambda \omega_N^{\lambda\kappa} = \frac{1}{\sqrt{N}} \sum_{\lambda=0}^{N-1} \sum_{\mu+\nu\equiv\ \lambda \pmod N} a_\mu b_\nu \omega_N^{\lambda\kappa}$$

and thus finally arrive at

$$\underline{W}_N(\underline{a} * \underline{b}) = \sqrt{N}\ (\underline{W}_N\underline{a} \circ \underline{W}_N\underline{b})$$

which is equivalent to (11).

Remark. Due to the fact that $z^N \equiv 1 \pmod N$ the components of the circular convolution vector $\underline{a} * \underline{b}$ may be alternatively obtained as the system of coefficients of the product of the polynomials $\sum_{\mu=0}^{N-1} a_\mu z^\mu$ and $\sum_{\nu=0}^{N-1} b_\nu z^\nu$ modulo $z^N - 1$.

This observation leads to an interpretation of convolutions and DFTs as operations defined over finite rings of polynomials. In connection with the Chinese remainder theorem it plays a key role in the development of the most recent algorithms for time-saving calculation of circular convolutions (Winograd [29], Nussbaumer [18], p. 151).

In many instances it is necessary to make use of multidimensional DFTs. If $\underline{x} = \{x_{\nu_1\nu_2}\}$, $-\infty<\nu_1,\nu_2<+\infty$, is a doubly infinite sequence of complex numbers with period N_1 in the first index ν_1 and period N_2 in the second index ν_2 then we define the two-dimensional DFT

$\underline{X} = \{X_{\mu_1\mu_2}\}$, $-\infty<\mu_1,\mu_2<+\infty$, of $\underline{x}$ by

$$X_{\mu_1\mu_2} = \frac{1}{\sqrt{N_1N_2}} \sum_{\nu_1=0}^{N_1-1} \sum_{\nu_2=0}^{N_2-1} x_{\nu_1\nu_2}\, \omega_{N_1}^{\mu_1\nu_1}\, \omega_{N_2}^{\mu_2\nu_2} \quad . \tag{12}$$

The generalization to arbitrary dimensions is obvious. Because (12) may be written as

$$X_{\mu_1\mu_2} = \frac{1}{\sqrt{N_1}} \sum_{\nu_1=0}^{N_1-1} \omega_{N_1}^{\mu_1\nu_1} \left(\frac{1}{\sqrt{N_2}} \sum_{\nu_2=0}^{N_2-1} x_{\nu_1\nu_2}\, \omega_{N_2}^{\mu_2\nu_2}\right) , \tag{13}$$

the evaluation of a two-dimensional DFT of length N_1N_2 is equivalent to the evaluation of N_1 one-dimensional DFTs of length N_2 corresponding to the N_1 distinct values of ν_1 and successively of N_2 one-dimensional DFTs of length N_1 corresponding to the N_2 distinct values of ν_2 .
A more intrinsic relation between one- and r-dimensional DFTs for the case that the periods $N_1,\ldots,N_r$ are pairwise mutually prime was set up by Good [8], [9] via the bijective mapping between the 1-index μ and the r-index $\mu_1\ldots\mu_r$ which is established for $0\le\mu<N_1\ldots N_r$ by the congruences

$$\mu_1 \equiv \mu \pmod{N_1}, \ldots, \mu_r \equiv \mu \pmod{N_r} \quad . \tag{14}$$

Note that μ can be reconstructed from $\mu_1,\ldots,\mu_r$ by means of Chinese remaindering (cf. Knuth [13], p. 270).

2.

We now describe some ideas which have been used for the construction of fast Fourier transform and convolution algorithms.

(i) The Cooley-Tukey algorithm (cf. [3])

Consider the problem of evaluating the N-point DFT

$$X_\mu = \frac{1}{\sqrt{N}} \sum_{\nu=0}^{N-1} x_\nu \omega_N^{\mu\nu} \quad , \mu = 0(1)N-1 \quad . \tag{15}$$

We assume that N = pq and start by establishing a bijective correspondence between the integers μ,ν in (15) and the pairs of integers (μ_1,μ_0), (ν_1, ν_0) defined by

$$\left.\begin{array}{l} \mu = \mu_1 q+\mu_0 \\ \\ \nu = \nu_1 p+\nu_0 \end{array}\right\} ;\mu_1,\nu_0 = 0(1)p-1 \; ; \; \mu_0,\nu_1 = 0(1)q-1 \quad . \tag{16}$$

This enables us to write

$$\begin{aligned} X_\mu &= X_{\mu_1 q+\mu_0} = \\ &= \frac{1}{\sqrt{N}} \sum_{\nu_0=0}^{p-1} \omega_N^{(\mu_1 q+\mu_0)\nu_0} \sum_{\nu_1=0}^{q-1} x_{\nu_1 p+\nu_0} \omega_N^{\mu_0\nu_1 p} = \\ &= \frac{1}{\sqrt{p}} \sum_{\nu_0=0}^{p-1} \omega_p^{\mu_1\nu_0} \omega_N^{\mu_0\nu_0} \left(\frac{1}{\sqrt{q}} \sum_{\nu_1=0}^{q-1} x_{\nu_1 p+\nu_0} \omega_q^{\mu_0\nu_1}\right) = \\ &= \frac{1}{\sqrt{q}} \sum_{\nu_1=0}^{q-1} \omega_q^{\mu_0\nu_1} \left[\frac{1}{\sqrt{p}} \sum_{\nu_0=0}^{p-1} (x_{\nu_1 p+\nu_0} \omega_N^{\mu_0\nu_0}) \, \omega_p^{\mu_1\nu_0}\right] \quad . \end{aligned} \tag{17}$$

We conclude that the one-dimensional DFT (15) of length N=pq apart from the "twiddle factors"

$\omega_N^{\mu_0\nu_0}$ may be looked upon as a two-dimensional DFT of length pq and therefore according to (13) can be performed in two basic steps of p DFTs of length q which require pq^2 multiplications and successively q DFTs of length p which need qp^2 multiplications. Because the multiplications by the twiddle factors need not be counted we arrive at a total of N(p+q) multiplications which usually is considerably less than the N^2 multiplications necessary for the direct calculation of (15).

Remarks. 1. Note that the order of the input and output (.,.)-indices in (17) is permuted. Thus a final permutation step has to be added after the two basic steps just described in order to complete the FFT prodecure.

2. Each of the two possibilities for the actual performance of the FFT algorithm defined by (17) requires the same amount of computational work.

3. Significant additional savings result by noting that a number of the multiplications in the FFT procedure may be trivial multiplications by ± 1 or $\pm i$.

4. If we define in the case that p and q are mutually prime according to Good [8], [9] a bijective correspondence between the indices μ,ν in (15) and the pairs of integers (μ_1,μ_0), (ν_1,ν_0) by (cf. 14))

$$\mu_0 \equiv \mu (\mathrm{mod}\ q) \ , \qquad \mu_0 = 0(1)q-1 \ ,$$

$$\mu_1 \equiv \mu (\mathrm{mod}\ p) \ , \qquad \mu_1 = 0(1)p-1 \ ,$$

and

$$\nu \equiv \nu_1 p + \nu_0 q (\mathrm{mod}\ N) \ , \quad \nu_0 = 0(1)p-1 \ , \quad \nu_1 = 0(1)q-1 \ ,$$

then there is no further need for twiddle factors and (17) may be replaced by

$$X_\mu = X_{(\mu_1,\mu_0)} = \frac{1}{\sqrt{p}} \sum_{\nu_0=0}^{p-1} \omega_p^{\mu_1\nu_0} \left(\frac{1}{\sqrt{q}} \sum_{\nu_1=0}^{q-1} x_{(\nu_1,\nu_0)} \omega_q^{\mu_0\nu_1} \right) . \qquad (18)$$

5. In the case that N is highly composite, e.g. $N = N_1 \ldots N_k$, the Cooley-Tukey algorithm may be used iteratively, thus yielding a total of $N(N_1 + \ldots + N_k)$ multiplications. If, for example, $N = 2520 = 5 \cdot 7 \cdot 8 \cdot 9$, the numerical work for the calculation of a DFT of length N is cut in this way by a factor of over 80.

6. The practical need for simplicity of the computational process connected with the FFT procedure yields DFT-length which are the power of an integer to be favoured about other possibilities. The most frequent use of N is a power of 2 (cf. (ii) below), although a base-3-algorithm would formally result to be sligthly more efficient.

7. Hints for the implementation of FFT procedures are given in Gander-Mazzario [6] and Nussbaumer [18], p. 94.

8. Computational problems which arise from accumulation of round-off errors in FFT algorithms are discussed by Kaneko-Liu [12] and Segeth [28].

(ii) The case $N = 2^k$

Let $N = 2^k$ and put $\hat{X}_\mu = \sqrt{N}\ X_\mu$. Then for $\mu = 0(1)2^k-1$

$$\hat{X}_\mu = \sum_{\nu=0}^{2^k-1} x_\nu \omega_N^{\mu\nu} = \sum_{\nu=0}^{2^{k-1}-1} x_{2\nu} \omega_N^{2\nu\mu} + \sum_{\nu=0}^{2^{k-1}-1} x_{2\nu+1} \omega_N^{(2\nu+1)\mu} .$$

From this we deduce

$$\hat{X}_\mu = \sum_{\nu=0}^{2^{k-1}-1} x_{2\nu} \omega_N^{2\mu\nu} + \omega_N^{\mu} \sum_{\nu=0}^{2^{k-1}-1} x_{2\nu+1}\ \omega_N^{2\mu\nu} \tag{19}$$

and - because $\omega_N^{2^{k-1}} = -1$ -

$$\hat{X}_{\mu+2^{k-1}} = \sum_{\nu=0}^{2^{k-1}-1} x_{2\nu} \omega_N^{2\mu\nu} - \omega_N^{\mu} \sum_{\nu=0}^{2^{k-1}-1} x_{2\nu+1}\ \omega_N^{2\mu\nu} \tag{20}$$

where now $\mu=0(1)2^{k-1}-1$. This so-called decimation in time procedure replaces the calculation of one 2^k-point DFT by that of two DFTs of length 2^{k-1} . Repetitive application of this bisection formalism results in a FFT procedure which needs only $(N/2)\log_2 N$ multiplications per step. Thus, for $N = 2^{10}$, there are about 200 times fewer multiplications necessary than with the ordinary DFT method.

A second (computationally equivalent) approach, called decimation in frequency starts from

$$\hat{X}_\mu = \sum_{\nu=0}^{2^k-1} x_\nu \omega_N^{\mu\nu} = \sum_{\nu=0}^{2^{k-1}-1} (x_\nu + \omega_N^{2^{k-1}\mu}\ x_{\nu+2^{k-1}}) \omega_N^{\mu\nu}$$

and leads to

$$\hat{X}_{2\mu} = \sum_{\nu=0}^{2^{k-1}-1} (x_\nu + x_{\nu+2^{k-1}})\ \omega_N^{2\mu\nu} \tag{21}$$

and

$$\hat{X}_{2\mu+1} = \sum_{\nu=0}^{2^{k-1}-1} [(x_\nu - x_{\nu+2^{k-1}})\omega_N^\nu]\ \omega_N^{2\nu\mu} \quad , \tag{22}$$

$\mu = 0(1)2^{k-1}-1$.

For N=8, the 3 steps of the decimation in frequency FFT algorithm run as follows (for an illustration by a signal flow graph see Nussbaumer [18], p. 90):

	1st step	2nd step	3rd step
x_0	$X_0^{(1)} = x_0+x_4$	$X_0^{(2)} = X_0^{(1)}+X_2^{(1)}$	$\hat{X}_0 = X_0^{(2)}+X_1^{(2)}$
x_1	$X_1^{(1)} = x_1+x_5$	$X_1^{(2)} = X_1^{(1)}+X_3^{(1)}$	$\hat{X}_4 = X_0^{(2)}-X_1^{(2)}$
x_2	$X_2^{(1)} = x_2+x_6$	$X_2^{(2)} = X_0^{(1)}-X_2^{(1)}$	$\hat{X}_2 = X_2^{(2)}+X_3^{(2)}$
x_3	$X_3^{(1)} = x_3+x_7$	$X_3^{(2)} = (X_1^{(1)}-X_3^{(1)})\ \omega_8^2$	$\hat{X}_6 = X_2^{(2)}-X_3^{(2)}$
x_4	$X_4^{(1)} = x_0-x_4$	$X_4^{(2)} = X_4^{(1)}+X_6^{(1)}$	$\hat{X}_1 = X_4^{(2)}+X_5^{(2)}$
x_5	$X_5^{(1)} = (x_1-x_5)\ \omega_8$	$X_5^{(2)} = X_5^{(1)}+X_7^{(1)}$	$\hat{X}_5 = X_4^{(2)}-X_5^{(2)}$
x_6	$X_6^{(1)} = (x_2-x_6)\ \omega_8^2$	$X_6^{(2)} = X_4^{(1)}-X_6^{(1)}$	$\hat{X}_3 = X_6^{(2)}+X_7^{(2)}$
x_7	$X_7^{(1)} = (x_3-x_7)\omega_8^3$	$X_7^{(2)} = (X_5^{(1)}-X_7^{(1)})\ \omega_8^2$	$\hat{X}_7 = X_6^{(2)}-X_7^{(2)}$

Note that the indices of the input sequence $\{x_\nu\}$ and the output sequence $\{\hat{X}_\mu\}$ correspond by virtue of a digit reversion in their respective dual number representations ("bit reversal")

ν	Dual number representation	Bit reserved dual number	μ
0	OOO	OOO	0
1	OOL	LOO	4
2	OLO	OLO	2
3	OLL	LLO	6
4	LOO	OOL	1
5	LOL	LOL	5
6	LLO	OLL	3
7	LLL	LLL	7

Remarks. (i) From (21), (22) a representation of $\underline{W}_N = \underline{W}_{2^k}$ as a product of sparse matrices can be deduced in the following way:
For $\kappa = 1(1)k$ define the matrices $\underline{S}_\kappa$ of type $(2^\kappa, 2^\kappa)$ by

$$\underline{S}_1 = \begin{pmatrix} 1 & 1 \\ 1 & -1 \end{pmatrix}$$

and - with the abbreviation $\tilde{\omega}_\kappa = \omega_{2^\kappa} = \omega_N^{2^{k-\kappa}}$ -

$$\underline{S}_\kappa = \left(\begin{array}{cccc|cccc} 1 & & & & 1 & & & \\ & 1 & & & & 1 & & \\ & & \ddots & & & & \ddots & \\ & & & 1 & & & & 1 \\ \hline 1 & & & & -1 & & & \\ & \tilde{\omega}_\kappa & & & & -\tilde{\omega}_\kappa & & \\ & & \tilde{\omega}_\kappa^2 \ddots & & & & -\tilde{\omega}_\kappa^2 \ddots & \\ & & & \tilde{\omega}_\kappa^{2^{\kappa-1}-1} & & & & -\tilde{\omega}_\kappa^{2^{\kappa-1}-1} \end{array}\right).$$

If we now put for $\kappa = 1(1)k$

$$\underline{R}_k^{(k-\kappa)} = \underbrace{\underline{S}_\kappa \otimes \underline{S}_\kappa \otimes \ldots \otimes \underline{S}_\kappa}_{2^{k-\kappa} \text{ times}} \quad \text{where } \underline{S}_\kappa \otimes \underline{S}_\kappa = \left[\begin{array}{c|c} \underline{S}_\kappa & \underline{0} \\ \hline \underline{0} & \underline{S}_\kappa \end{array}\right]$$

and finally take the $(2^k, 2^k)$ - permutation matrix $\underline{Q}_k$ to be given by

$$\underline{Q}_k = ((\alpha_{\mu\nu})) = \begin{cases} 1 & \text{if the index } \mu \text{ is transformed into the index } \nu \text{ through bit reversal in the respective dual number representations of } \mu \text{ and } \nu \\ 0 & \text{otherwise} \end{cases}$$

then we arrive at

$$\underline{W}_N = \underline{W}_{2^k} = \frac{1}{\sqrt{N}}\, \underline{Q}_k\, \underline{R}_k^{(k-1)}\, \underline{R}_k^{(k-2)} \ldots \underline{R}_k^{(0)} \quad . \tag{23}$$

For $N = 8$ we thus obtain

$$\underline{W}_8 = \frac{1}{2\sqrt{2}}\, \underline{Q}_3\, \underline{R}_3^{(2)}\, \underline{R}_3^{(1)}\, \underline{R}_3^{(0)} \quad ;$$

here

$$\underline{R}_3^{(0)} = \left(\begin{array}{cccc|cccc} 1 & & & & 1 & & & \\ & 1 & & & & 1 & & \\ & & 1 & & & & 1 & \\ & & & 1 & & & & 1 \\ \hline 1 & & & & -1 & & & \\ & \omega_8 & & & & -\omega_8 & & \\ & & \omega_8^2 & & & & -\omega_8^2 & \\ & & & \omega_8^3 & & & & -\omega_8^3 \end{array}\right) = \underline{S}_3 \quad ,$$

$$\underline{R}_3^{(1)} = \left(\begin{array}{cc:cc|cccc} 1 & 0 & 1 & 0 & & & & \\ 0 & 1 & 0 & 1 & & & & \\ \hdashline 1 & 0 & -1 & 0 & & \multicolumn{2}{c}{0} & \\ 0 & \omega_8^2 & 0 & -\omega_8^2 & & & & \\ \hline & & & & 1 & 0 & 1 & 0 \\ & \multicolumn{2}{c}{0} & & 0 & 1 & 0 & 1 \\ & & & & 1 & 0 & -1 & 0 \\ & & & & 0 & \omega_8^2 & 0 & -\omega_8^2 \end{array}\right) = \underline{S}_2 \otimes \underline{S}_2 \quad ,$$

$$\underline{R}_3^{(2)} = \left[\begin{array}{c|c} \begin{array}{cc|cc} 1 & 1 & & \\ 1 & -1 & \multicolumn{2}{c}{0} \\ \hline & & 1 & 1 \\ \multicolumn{2}{c|}{0} & 1 & -1 \end{array} & 0 \\ \hline 0 & \begin{array}{cc|cc} 1 & 1 & & \\ 1 & -1 & \multicolumn{2}{c}{0} \\ \hline & & 1 & 1 \\ \multicolumn{2}{c|}{0} & 1 & -1 \end{array} \end{array}\right] = \underline{S}_1 \otimes \underline{S}_1 \otimes \underline{S}_1 \otimes \underline{S}_1 \;,$$

$$\underline{Q}_3 = \begin{bmatrix} 1 & 0 & 0 & 0 & 0 & 0 & 0 & 0 \\ 0 & 0 & 0 & 0 & 1 & 0 & 0 & 0 \\ 0 & 0 & 1 & 0 & 0 & 0 & 0 & 0 \\ 0 & 0 & 0 & 0 & 0 & 0 & 1 & 0 \\ 0 & 1 & 0 & 0 & 0 & 0 & 0 & 0 \\ 0 & 0 & 0 & 0 & 0 & 1 & 0 & 0 \\ 0 & 0 & 0 & 1 & 0 & 0 & 0 & 0 \\ 0 & 0 & 0 & 0 & 0 & 0 & 0 & 1 \end{bmatrix} .$$

The splitting (23) of the DFT-matrix $\underline{W}_N$ for $N = 2^k$ is due to Schwarz [26]. The recursive construction of the matrices $\underline{R}_k^{(k-\kappa)}$ presented here together with a lucid proof of (23) by induction has been given by Forst [5]. For the general case where N is not limited to a power of 2 we refer to Schwarz [27].

(ii) Rader and Brenner [20] have proposed a modification of the 2^k-point FFT procedure which replaces the <u>complex</u> multiplications needed in the decimation in time and decimation in frequency arrangements, respectively, by multiplications of complex numbers by either real or pure imaginary ones.

(iii) <u>The method of Fiduccia</u> (cf. [4])
The X_μ as defind by (15) may be interpreted as the values of the polynomial
$v(t) = \frac{1}{\sqrt{N}} \sum_{\nu=0}^{N-1} x_\nu t^\nu$ at the N-th roots of unity, i.e.

$$X_\mu = v(\omega_N^\mu) \quad , \quad \mu = 0\,(1)N-1 \quad . \tag{24}$$

The method of Fiduccia is based on the following two observations:
1. If $r_\rho(t)$ denotes the remainder when $v(t)$ is divided by a polynomial $w_\rho(t)$, then for all zeros $\alpha_\sigma^{(\rho)}$ of $w_\rho(t)$

$$v(\alpha_\sigma^{(\rho)}) = r_\rho(\alpha_\sigma^{(\rho)}) \quad . \tag{25}$$

2. Let $N = pq$. Then

$$v(t) = \frac{1}{\sqrt{N}} \sum_{\nu=0}^{N-1} x_\nu t^\nu = \frac{1}{\sqrt{N}} \sum_{\nu_0=0}^{p-1} t^{\nu_0} \Big(\sum_{\nu_1=0}^{q-1} x_{\nu_1 p+\nu_0} t^{\nu_1 p} \Big) \quad . \tag{26}$$

Now, if $N=pq$, then

$$t^N-1 = (t^p)^q-1 = \prod_{\rho=0}^{q-1} (t^p-\omega_N^{\rho p})$$

and according to (26) the remainders $r_\rho(t)$ of $v(t)$ after division by $w_\rho(t) = t^p-\omega_N^{\rho p}$, $\rho = 0(1)q-1$, are given by

$$r_\rho(t) = \frac{1}{\sqrt{N}} \sum_{\nu_0=0}^{p-1} t^{\nu_0} \Big[\sum_{\nu_1=0}^{q-1} x_{\nu_1 p+\nu_0} (\omega_N^{\rho p})^{\nu_1} \Big] \quad . \tag{27}$$

Evaluation of each $r_\rho(t)$ at the p zeros $\alpha_\sigma^{(\rho)} = \omega_N^{\sigma q+\rho}$, $\sigma= 0(1)p-1$, of $w_\rho(t)$ results in a set of $N = pq$ function values which by (17) and (25) coincide with the N DFT-values X_μ defined by (24). Obviously this construction of a DFT of length $N= pq$ costs $N(p+q)$ multiplications.

Take for example $N = 8$. Then the different possibilities for a factorization of t^8-1 can be arranged in the trees

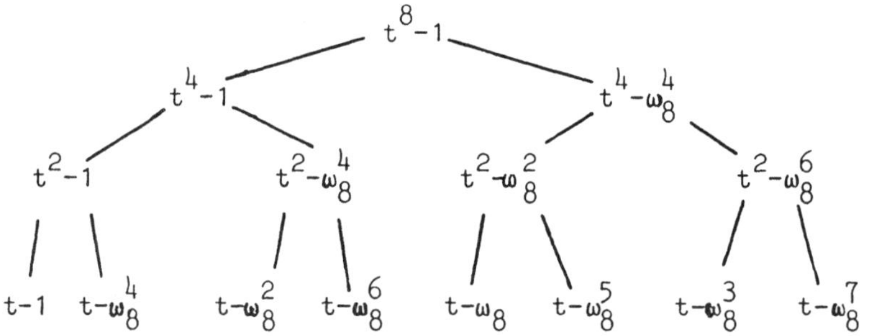

or

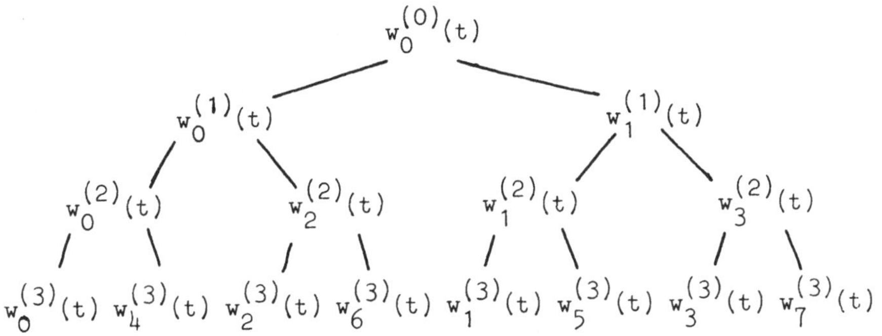

The respective remainders $r_\rho^{(\sigma)}(t)$ of $\sum_{\nu=0}^{7} x_\nu t^\nu$ after division by $w_\rho^{(\sigma)}(t)$ are

$$r_0^{(0)}(t) = x_0+x_1t+x_2t^2+\ldots+x_7t^7$$

$$r_0^{(1)}(t) = (x_0+x_4)+(x_1+x_5)t+(x_2+x_6)t^2+(x_3+x_7)t^3$$

$$r_1^{(1)}(t) = (x_0+x_4\omega_8^4)+(x_1+x_5\omega_8^4)t+(x_2+x_6\omega_8^4)t^2+(x_3+x_7\omega_8^4)t^3$$

$$r_0^{(2)}(t) = (x_0+x_2+x_4+x_6)+(x_1+x_3+x_5+x_7)t$$

$$r_1^{(2)}(t) = (x_0+x_2\omega_8^2+x_4\omega_8^4+x_6\omega_8^6)+(x_1+x_3\omega_8^2+x_5\omega_8^4+x_7\omega_8^6)t$$

$$r_2^{(2)}(t) = (x_0+x_2\omega_8^4+x_4+x_6\omega_8^4)+(x_1+x_3\omega_8^4+x_5+x_7\omega_8^4)t$$

$$r_3^{(2)}(t) = (x_0+x_2\omega_8^6+x_4\omega_8^4+x_6\omega_8^2)+(x_1+x_3\omega_8^6+x_5\omega_8^4+x_7\omega_8^2)t$$

$$r_\rho^{(3)}(t) = x_0+x_1\omega_8^\rho+x_2\omega_8^{2\rho}+\ldots+x_7\omega_8^{7\rho} \quad , \ \rho = 0(1)7 \ .$$

Remark. Implementations of this method are given by Kahaner [11] and in [0], p. 257.

(iv) The method of Meinardus (cf. [15])

This method is based upon a connection between polynomial interpolation in the N-th roots of unity and the DFT which manifests itself as follows:

Let $f_\nu \in \mathbb{C}$, $\nu = 0(1)N-1$, be arbitrarily prescribed numbers and denote by $B(z) = \sum_{\mu=0}^{N-1} \beta_\mu z^\mu$ that polynomial of degree $\leq N-1$ which is uniquely defined by the interpolation conditions

$$B(\omega_N^\nu) = f_\nu \quad , \quad \nu = 0(1)N-1. \tag{28}$$

If we put

$$\underline{b} = \begin{pmatrix} \beta_0 \\ \vdots \\ \beta_{N-1} \end{pmatrix} \quad \text{and} \quad \underline{f} = \begin{pmatrix} f_0 \\ \vdots \\ f_{N-1} \end{pmatrix} ,$$

then (28) implies $\sqrt{N}\ \underline{W}_N\ \underline{b} = \underline{f}$, i.e.

$$\sqrt{N}\ \underline{P}_N\ \underline{b} = \underline{W}_N\ \underline{f} \ . \tag{29}$$

Consequently, the computation of the DFT $\underline{W}_N\underline{f}$ is equivalent to the determination of the coefficient vector $\underline{b}$ of the interpolation polynomial $B(z)$ - represented by its values f_ν at the N-th roots of unity - and a subsequent shifting of the components of $\underline{b}$ corresponding to the multiplication of $\underline{b}$ by the permutation matrix $\underline{P}_N$ (cf. (6)).

For $N = 2^k$ the construction of $B(z)$ may be accomplished in a recursive process requiring a number of multiplications proportional to $N \log N$ as follows (for the case $N = 3^k$ see [17], p. 306):
If $B_{\iota,\kappa}(z)$ denotes that polynomial of degree $\leq 2^\kappa-1$ which for $\kappa = 0(1)k$ and $\iota = 0(1)2^{k-\kappa}-1$ is defined by the interpolation conditions

$$B_{\iota,\kappa}(\omega_N^{\iota+\lambda 2^{k-\kappa}}) = f_{\iota+\lambda 2^{k-\kappa}} \quad , \quad \lambda = 0(1)2^\kappa-1 \quad ,$$

then

$$B_{0,k}(z) = B(z) \quad .$$

Furthermore, for $\kappa = 0(1)k-1$ and $\iota = 0(1)2^{k-\kappa-1}-1$ the recursion formula

$$\begin{aligned} B_{\iota,\kappa+1}(z) &= \tfrac{1}{2}\,(1+\omega_N^{-\iota 2^\kappa} z^{2^\kappa})\, B_{\iota,\kappa}(z) + \\ &+ \tfrac{1}{2}\,(1-\omega_N^{-\iota 2^\kappa} z^{2^\kappa})\, B_{\iota+2^{k-\kappa-1},\kappa}(z) \end{aligned} \tag{30}$$

with starting values $B_{\iota,0}(z) \equiv f_\iota$, $\iota = 0(1)2^k-1$, holds.

For N = 8 formula (30) constitutes the following tree

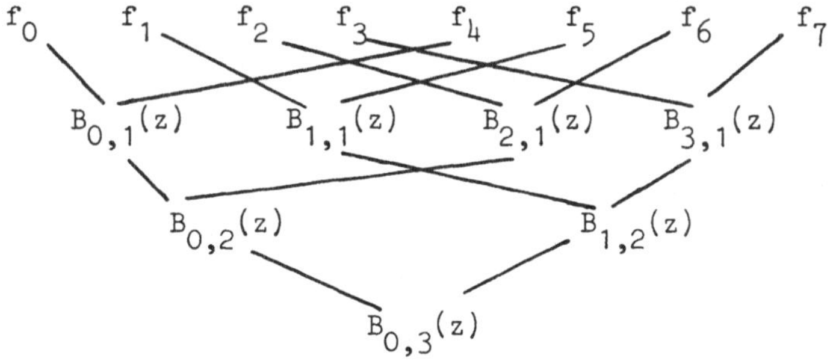

Proof of (30): Let Γ denote the positively oriented circle with radius R>1 and center at the origin. Because $B_{\iota,\kappa}(z)$ and $B_{\iota+2^{k-\kappa-1},\kappa}(z)$ interpolate at the 2^κ zeros of the polynomials $z^{2^\kappa}-\omega_N^{\iota 2^\kappa}$ and $z^{2^\kappa}+\omega_N^{\iota 2^\kappa}$, respectively, the contour integral representations

$$B_{\iota,\kappa}(z) = \frac{1}{2\pi i}\oint_\Gamma \frac{\zeta^{2^\kappa}-z^{2^\kappa}}{\zeta-z}\;\frac{B(\zeta)}{\zeta^{2^\kappa}-\omega_N^{\iota 2^\kappa}}\,d\zeta \tag{31}$$

and

$$B_{\iota+2^{k-\kappa-1},\kappa}(z) = \frac{1}{2\pi i}\oint_\Gamma \frac{\zeta^{2^\kappa}-z^{2^\kappa}}{\zeta-z}\;\frac{B(\zeta)}{\zeta^{2^\kappa}+\omega_N^{\iota 2^\kappa}}\,d\zeta$$

are valid. Thus the linear combination on the right hand side of (30) equals

$$\frac{1}{2\pi i}\oint_\Gamma \frac{\zeta^{2^{\kappa+1}}-z^{2^{\kappa+1}}}{\zeta-z}\;\frac{B(\zeta)}{\zeta^{2^{\kappa+1}}-\omega_N^{\iota 2^{\kappa+1}}}\,d\zeta$$

which by (31) is $B_{\iota,\kappa+1}(z)$.

Remarks. 1.Note that the relatively complicated bit reversal, which has to be performed as a final step in the algorithms of Cooley-Tukey and Fiduccia, is replaced in Meinardus' method by a simple shifting operation.

2. A numerical implementation of the recursion formaula (30) is given in [15] and [17] , p. 89.

3. A connection between the methods of Fiduccia and Meinardus may be established via modular polynomial arithmetic and the Chinese remainder theorem (cf. [0], Chapter 8).

(v) The method of Winograd (cf. [29]; see also [1], Chapter III) The convolution property (11) of the DFT provides a skilful instrument for the calculation of circular convolutions using FFT techniques. Contrary to this one of the basic concepts of Winograd's method manifests itself in the at a first sight perhaps somewhat strange suggestion to use circular convolutions for the construction of fast algorithms for DFT. Most surprisingly this idea, which is originally due to Rader [19], together with applications of modular arithmetic based on the use of (14) and of Chinese remaindering for polynomials, eventually results in a procedure which gets along with only about 20% of the multiplications needed for the FFT algorithms we have derived so long and from which the new algorithm differs fundamentally.

We illustrate the main ideas of Winograd's method for the case $N = 7$, where - due to the fact that N is prime - the Cooley-Tukey algorithm allows no saving of computational work.
After reduction of the respective powers of $\omega = \omega_7$ we have to calculate

$$\begin{pmatrix} \hat{X}_0 \\ \hat{X}_1 \\ \hat{X}_2 \\ \hat{X}_3 \\ \hat{X}_4 \\ \hat{X}_5 \\ \hat{X}_6 \end{pmatrix} = \begin{pmatrix} 1 & 1 & 1 & 1 & 1 & 1 & 1 \\ 1 & \omega & \omega^2 & \omega^3 & \omega^4 & \omega^5 & \omega^6 \\ 1 & \omega^2 & \omega^4 & \omega^6 & \omega & \omega^3 & \omega^5 \\ 1 & \omega^3 & \omega^6 & \omega^2 & \omega^5 & \omega & \omega^4 \\ 1 & \omega^4 & \omega & \omega^5 & \omega^2 & \omega^6 & \omega^3 \\ 1 & \omega^5 & \omega^3 & \omega & \omega^6 & \omega^4 & \omega^2 \\ 1 & \omega^6 & \omega^5 & \omega^4 & \omega^3 & \omega^2 & \omega \end{pmatrix} \begin{pmatrix} x_0 \\ x_1 \\ x_2 \\ x_3 \\ x_4 \\ x_5 \\ x_6 \end{pmatrix} . \quad (32)$$

Note that due to the omission of the factor $\frac{1}{\sqrt{7}}$ before the matrix we have replaced the usual X_μ on the left hand side of (32) by $\hat{X}_\mu$.
The framed part of the matrix in (32) can be interpreted as the group table of the (non-zero) residue classes modulo 7 under multiplication. Because this group is cyclic of order 6 it must be isomorphic to the additive group of residue classes modulo 6. Therefore it is possible to rearrange the "main part" of (32) as

$$\begin{pmatrix} \hat{X}_1 \\ \hat{X}_3 \\ \hat{X}_2 \\ \hat{X}_6 \\ \hat{X}_4 \\ \hat{X}_5 \end{pmatrix} = \begin{pmatrix} \omega & \omega^5 & \omega^4 & \omega^6 & \omega^2 & \omega^3 \\ \omega^3 & \omega & \omega^5 & \omega^4 & \omega^6 & \omega^2 \\ \omega^2 & \omega^3 & \omega & \omega^5 & \omega^4 & \omega^6 \\ \omega^6 & \omega^2 & \omega^3 & \omega & \omega^5 & \omega^4 \\ \omega^4 & \omega^6 & \omega^2 & \omega^3 & \omega & \omega^5 \\ \omega^5 & \omega^4 & \omega^6 & \omega^2 & \omega^3 & \omega \end{pmatrix} \begin{pmatrix} x_1 \\ x_5 \\ x_4 \\ x_6 \\ x_2 \\ x_3 \end{pmatrix} \quad . \tag{33}$$

Equation (33) exhibits - apart from the "wrong" arrangement of the components $\hat{X}_\mu$ which may be corrected through a multiplication of $(\hat{X}_1,\hat{X}_3,\hat{X}_2,\hat{X}_6,\hat{X}_4,\hat{X}_5)^T$ by the permutation matrix P_6 defined in (6) - the structure of a circular convolution (cf.(9),(10)). As a consequence of the isomorphism $Z_n \cong Z_{n_1} \times Z_{n_2}$ for cyclic groups of composite order $n = n_1 n_2$, $(n_1, n_2) = 1$, and its possible realization through application of modular arithmetic, the matrix of the circular convolution(33) can eventually be partitioned into four blocks of circulant (3,3)-matrices which together form a circulant block matrix of type (2,2). Thus we obtain

$$\left(\begin{array}{c} \hat{X}_1 \\ \hat{X}_2 \\ \hat{X}_4 \\ \hline \hat{X}_6 \\ \hat{X}_5 \\ \hat{X}_3 \end{array}\right) = \left(\begin{array}{ccc|ccc} \omega & \omega^4 & \omega^2 & \omega^6 & \omega^3 & \omega^5 \\ \omega^2 & \omega & \omega^4 & \omega^5 & \omega^6 & \omega^3 \\ \omega^4 & \omega^2 & \omega & \omega^3 & \omega^5 & \omega^6 \\ \hline \omega^6 & \omega^3 & \omega^5 & \omega & \omega^4 & \omega^2 \\ \omega^5 & \omega^6 & \omega^3 & \omega^2 & \omega & \omega^4 \\ \omega^3 & \omega^5 & \omega^6 & \omega^4 & \omega^2 & \omega \end{array}\right) \left(\begin{array}{c} x_1 \\ x_4 \\ x_2 \\ \hline x_6 \\ x_3 \\ x_5 \end{array}\right) \quad . \tag{34}$$

Finally, we take advantage of this structure by combining (34) from two single fast algorithms for circular convolutions of length 2 and 3, respectively. These algorithms were constructed via the interpretation of circular convolutions as generated through multiplication of polynomials modulo Z^N-1 (cf. Part 1 of this paper) and application of the Chinese remainder theorem for polynomials. They run as follows:

1. In order to calculate the 3-term circular convolution

$$\begin{pmatrix} c_0 \\ c_1 \\ c_2 \end{pmatrix} = \begin{pmatrix} a_0 & a_2 & a_1 \\ a_1 & a_0 & a_2 \\ a_2 & a_1 & a_0 \end{pmatrix} \begin{pmatrix} b_0 \\ b_1 \\ b_2 \end{pmatrix}$$

let

$$m_0 = \frac{1}{3}(a_0+a_1+a_2)(b_0+b_1+b_2), \quad m_1 = \frac{1}{3}(a_0+a_1-2a_2)(b_0-b_1),$$
$$m_2 = \frac{1}{3}(-2a_0+a_1+a_2)(b_2-b_1), \quad m_3 = \frac{1}{3}(a_0-2a_1+a_2)(b_0-b_2). \tag{35}$$

Then

$$\begin{aligned} c_0 &= m_0 + m_1 + m_3 \\ c_1 &= m_0 + m_2 - m_3 \\ c_2 &= m_0 - m_1 - m_2 . \end{aligned} \tag{36}$$

2. Let

$$\underline{\Omega}_0 = \begin{pmatrix} \omega & \omega^4 & \omega^2 \\ \omega^2 & \omega & \omega^4 \\ \omega^4 & \omega^2 & \omega \end{pmatrix}, \quad \underline{\Omega}_1 = \begin{pmatrix} \omega^6 & \omega^3 & \omega^5 \\ \omega^5 & \omega^6 & \omega^3 \\ \omega^3 & \omega^5 & \omega^6 \end{pmatrix},$$

$$\underline{\psi}_0 = \begin{pmatrix} \hat{x}_1 \\ \hat{x}_2 \\ \hat{x}_4 \end{pmatrix}, \quad \underline{\psi}_1 = \begin{pmatrix} \hat{x}_6 \\ \hat{x}_5 \\ \hat{x}_3 \end{pmatrix}, \quad \underline{\varphi}_0 = \begin{pmatrix} x_1 \\ x_4 \\ x_2 \end{pmatrix}, \quad \underline{\varphi}_1 = \begin{pmatrix} x_6 \\ x_3 \\ x_5 \end{pmatrix}.$$

Then(34) is the circular block-convolution

$$\begin{pmatrix} \underline{\psi}_0 \\ \underline{\psi}_1 \end{pmatrix} = \begin{pmatrix} \underline{\Omega}_0 & \underline{\Omega}_1 \\ \underline{\Omega}_1 & \underline{\Omega}_0 \end{pmatrix} \begin{pmatrix} \underline{\varphi}_0 \\ \underline{\varphi}_1 \end{pmatrix}$$

which can be performed according to

$$\underline{M}_1 = \frac{1}{2}(\underline{\Omega}_0+\underline{\Omega}_1)(\underline{\varphi}_0+\underline{\varphi}_1), \quad \underline{M}_2 = \frac{1}{2}(\underline{\Omega}_0-\underline{\Omega}_1)(\underline{\varphi}_0-\underline{\varphi}_1),$$

$$\underline{\psi}_0 = \underline{M}_1+\underline{M}_2, \quad \underline{\psi}_1 = \underline{M}_1-\underline{M}_2 .$$

$\underline{M}_1$, $\underline{M}_2$ are obtained by use of (35), (36). The result is a procedure for the calculation of (34) which needs a total of only 8 multiplications.

Further investigations in this direction which lie beyond the scope of this short survey have led to algorithms which require an even more reduced amount of computational work (cf. Nussbaumer [18], Chapter 7).

3.

We point out now some of the numerous possibilities for the application of fast DFT and convolution techniques. For a rather complete account of this topic we refer to Henrici [10].

(i) Numerical calculation of Fourier coefficients

For any 2π - periodic R-integrable function $f(e^{it})$ of the real variable t the coefficients of its formal Fourier series

$$\sum_{\mu=-\infty}^{\infty} b_\mu e^{i\mu t}$$

are defined by

$$b_\mu = b_\mu(f) = \frac{1}{2\pi} \int_0^{2\pi} f(e^{it}) e^{-i\mu t}\, dt \ , \ \mu \in \mathbb{Z} \ .$$

If we divide the interval $[0,2\pi]$ into N subintervals of equal length $\frac{2\pi}{N}$ and apply the trapezoidal rule for the numerical calculation of b_μ we arrive at the approximations

$$\beta_\mu = \beta_\mu(f) = \frac{1}{N} \sum_{\nu=0}^{N-1} f_\nu \omega_N^{-\mu\nu} \ , \mu = 0(1)N-1 \ , \tag{37}$$

where

$$f_\nu = f(\omega_N^\nu) \quad , \quad \nu = 0(1)N-1 \quad .$$

If we let

$$\underline{b} = \begin{pmatrix} \beta_0 \\ \vdots \\ \vdots \\ \beta_{N-1} \end{pmatrix} , \quad \underline{f} = \begin{pmatrix} f_0 \\ \vdots \\ \vdots \\ f_{N-1} \end{pmatrix} ,$$

this may be written in a vectorial form as

$$\underline{b} = \frac{1}{\sqrt{N}} \underline{W}_N^{-1} \ \underline{f} = \frac{1}{\sqrt{N}} \underline{P}_N \underline{W}_N \underline{f} \quad . \tag{38}$$

Obviously, these relations call for the application of FFT procedures in connection with their numerical realization.

If f(z) has a Laurent series expansion

$$f(z) = \sum_{\kappa=-\infty}^{\infty} b_\kappa z^\kappa$$

which is convergent for $0<r<1<R$, the Fourier coefficients of $f(e^{it})$ and the Laurent coefficients of f(z) coincide. In this case

$$\beta_\mu = \sum_{\rho=-\infty}^{\infty} b_{\mu+\rho N}, \quad \mu = 0(1)N-1 \quad . \tag{39}$$

Proof: The polynomial $B(z) = \sum_{\mu=0}^{N-1} \beta_\mu z^\mu$ with coefficients defined by (39) takes at the N-th roots of unity the values

$$B(\omega_N^\nu) = \sum_{\mu=0}^{N-1} \omega_N^{\mu\nu} \left(\sum_{\rho=-\infty}^{\infty} b_{\mu+\rho N}\right) =$$

$$= \sum_{\mu=0}^{N-1} \sum_{\rho=-\infty}^{\infty} b_{\mu+\rho N}\, \omega_N^{(\mu+\rho N)\nu} = \sum_{\kappa=-\infty}^{\infty} b_\kappa \omega_N^{\kappa\nu} = f(\omega_N^\nu) = f_\nu \quad ,$$

i.e. by (29)

$$\sqrt{N}\, \underline{P}_N\, \underline{b} = \underline{W}_N\, \underline{f}$$

and this is equivalent with (38).
As an immediate consequence of (39) we arrive at the error estimate

$$\beta_\mu - b_\mu = \sum_{\substack{\rho=-\infty \\ \rho\neq 0}}^{\infty} b_{\mu+\rho N} \quad .$$

Remarks. 1. If f is absolutely continuous on the real line then $\lim_{\mu\to\pm\infty} b_\mu = 0$, whereas the approximate Fourier coefficients β_μ according to (37) are N-periodic and therefore do not reproduce the asymptotic behavior of the b_μ. In order to correct this deficiency one may use the following approach: Approximate f by a function g and then take the Fourier coefficients $b_\mu(g)$ of g as approximations for $b_\mu(f)$. By an appropriate choice of g (cf. Gautschi [7]) it is often possible to deduce a relation of the form

$$b_\mu(g) = \tau_\mu\, \beta_\mu(f), \quad \mu \in \mathbb{Z}, \tag{40}$$

where the so-called attenuation factors τ_μ are independent of f and go to zero at a rate comparable to that of the $b_\mu(f)$. Note that (40) still permits the use of FFT procedures.

2. A FFT-like algorithm which yields the values (37) of the trapezoidal rule for all stepsizes needed in the use of convergence accelerating procedures like Romberg's method without additional computational work is due to Meinardus [15].

(ii) Multiplication of polynomials and of large integers

1. Multiplication of polynomials

Let

$$p(z) = \sum_{\mu=0}^{N-1} a_\mu z^\mu , \quad q(z) = \sum_{\nu=0}^{N-1} b_\nu z^\nu . \tag{41}$$

Then

$$p(z)\, q(z) = \sum_{\kappa=0}^{2N-2} c_\kappa z^\kappa \tag{42}$$

and

$$c_\kappa = \sum_{\iota=0}^{N-1} a_\iota b_{\kappa-\iota} , \quad \kappa = 0(1)2N-2 , \tag{43}$$

with $a_\lambda = b_\lambda = 0$ if $\lambda<0$ or $\lambda \geq N$.

In order to make the convolution theorem for circular convolutions and thereby FFT techniques applicable for the construction of the coefficients c_κ of the product polynomial we extend the coefficient vectors of p and q by adding N zeros to each of them

$$\underline{a} = (a_0,\dots,a_{N-1},\underbrace{0,\dots,0}_{N \text{ zeros}})^T , \quad \underline{b} = (b_0,\dots,b_{N-1},\underbrace{0,\dots,0}_{N \text{ zeros}})^T .$$

Then the first 2N-1 components of the vector $\underline{c} = \underline{a} * \underline{b} = (c_0,\dots,c_{2N-1})^T$ coincide with the values c_κ prescribed by (43) and additionally $c_{2N-1} = 0$.

2. Multiplication of large integers

If z in (41) is taken equal to a (small) integer $\gamma \geq 2$ the method just described provides a tool for the time-saving multiplication of (large) integers. However, c_κ in (43) needs not satisfy $0 \leq c_\kappa < \gamma$ so that (42) with $z = \gamma$ does not necessarily give the correct base - γ representation of the product. But we can assure $0 \leq c_\kappa < N\gamma^2$ and hence

$$c_\kappa = \sum_{\mu=0}^{m} c_{\kappa,\mu}\gamma^\mu \tag{44}$$

with $m = [\log_\gamma N] + 2$. The correct base-γ digits of the product are now obtained by inserting (44) into (42).
A much more sophisticated algorithm for the fast multiplication of large integers which also makes important use of the FFT was developed by Schönhage and Strassen [24] (see also Knuth [13], p. 278).

(iii) Circulant matrices

We define a circulant (N,N) - matrix $\underline{C}_N$ by

$$\underline{C}_N = \begin{pmatrix} \gamma_0 & \gamma_1 & \gamma_2 & \cdots & \gamma_{N-1} \\ \gamma_{N-1} & \gamma_0 & \gamma_1 & \cdots & \gamma_{N-2} \\ \gamma_{N-2} & \gamma_{N-1} & \gamma_0 & \cdots & \gamma_{N-3} \\ \cdot & \cdot & \cdot & \cdots & \cdot \\ \gamma_1 & \gamma_2 & \gamma_3 & \cdots & \gamma_0 \end{pmatrix}$$

and put $\rho(z) = \sum_{\kappa=0}^{N-1} \gamma_\kappa z^\kappa$. Because the eigenvalues λ_ν of $\underline{C}_N$ are given by

$$\lambda_\nu = \rho(\omega_N^\nu), \quad \nu = 0(1)N-1 , \tag{45}$$

and the corresponding eigenvectors $\underline{\xi}_\nu$ result from

$$\underline{\xi}_\nu = \frac{1}{\sqrt{N}} \begin{pmatrix} 1 \\ \omega_N^\nu \\ \omega_N^{2\nu} \\ \vdots \\ \omega_N^{(N-1)\nu} \end{pmatrix} , \quad \nu = 0(1)N-1 ,$$

we get

$$\underline{W}_N^* \underline{C}_N \underline{W}_N = ((\lambda_\nu \delta_{\mu\nu}))_{\mu,\nu=0}^{N-1} =: \underline{\Delta}_N . \tag{46}$$

As an application of (46) one may give another proof of the convolution theorem (11) (cf. [25], p. 176). Here we concentrate on the calculation of the solution of a system of linear equations

$$\underline{C}_N \underline{u} = \underline{v} \tag{47}$$

with

$$\underline{u} = \begin{pmatrix} u_0 \\ \vdots \\ u_{N-1} \end{pmatrix}, \quad \underline{v} = \begin{pmatrix} v_0 \\ \vdots \\ v_{N-1} \end{pmatrix} \tag{48}$$

and the circulant matrix $\underline{C}_N$ supposed to be non-singular, i.e.

$$\rho(\omega_N^{\nu}) \neq 0 \ , \quad \nu = 0(1)N-1 \quad .$$

It follows from (46) that

$$\underline{C}_N^{-1} = \underline{W}_N \underline{\Delta}_N^{-1} \ \underline{P}_N \underline{W}_N$$

and this permits the calculation of the solution vector $\underline{u}$ in (47) by means of two FFT applications.

Examples. 1. In cubic periodic spline interpolation with equally spaced knots the circulant (N,N)-matrix

$$\underline{\Gamma}_N^{(1)} = \begin{pmatrix} 4 & 1 & 0 & \cdots & & 1 \\ 1 & 4 & 1 & & & 0 \\ & \ddots & \ddots & \ddots & & \\ & & \ddots & \ddots & \ddots & \\ 0 & & & 1 & 4 & 1 \\ 1 & 0 & \cdots & 0 & 1 & 4 \end{pmatrix} \tag{49}$$

plays an important role (cf. [17], p.154). According to (45), the eigenvalues of $\underline{\Gamma}_N^{(1)}$ are

$$\lambda_\nu = 4+2\cos\frac{2\pi\nu}{N} \ , \quad \nu = 0(1)N-1 \quad ,$$

and thus the existence problem for cubic interpolating spline functions is completely settled in this case for arbitrary N.

2. In [16] a constructive solution of the interpolation problem with periodic splines of arbitrary degree for the case of equally spaced knots and nodes is described. The relevant system of cyclic difference equations is solved by means of two DFT appli-

cations.

3. If in 1. the periodic boundary conditions are replaced by essential ones, the matrix of the respective system of linear equations becomes the (N-2, N-2)-matrix

$$\underline{\Gamma}_{N-2}^{(2)} = \begin{pmatrix} 4 & 1 & & & & 0 \\ 1 & 4 & 1 & & & \\ & \ddots & \ddots & \ddots & & \\ & & & 1 & 4 & 1 \\ 0 & & & & 1 & 4 \end{pmatrix} .$$

$\Gamma_{N-2}^{(2)}$ is no longer of circulant type, but may be embedded in a circulant matrix by the following procedure:

Starting from the linear system

$$\underline{\Gamma}_{N-2}^{(2)} \; \underline{\tilde{u}} = \underline{\tilde{v}} \tag{50}$$

with

$$\underline{\tilde{u}} = \begin{pmatrix} u_1 \\ \vdots \\ u_{N-2} \end{pmatrix}, \qquad \underline{\tilde{v}} = \begin{pmatrix} v_1 \\ \vdots \\ v_{N-2} \end{pmatrix}$$

we enlarge $\underline{\Gamma}_{N-2}^{(2)}$ to the matrix $\Gamma_N^{(1)}$ in (49) and consider together with (50) the linear system

$$\underline{\Gamma}_N^{(1)} \underline{u} = \underline{v} \tag{51}$$

with $\underline{u}$ and $\underline{v}$ as in (48).

Let us now take v_0 and v_{N-1} in (51) as uniquely determined from the equations

$$\begin{aligned} v_0 + v_{N-1}\lambda_1^{N-1} + \sum_{\nu=1}^{N-2} v_\nu \lambda_1^\nu &= 0 \\ v_0 + v_{N-1}\lambda_2^{N-1} + \sum_{\nu=1}^{N-2} v_\nu \lambda_2^\nu &= 0 \end{aligned} \tag{52}$$

($\lambda_{1,2} = -2 \pm \sqrt{3}$ denote the roots of the polynomial z^2+4z+1). Then the components u_0 and u_{N-1} in the solution vector $\underline{u}$ of (51) become equal to zero. Hence, if we solve (51) with v_0, v_{N-1} according to (52) - which may be done by use of a FFT technique - we arrive automatically at the solution $\underline{\tilde{u}}$ of (50).

4. In connection with cubic spline interpolation for a function $f(t)$ which has a pole of order p at a previously known point $t = t_1$ the following simple method may be of interest: In order to find a good approximation for $f(t)$ first interpolate the auxiliary function $g(t) = (t - t_1)^p f(t)$ by a cubic spline function $s(t)$ and then take $s(t)/(t-t_1)^p$ as final approximation for $f(t)$.

Let us for example interpolate $f(t) = \tan(\pi t/4)$ at four equally spaced points in the interval $0 \le t \le 1.7$
(α) by a cubic spline $s_1(t)$ with boundary conditions $s_1'(o) = f'(0)$, $s_1'(1.7) = f'(1.7)$
(β) by a "pseudorational" spline $(t-2)^{-1} s_2(t)$ where the cubic spline $s_2(t)$ is obtained through interpolation of $g(t) = (t-2)\tan(\pi t/4)$ with boundary conditions $s_2'(0) = g'(0)$, $s_2'(1.7) = g'(1.7)$.

Then

$$\max_{0 \le t \le 1.7} \left| f(t) - s_1(t) \right| \sim 0.28$$

$$\max_{0 \le t \le 1.7} \left| f(t) - \frac{s_2(t)}{t-2} \right| \sim 4.2(-4) \quad .$$

If we consider the same problem in the enlarged interval $0 \le t \le 1.9$ we obtain for the respective error bounds in polynomial and pseudorational cubic spline interpolation the still more surprising values 7.6 and 6.3(-4).

With this last example we have left the actual topic of this paper. But we have continued to do numerical mathematics and hence this slip hopefully calls only for slight apologies.

References

0. A.V. Aho, J.E. Hopcroft, J.D. Ullman:
The design and analysis of computer algorithms
Reading, Mass.: Addison-Wesley 1974

1. L. Auslander and R. Tolmieri:
Is computing with the finite Fourier transform pure or applied mathematics?
Bull. Amer. Math. Soc. (New Series) 1(1979), 847-897

2. J.W. Cooley, P.A.W. Lewis, P.D. Welch:
The finite Fourier transform
IEEE Trans. AU-17(1969), 77 - 85

3. J.W. Cooley, J.W. Tukey: An algorithm for the machine calculation of complex Fourier series
Math. Comput. 19(1965), 297 - 301

4. C.M. Fiduccia:
Polynomial evaluation via the division algorithm. The fast Fourier transform revisited, in:
Proc. 4th Ann. ACM Symp. on Theory of Computing
1972, p. 88 - 93

5. W. Forst: Private communication to G. Meinardus 1978

6. W. Gander, A. Mazzario:
Numerische Prozeduren I (in memoriam Heinz Rutishauser)
Berichte der Fachgruppe Computer-Wissenschaften 4
Zürich: Eidgenössische Technische Hochschule, Oktober 1972

7. W. Gautschi:
Attenuation factors in practical Fourier analysis
Numer. Math. 18(1972), 373 - 400

8. I.J. Good:
The interaction algorithm and practical Fourier analysis
J. Roy. Stat. Soc. B-20(1958), 361 - 372, B-22(1960), 372 - 375

9. I.J. Good:
The relationship between two fast Fourier transforms
IEEE Trans. C-20(1971), 310 - 317

10. P. Henrici: Fast Fourier methods in computational complex analysis
SIAM Review 21(1979), 481 - 527

11. D. Kahaner:
The fast Fourier Transform by polynomial evaluation
Z. Angew. Math. Phys. 29(1978), 387 - 394

12. T. Kaneko, B. Liu:
Accumulation of round-off error in fast Fourier transforms
J. Assoc. Comput. Mach. 17(1970), 637 - 654

13. D.E. Knuth:
The art of computer programming,
vol. 2: Seminumerical algorithms, 2nd edition
Reading, Mass.: Addison-Wesley 1981

14. J.H. Mc Clellan, T.W. Parks:
Eigenvalue and eigenvector decomposition of the discrete Fourier transform
IEEE Trans. AU-20(1972), 66 - 74

15. G. Meinardus:
Schnelle Fourier-Transformation, in: L. Collatz et al.(eds.): Numerische Methoden der Approximationstheorie Band 4,
ISNM 42, p. 192 - 203
Basel: Birkhäuser 1978

16. G. Meinardus, G. Merz:
Zur periodischen Spline-Interpolation, in: K. Böhmer et al. (eds.): Spline-Funktionen
Mannheim: Bibliographisches Institut 1974, p. 177 - 195

17. G. Meinardus, G. Merz:
Praktische Mathematik I
Mannheim: Bibliographisches Institut 1979

18. H.J. Nussbaumer:
Fast Fourier transform and convolution algorithms
Berlin: Springer 1981

19. C.M. Rader:
Discrete Fourier transforms when the number of data samples is prime
Proc. IEEE 56(1968), 1107 - 1108

20. C.M. Rader, N.M. Brenner:
A new principle for fast Fourier transformations
IEEE Trans. ASSP-24(1976), 264 - 266

21. C. Runge:
Über die Zerlegung empirisch gegebener periodischer Funktionen in Sinuswellen
Z. Math. Phys. 48(1903), 443 - 456

22. C. Runge:
Über die Zerlegung einer empirischen Funktion in Sinuswellen
Z. Math. Phys. 53(1905), 117 - 123

23. C. Runge, H. König:
Vorlesungen über numerisches Rechnen
Berlin: Springer 1924

24. A. Schönhage, V. Strassen:
Schnelle Multiplikation großer Zahlen
Computing 7(1971), 281 - 292

25. H.W. Schüßler:
Digitale Systeme zur Signalverarbeitung
Berlin: Springer 1973

26. H.R. Schwarz:
Elementare Darstellung der schnellen Fourier-Transformation
Computing 18(1977), 107 - 116

27. H.R. Schwarz:
The fast Fourier transform for general order
Computing 19(1978), 341 - 350

28. K. Segeth:
Roundoff errors in the fast computation of discrete convolutions
Apl. Mat. 26(1981), 241 - 262

29. S. Winograd:
On computing the discrete Fourier transform
Math. Comput. 32(1978), 175 - 199

ERROR ANALYSIS OF COMPLEX ARITHMETIC

F.W.J. Olver

The University of Maryland and
The National Bureau of Standards

This lecture begins with a brief account of recent work on unrestricted algorithms for computing mathematical functions, especially the development of error analysis based on a non-traditional definition of relative error. The main part of the talk describes the application of this analysis to real and complex arithmetic, and concludes with some new extensions that have been made in complex arithmetic.

1. INTRODUCTION

1.1. Unrestricted algorithms

Since 1974 C. W. Clenshaw at the University of Lancaster and the present speaker have been studying the problem of constructing unrestricted algorithms for generating mathematical functions, beginning with the elementary functions. By an unrestricted algorithm we mean one in which the user may demand any guaranteed accuracy in the computed value of the function for any value(s) of the argument(s). For example, we may wish to compute the Bessel function $J_\nu(x)$ to a thousand significant figures of accuracy for real or complex values of x and ν of the order of a million - or the reciprocal of a million. In practice, of course, there will always be limits on the extent to which such aspirations can be fulfilled, governed by (if nothing else) the number of particles in the universe and the speed of electricity. However, this consideration does not prevent our seeking algorithms that cover all cases in theory, and are implementable in practice for very high accuracy requirements and vast ranges of the arguments. Successful

H. Werner et al. (eds.), Computational Aspects of Complex Analysis, 279–292.

implementations hinge on the availability of comprehensive and reliable multiple-precision software packages such as that developed at the Australian National University by R. P. Brent [1], [2]. Currently D. W. Lozier of the National Bureau of Standards is extending Brent's MP package to increase its usefulness for unrestricted algorithms, for example, by permitting the use of an arbitrary number of word lengths to represent the exponent of floating-point numbers.

Why should anyone wish to be able to compute Bessel functions, or even elementary functions, to thousands of significant figures for arguments that may be of enormous size? There are various answers to this question, but I shall mention only one. Clenshaw and I felt instinctively that in the course of solving analytical and numerical problems that may arise in the construction of unrestricted algorithms new light would be shed upon processes of numerical approximation, and perhaps also on other branches of numerical analysis. More specifically, by passing to the limiting or "infinite" case we ought to be able to learn something new about the finite case. There are ample precedents for this point of view. For example, it is easier to approach parts of the analytic theory of difference equations via the corresponding theory of differential equations.

1.2. Error analysis

The expectations expressed in the preceding paragraph have already been realized in one way. A new form of error analysis has been developed for floating-point arithmetic that offers certain advantages over existing procedures, including interval arithmetic and the theories of J. H. Wilkinson and F. Stummel.

For the purpose of constructing unrestricted algorithms, interval arithmetic, and its complex analogues such as disc arithmetic, have the following disadvantages. First, interval arithmetic is a posteriori in nature and does not adapt easily to a priori requirements. Secondly, interval arithmetic is a rather crude arithmetic tool that usually fails to take account of interactive effects of errors at successive steps of a computing algorithm. In consequence, it may yield error bounds that exceed the minimal error bounds by an enormous factor, or worse still, it may fail altogether owing to the need to divide by an interval (or disc) that contains zero; see for example [3].

Wilkinson's theory, see for example [10], is analytic rather than arithmetic and is free from the disadvantages mentioned in the preceding paragraph. However, the mathematical formulation is less efficient, especially for complex arithmetic. Furthermore, backward error analysis serves only to reflect the effect of the abbreviation errors, that is, chopping or rounding errors, into the effect of the inherent errors: it does not determine directly the effect of the inherent (and abbreviation) errors on the solution.

Stummel's theory, see for example [8], [9], is analytic, and like that of Wilkinson it is a powerful tool for assessing stability and efficacy of computational algorithms. Stummel's theory traces the effect of all errors, inherent and abbreviation, to the final solution, but only within a linearized framework. Thus it yields error estimates rather than strict error bounds.

The new form of error analysis was introduced four years ago [4] and has advanced considerably in the meantime; see [5]. Currently it is being applied in ways quite different from the original purpose of constructing unrestricted algorithms. One such application is the computation of rigorous and realistic error bounds of a posteriori type for the solution of systems of linear algebraic equations by direct methods, especially Gaussian elimination; this is described in a forthcoming paper by J. H. Wilkinson and the speaker [7]. Another application that will be reported in due course concerns the evaluation of real or complex polynomials.

The purpose of the present lecture is to describe the basics of the new form of error analysis, especially with a view to arithmetic operations with complex numbers.

2. ABSOLUTE AND RELATIVE PRECISION

2.1. Absolute precision

Let $\bar{a}$ be an approximation to a number a such that

$$a = \bar{a} + u, \quad \text{where} \quad |u| \le \alpha,$$

α being a known nonnegative real number. Then we say that a is approximated by $\bar{a}$ to *absolute precision* α, and express this relation symbolically in the form

$$a \simeq \bar{a}\ ;\ \text{ap}(\alpha). \tag{2.1}$$

This is, of course, nothing more than the ordinary definition of absolute error, and it applies equally to the real and complex number fields.

The following properties are simple consequences of the definition:

I. (Symmetry). $\bar{a} \simeq a\ ;\ \text{ap}(\alpha)$.

II. (Inclusion). $a \simeq \bar{a}\ ;\ ap(\delta),\ \ \forall\delta \geq \alpha.$

III. $a + k \simeq \bar{a} + k\ ;\ ap(\alpha),\ \ \forall k.$

IV. $ka \simeq k\bar{a}\ ;\ ap(|k|\alpha),\ \ \forall k.$

V. (Addition and subtraction). If, also,

$$b \simeq \bar{b}\ ;\ ap(\beta), \tag{2.2}$$

then

$$a + b \simeq \bar{a} + \bar{b}\ ;\ ap(\alpha+\beta), \tag{2.3}$$

$$a - b \simeq \bar{a} - \bar{b}\ ;\ ap(\alpha+\beta). \tag{2.4}$$

VI. (Repeated approximation). If, also,

$$\bar{a} \simeq \bar{\bar{a}}\ ;\ ap(\delta),$$

then

$$a \simeq \bar{\bar{a}}\ ;\ ap(\alpha+\delta).$$

Multiplication and division are more cumbersome. If (2.1) and (2.2) hold, then we have

$$ab \simeq \bar{a}\bar{b}\ ;\ ap(|\bar{a}|\beta+|\bar{b}|\alpha+\alpha\beta),$$

$$\frac{a}{b} \simeq \frac{\bar{a}}{\bar{b}}\ ;\ ap\left\{\frac{|\bar{a}|\beta+|\bar{b}|\alpha}{|\bar{b}|(|\bar{b}|-\beta)}\right\},$$

provided that $|\bar{b}| > \beta$ in the latter case.

2.2. Relative precision

The definition and properties given in the preceding subsection are natural for use when arithmetic operations are carried out in a fixed-point mode. In effect, they form the basis of interval arithmetic (real variables) or disc arithmetic (complex variables). Nowadays, however, most computations are performed in a floating-point mode and it is natural to work in terms of relative error rather than absolute error. Unfortunately the usual definition

$$\left|\frac{\bar{a}}{a} - 1\right| \leq \alpha$$

does not lead to properties that are analogous to I to VI of §2.1. For example, it certainly does not follow that

$$\left|\frac{a}{\bar{a}} - 1\right| \leq \alpha.$$

The discrepancy is of only the second order when α is small, but it cannot be ignored if strict error bounds are required.

Instead of relative error we employ another definition. Let a and $\bar{a}$ be nonzero numbers such that

$$a = \bar{a}e^{u}, \quad \text{where} \quad |u| \leq \alpha,$$

α again denoting a known nonnegative real number. Then we say that a is approximated by $\bar{a}$ to *relative precision* α, and write

$$a \simeq \bar{a}\ ;\ \text{rp}(\alpha). \tag{2.5}$$

Clearly (2.5) is equivalent to

$$\ell n\ a \simeq \ell n\ \bar{a}\ ;\ \text{ap}(\alpha), \tag{2.6}$$

with an appropriate choice of branches for the natural logarithms when a and $\bar{a}$ are other than real and positive. It is also clear that when α is small relative precision and relative error agree to within $O(\alpha^2)$.

With the new definition, Properties I to VI of §2.1 have the following analogues:

I. (Symmetry). $\bar{a} \simeq a\ ;\ \text{rp}(\alpha)$.

II. (Inclusion). $a \simeq \bar{a}\ ;\ \text{rp}(\delta),\quad \forall\delta \geq \alpha$.

III. $ka \simeq k\bar{a}\ ;\ \text{rp}(\alpha),\quad k \neq 0$.

IV. $a^k \simeq \bar{a}^k\ ;\ \text{rp}(|k|\alpha)$,

with an appropriate choice of branches when k is not an integer or zero.

V. (Multiplication and division). If, also,

$$b \simeq \bar{b}\ ;\ \text{rp}(\beta), \tag{2.7}$$

then

$$ab \simeq \bar{a}\bar{b}\ ;\ \text{rp}(\alpha+\beta), \tag{2.8}$$

$$a/b \simeq \bar{a}/\bar{b}\ ;\ \text{rp}(\alpha+\beta). \tag{2.9}$$

VI. (Repeated approximation). If, also,

$$\bar{a} \simeq \bar{\bar{a}}\ ;\quad rp(\delta),$$

then

$$a \simeq \bar{\bar{a}}\ ;\quad rp(\alpha+\delta).$$

It should be observed that of this set of properties only II and III hold with the conventional definition of relative error.

Addition and subtraction are more cumbersome. Generally, however, we use the ap rules (2.3) and (2.4) for these operations, preferring the rp rules (2.8) and (2.9) for multiplication and division. In consequence, conversion rules are needed for the ap and rp formats. These are supplied in the next subsection.

There is one frequent exception to the procedures suggested in the preceding paragraph. This occurs with the addition of real positive numbers having the rp forms (2.5) and (2.7). In this case it is easily verified that

$$a + b \simeq \bar{a} + \bar{b}\ ;\quad rp\left\{\ell n\left(\frac{\bar{a}e^{\alpha}+\bar{b}e^{\beta}}{\bar{a}+\bar{b}}\right)\right\}.$$

Furthermore, as a consequence of the inclusion property II, this result can be replaced by the simpler (but generally weaker) form

$$a + b \simeq \bar{a} + \bar{b}\ ;\quad rp\{\max(\alpha,\beta)\}. \tag{2.10}$$

2.3. Conversion rules

I. (ap to rp). Suppose that (2.1) holds and $|\bar{a}| > \alpha$. Then

$$a \simeq \bar{a}\ ;\quad rp\left\{-\ell n\left(1 - \frac{\alpha}{|\bar{a}|}\right)\right\}. \tag{2.11}$$

II. (rp to ap). Suppose that (2.5) holds. Then

$$a \simeq \bar{a}\ ;\quad ap\{|\bar{a}|(e^{\alpha}-1)\}. \tag{2.12}$$

Both rules apply whether the numbers a and $\bar{a}$ are real or complex.

Perhaps at this stage it should be pointed out that although exponential and logarithmic functions appear frequently during the analytic derivation of error bounds, in applications there is never a need to compute exponentials and logarithms in evaluating the final bounds. This important feature is explained in [5].

3. REAL COMPUTER ARITHMETIC

3.1. Notation

In practical computations we have to contend with three types of error: inherent errors in the data, truncation errors arising from the use of analytic approximations (such as finite-difference representations), and abbreviation errors incurred in implementing arithmetic operations. In this lecture we shall be concerned only with the first and third types of error.

For real floating-point arithmetic we denote by γ the maximum abbreviation error, in rp form, of all internal arithmetic operations. These operations include addition, subtraction, multiplication and division. We call γ the *working relative precision* (wrp) associated with the computing facility under consideration.

For example, if the abbreviation mode is chopping and the accumulator register has at least one guard digit, then $\gamma = r^{1-d}$ where r is the internal radix and d is the number of digits (in base r) in the floating-point mantissae. Similarly, if the abbreviation mode is properly executed rounding, then $\gamma = \frac{1}{2} r^{1-d}$. These values of γ are exactly what might be expected, and a rigorous proof is easily supplied [4].

On certain computers, especially ones having no form of guard digit in the accumulator register, it is not possible to assign a realistic value to γ to cover all operations. The analysis that follows can be modified easily to cover these cases; this will be described elsewhere [6].

So far we have added bars to symbols simply to indicate "an approximation". Henceforth we restrict the use of bars to denote stored values. For example, the stored approximations to given numbers a and b are denoted by $\bar{a}$ and $\bar{b}$, respectively, and the stored approximation to ab is denoted by $\overline{ab}$. Usually, of course, $\overline{ab}$ differs from $\bar{a}\bar{b}$.

3.2. Products and quotients

Assume that

$$a_j \simeq \bar{a}_j \; ; \; rp(\alpha_j), \quad j = 1, 2, \ldots, n, \tag{3.1}$$

where a_j and $\bar{a}_j$ are nonzero. Then by V of §2.2 we have

$$a_1 a_2 \simeq \bar{a}_1 \bar{a}_2 \; ; \; rp(\alpha_1 + \alpha_2).$$

From the definition of γ it follows that

$$\bar{a}_1\bar{a}_2 \simeq \overline{a_1a_2} \ ; \quad rp(\gamma).$$

Hence by use of VI of §2.2 we have

$$a_1a_2 \simeq \overline{a_1a_2} \ ; \quad rp(\alpha_1+\alpha_2+\gamma).$$

Continuing this process we arrive at the desired relation between the product $a_1a_2 \ldots a_n$ and its stored value $\overline{a_1a_2 \ldots a_n}$, given by

$$a_1a_2 \ldots a_n \simeq \overline{a_1a_2 \ldots a_n} \ ; \quad rp\{\alpha_1+\alpha_2+\ldots+\alpha_n+(n-1)\gamma\}. \tag{3.2}$$

If any multiplication operation is replaced by a division, then the error bound remains the same:

$$a_1a_2^{\pm 1} \ldots a_n^{\pm 1} \simeq \overline{a_1a_2^{\pm 1} \ldots a_n^{\pm 1}} \ ; \quad rp\{\alpha_1+\alpha_2+\ldots+\alpha_n+(n-1)\gamma\}. \tag{3.3}$$

3.3. Sums

In this subsection it is more convenient to assume that the data are supplied in ap form; thus

$$a_j \simeq \bar{a}_j \ ; \quad ap(\alpha_j), \quad j = 1,2,\ldots,n. \tag{3.4}$$

(Of course, if we were given (3.1) instead, then we would apply the conversion rule (2.12) to find the corresponding ap form.)

The effect of the inherent errors on the sum is expressed by the simple formula

$$a_1+a_2+\ldots+a_n \simeq \bar{a}_1+\bar{a}_2+\ldots+\bar{a}_n \ ; \quad ap(\alpha_1+\alpha_2+\ldots+\alpha_n). \tag{3.5}$$

The cumulative effect of the abbreviation errors depends on the manner in which the sum is computed. We shall assume that the terms are summed in the given order, each partial sum being abbreviated to $rp(\gamma)$ as soon as it is formed. On the first step, the stored value $\overline{a_1+a_2}$ is related to $\bar{a}_1 + \bar{a}_2$ by the formula

$$\bar{a}_1 + \bar{a}_2 \simeq \overline{a_1+a_2} \ ; \quad rp(\gamma).$$

From the conversion rule (2.12) we derive

$$\bar{a}_1 + \bar{a}_2 \simeq \overline{a_1+a_2} \ ; \quad ap\{|\overline{a_1+a_2}|(e^{\gamma}-1)\}.$$

The next step is to add $\bar{a}_3$ and then abbreviate. This increases the rp error by γ and hence the ap error by $|\overline{a_1+a_2+a_3}| \times (e^\gamma-1)$. Continuing the process for subsequent terms, we arrive at

$$\bar{a}_1 + \bar{a}_2 + \ldots + \bar{a}_n \simeq \overline{a_1+a_2+\ldots+a_n}; \quad \mathrm{ap}\{(|\bar{s}_2|+|\bar{s}_3|+\ldots+|\bar{s}_n|)(e^\gamma-1)\}, \tag{3.6}$$

where

$$s_j = a_1 + a_2 + \ldots + a_j.$$

Combination of (3.5) and (3.6) yields the required error bound in the form

$$s_n \simeq \bar{s}_n ; \quad \mathrm{ap}(\rho_n), \tag{3.7}$$

where

$$\rho_n = \alpha_1+\alpha_2+\ldots+\alpha_n+ (|\bar{s}_2|+|\bar{s}_3|+\ldots+|\bar{s}_n|)(e^\gamma-1). \tag{3.8}$$

3.4. Inner products

These can be handled by combining the results for products and sums. For $j = 1,2,\ldots,n$ assume that

$$a_j \simeq \bar{a}_j ; \quad \mathrm{rp}(\alpha_j), \qquad b_j \simeq \bar{b}_j ; \quad \mathrm{rp}(\beta_j), \tag{3.9}$$

and write

$$s_j = a_1b_1 + a_2b_2 + \ldots + a_jb_j.$$

In computing s_n we assume that each product a_jb_j and also each partial inner product s_j is abbreviated to $\mathrm{rp}(\gamma)$ as soon as it is formed. The desired relation between s_n and $\bar{s}_n$ is then seen to be

$$s_n \simeq \bar{s}_n ; \quad \mathrm{ap}(\rho_n), \tag{3.10}$$

where

$$\rho_n = \sum_{j=1}^{n} |\overline{a_jb_j}|(e^{\alpha_j+\beta_j+\gamma}-1) + (\sum_{j=2}^{n} |\bar{s}_j|)(e^\gamma-1). \tag{3.11}$$

4. COMPLEX COMPUTER ARITHMETIC

All of the results for products, quotients, sums and inner products that were obtained for real numbers in §§3.2 to 3.4 carry over to complex numbers, provided that we can assign a meaning to γ, the working relative precision. To solve this problem we first replace the symbol γ for the wrp of real floating-point arithmetic operations by γ_0. We continue to use bars to denote stored values (and not complex conjugates).

4.1. Addition

Let

$$\bar{z}_1 = \bar{x}_1 + i\bar{y}_1, \tag{4.1}$$

$$\bar{z}_2 = \bar{x}_2 + i\bar{y}_2, \tag{4.2}$$

where $\bar{x}_1$, $\bar{y}_1$, $\bar{x}_2$ and $\bar{y}_2$ are real numbers in floating-point form. Then

$$\bar{z}_1 + \bar{z}_2 = (\bar{x}_1+\bar{x}_2) + i(\bar{y}_1+\bar{y}_2).$$

By hypothesis

$$\bar{x}_1 + \bar{x}_2 \simeq \overline{x_1+x_2} \;;\; rp(\gamma_0),$$

$$\bar{y}_1 + \bar{y}_2 \simeq \overline{y_1+y_2} \;;\; rp(\gamma_0).$$

From these relations it follows that

$$\bar{z}_1 + \bar{z}_2 \simeq \overline{z_1+z_2} \;;\; rp(\gamma_0). \tag{4.3}$$

The last step is not trivial. It is justified in [4], pp. 391-392 by conformal mapping, and the condition $\gamma_0 < \frac{1}{2}\pi$ is imposed.

The relation (4.3) represents the desired result. It shows that for addition in $\mathbb{C}$ we may take $\gamma = \gamma_0$, provided that $\gamma_0 < \frac{1}{2}\pi$.

4.2. Multiplication

From (4.1) and (4.2) we have

$$\bar{z}_1\bar{z}_2 = (\bar{x}_1\bar{x}_2-\bar{y}_1\bar{y}_2) + i(\bar{x}_1\bar{y}_2+\bar{x}_2\bar{y}_1).$$

If each of the inner products $\bar{x}_1\bar{x}_2 - \bar{y}_1\bar{y}_2$ and $\bar{x}_1\bar{y}_2 + \bar{x}_2\bar{y}_1$ is accumulated exactly and subsequently abbreviated to $\mathrm{rp}(\gamma_0)$, then evidently the same error bound applies as for addition, that is,

$$\bar{z}_1\bar{z}_2 \simeq \overline{z_1 z_2} \ ; \quad \mathrm{rp}(\gamma_0), \tag{4.4}$$

provided that $\gamma_0 < \frac{1}{2}\pi$.

More usually, $\bar{x}_1\bar{x}_2 - \bar{y}_1\bar{y}_2$ is computed by abbreviating $\bar{x}_1\bar{x}_2$ and $\bar{y}_1\bar{y}_2$ to form $\overline{x_1x_2}$ and $\overline{y_1y_2}$, respectively, and then abbreviating the difference $\overline{x_1x_2} - \overline{y_1y_2}$ to form $\overline{x_1x_2 - y_1y_2}$. From the result of §3.4 we deduce that

$$\bar{x}_1\bar{x}_2 - \bar{y}_1\bar{y}_2 \simeq \overline{x_1x_2 - y_1y_2} \ ; \quad \mathrm{ap}\{(|\overline{x_1x_2}| + |\overline{y_1y_2}| + |\overline{x_1x_2 - y_1y_2}|)(e^{\gamma_0} - 1)\}.$$

Similarly

$$\bar{x}_1\bar{y}_2 + \bar{x}_2\bar{y}_1 \simeq \overline{x_1y_2 + x_2y_1} \ ; \quad \mathrm{ap}\{(|\overline{x_1y_2}| + |\overline{x_2y_1}| + |\overline{x_1y_2 + x_2y_1}|)(e^{\gamma_0} - 1)\}.$$

It will be shown elsewhere that these two relations imply

$$\bar{z}_1\bar{z}_2 \simeq \overline{z_1 z_2} \ ; \quad \mathrm{rp}(\gamma_1), \tag{4.5}$$

where

$$\gamma_1 = \ell n \left\{ 1 - \frac{4}{\sqrt{3}} e^{\gamma_0} (e^{\gamma_0} \ 1) \right\}, \tag{4.6}$$

provided that

$$4e^{\gamma_0}(e^{\gamma_0} - 1) < \sqrt{3}, \tag{4.7}$$

that is, $\gamma_0 < 0.2825\ldots$.

Usually, of course, γ_0 is small, and in these circumstances we have

$$\gamma_1 \sim \frac{4}{\sqrt{3}} \gamma_0 = (2.309\ldots)\gamma_0. \tag{4.8}$$

Therefore if the real and imaginary points of a complex product are computed in the manner indicated at the beginning of the preceding paragraph, then the rp error bound for the product is about 2.3 times the corresponding wrp for real arithmetic. Bearing in mind that a total of six abbreviation errors are incurred in computing the complex product in this manner, we perceive that this is a very satisfactory result.

Another way to carry out complex multiplication would be to use polar forms. Since this procedure depends on the evaluation of trigonometric and inverse trigonometric functions we shall not analyze it here.

4.3. Division

We suppose that the quotient $\bar{z}_1/\bar{z}_2$ is computed from the formula

$$\frac{\bar{z}_1}{\bar{z}_2} \simeq \frac{(\bar{x}_1+i\bar{y}_1)(\bar{x}_2-i\bar{y}_2)}{\bar{x}_2^2+\bar{y}_2^2}.$$

By the multiplication rule of §4.2 we have

$$(\bar{x}_1+i\bar{y}_1)(\bar{x}_2-i\bar{y}_2) \simeq \overline{(x_1+iy_1)(x_2-iy_2)}\ ;\quad rp(\gamma_1), \tag{4.9}$$

where γ_1 is defined by (4.6). Also, by real rp analysis it follows that

$$\bar{x}_2^2 + \bar{y}_2^2 \simeq \overline{x_2^2+y_2^2}\ ;\quad rp(2\gamma_0).$$

The next step is to divide the real and imaginary parts of $\overline{(x_1+iy_1)(x_2-iy_2)}$ by $\overline{x_2^2+y_2^2}$. This introduces a further error of $rp(\gamma_0)$ in each part. Another application of the result of [4], p.391 yields

$$\frac{\overline{(x_1+iy_1)(x_2-iy_2)}}{\bar{x}_2^2+\bar{y}_2^2} \simeq \overline{\left\{\frac{(x_1+iy_1)(x_2-iy_2)}{x_2^2+y_2^2}\right\}}\ ;\quad rp(3\gamma_0), \tag{4.10}$$

provided that $3\gamma_0 < \frac{1}{2}\pi$ — which is certainly the case when (4.7) is satisfied. Dividing both sides of (4.9) by $\bar{x}_2^2 + \bar{y}_2^2$ (§2.2, Property III) and combining the result with (4.10) (§2.2, Property VI) we obtain the required result in the form

$$\frac{\bar{z}_1}{\bar{z}_2} \simeq \overline{\left(\frac{z_1}{z_2}\right)}\ ;\ rp(3\gamma_0+\gamma_1). \tag{4.11}$$

If γ_0 is small, then we have

$$3\gamma_0 + \gamma_1 \sim (5.309\ldots)\gamma_0 \ ;$$

compare (4.8). To appreciate this result it should be observed that a total of eleven abbreviation errors are incurred in computing the complex quotient.

In the special case $\bar{z}_1 = 1$, no complex multiplication need take place. In consequence we have

$$\frac{1}{\bar{z}_2} \simeq \overline{\left(\frac{1}{z_2}\right)} \ ; \ \mathrm{rp}(3\gamma_0), \tag{4.12}$$

provided that $3\gamma_0 < \frac{1}{2}\pi$.

5. CONCLUSIONS

We have determined explicit and strict error bounds for products, quotients, sums and inner products of real floating-point numbers, based on the concept of a working relative precision associated with each computing facility. In order to extend these results to complex numbers whose real and imaginary parts are stored in floating-point form, we showed how to determine the corresponding working relative precision for complex arithmetic operations. As a consequence, some recent error analyses can be extended immediately to the complex number field. This includes, for example, an analysis of the solution of systems of linear algebraic equations by Gaussian elimination.

There has not been time in this lecture to describe the conversion of the error bounds we have obtained into computable form. Listeners who are interested in the procedures should consult [5].

REFERENCES

[1] Brent, R.P. "A Fortran multiple-precision arithmetic package", *ACM Trans. Math. Software* 4, pp. 57-70, 1978.

[2] Brent, R.P. "Algorithm 524. MP, A Fortran multiple-precision arithmetic package", *ACM Trans. Math. Software* 4, pp. 71-81, 1978.

[3] Nickel, K. "Interval-analysis". In *The State of the Art in Numerical Analysis* (D. Jacobs ed.), pp. 193-225. Academic Press, London, 1977.

[4] Olver, F.W.J. "A new approach to error arithmetic", *SIAM J. Numer. Anal.* 15, pp. 368-393, 1978.

[5] Olver, F.W.J. "Further developments of rp and ap error analysis", *IMA J. Numer. Anal.* [In press.]

[6] Olver, F.W.J. "Error bounds for arithmetic operations on computers without guard digits". [Manuscript.]

[7] Olver, F.W.J. and Wilkinson, J.H. "A posteriori error bounds for Gaussian elimination", *IMA J. Numer. Anal.* [In press.]

[8] Stummel, F. "Rounding error analysis of elementary numerical algorithms", *Computing, Suppl.* 2, pp. 169-195, 1980.

[9] Stummel, F. "Perturbation theory for evaluation algorithms of arithmetic expressions", *Math. Comp.* 37, pp. 435-473, 1981.

[10] Wilkinson, J.H. *Rounding Errors in Algebraic Processes.* N.P.L. Notes on Applied Science No. 32, Her Majesty's Stationery Office, London, 1963.

This work was supported by the U.S. Army Research Office, Durham, under Contract DAAG 29-80-C-0032, and by the National Science Foundation under Grant MCS 81-11725.

THE USE OF GREEN'S BOUNDARY FORMULA IN POTENTIAL THEORY

George T. Symm

National Physical Laboratory, Teddington, U.K.

INTRODUCTION

The solution ϕ of Laplace's equation in two dimensions:

$$\nabla^2\phi \equiv \frac{\partial^2\phi}{\partial x^2} + \frac{\partial^2\phi}{\partial y^2} = 0 \tag{1}$$

in a plane domain D, bounded internally or externally by one or more closed contours C, also satisfies Green's third identity:

$$\int_C \phi'(q)\log|q-p|dq - \int_C \phi(q)\log'|q-p|dq = \theta(p)\phi(p), \tag{2}$$

where p and q are vector variables specifying points of the plane and points on C respectively, the prime denotes differentiation at the point q along the normal to C directed <u>into</u> the domain D and dq is the differential increment of C at q. When $p \in D$, the parameter θ has the value 2π and ϕ is represented as the sum of two logarithmic potentials - a simple-layer potential with source density $\phi'/2\pi$ and a double-layer potential with source density $-\phi/2\pi$. When $p \in C$, the first of these potentials remains continuous but the second has a jump discontinuity such that θ takes the value of the "internal" angle of C at the point p, i.e. the angle <u>in D</u> between the two tangents to C which meet at p. Denoting this angle by $\Omega(p)$, which has the value π if C is smooth at p, equation (2) becomes

$$\int_C \phi'(q)\log|q-p|dq - \int_C \phi(q)\log'|q-p|dq = \Omega(p)\phi(p), \quad p \in C, \tag{3}$$

H. Werner et al. (eds.), Computational Aspects of Complex Analysis, 293–308.

which we call "Green's boundary formula" (Jaswon and Symm, 1977).

In this paper we show how each of the common boundary-value problems for Laplace's equation (1) can be formulated in terms of integral equations by means of Green's boundary formula (3) and how they may thereby be solved numerically.

When the domain D is multiply-connected, we denote its outer boundary by C_0 and its internal boundaries, supposing there are m of them, by $C_1, C_2, \ldots, C_m$. Then the complete boundary is given by

$$C = C_0 + C_1 + \ldots + C_m, \tag{4}$$

from which C_0 may be omitted if the domain is unbounded.

THE DIRICHLET PROBLEM

In the Dirichlet problem for Laplace's equation, the function ϕ is prescribed everywhere on the boundary C. In this case, Green's boundary formula (3) becomes an integral equation of the first kind for ϕ':

$$\int_C \phi'(q)\log|q-p|dq = f(p), \quad p \in C, \tag{5}$$

where

$$f(p) = \Omega(p)\phi(p) + \int_C \phi(q)\log'|q-p|dq. \tag{6}$$

If the domain D is finite (an interior problem), the integral equation (5) has generally (for sufficiently smooth C and ϕ) a unique solution $\phi'(q)$, which satisfies the Gauss condition

$$\int_C \phi'(q)dq = 0. \tag{7}$$

Substitution of this solution, together with the prescribed boundary values $\phi(q)$, back into Green's third identity (2), with $\theta(p) = 2\pi$, yields the unique solution of the Dirichlet problem at any point $p \in D$.

The only exceptional case arises when the outer boundary C_0 is a so-called Γ-contour, i.e. a contour which has unit transfinite diameter. In this case, the equation

$$\int_{C_0} \lambda(q)\log|q-p|dq = 0, \quad p \in C_0, \tag{8}$$

has a non-trivial solution λ, which, as a source density, generates a logarithmic potential with the constant value zero everywhere inside C_0. In particular, this potential is zero on any internal boundaries $C_1, C_2, \ldots, C_m$, whence equation (8) is valid not only on C_0 but for all $p \in C$. Hence an arbitrary multiple of λ (on C_0) may be added to any solution ϕ' of equation (5). That equation (5) still has a solution in this case is easily shown (Jaswon and Symm, 1977) and this solution may be made unique by imposing the Gauss condition (7) upon it (cf. Christiansen, 1975; Wendland, 1980). Alternatively, since there is only one Γ-contour in any set of geometrically similar contours (Jaswon, 1963), this case may be completely avoided by making a simple change of scale.

If the domain D is unbounded (an exterior problem), equation (5) again has a unique solution $\phi'(q)$, in general, but this solution need not satisfy the Gauss condition (7). Indeed, substitution of this solution, together with the prescribed boundary values $\phi(q)$, back into Green's third identity (2), yields a potential ϕ such that

$$\phi(p) \to A \log r + O(r^{-1}) \text{ as } r = |p| \to \infty, \tag{9}$$

where

$$A = \frac{1}{2\pi}\int_C \phi'(q)dq. \tag{10}$$

If, as is more usually the case, ϕ is required to be bounded (i.e. to have O(1) behaviour) at infinity, then a constant term k must be added to the left-hand side of equation (5), thus:

$$\int_C \phi'(q)\log|q-p|dq + k = f(p), \quad p \in C, \tag{11}$$

and this equation solved for ϕ' and k subject to the condition

$$\int_C \phi'(q)dq = 0. \tag{12}$$

Correspondingly, Green's third identity (2) becomes

$$\int_C \phi'(q)\log|q-p|dq - \int_C \phi(q)\log'|q-p|dq + k = \theta(p)\phi(p), \tag{13}$$

from which, once equations (11) and (12) have been solved, the required solution ϕ may be obtained at any point $p \in D$. Note that this latter formulation is valid even when the internal boundary C is a Γ-contour, in which case equation (5) could fail to have a solution.

THE NEUMANN PROBLEM

In the Neumann problem, the normal derivative ϕ' is prescribed everywhere on the boundary C, except possibly at a finite number of discrete points such as corners. In this case, Green's boundary formula (3) becomes an integral equation of the second kind for ϕ:

$$\Omega(p)\phi(p) + \int_C \phi(q)\log'|q-p|dq = g(p), \quad p \in C, \tag{14}$$

where

$$g(p) = \int_C \phi'(q)\log|q-p|dq. \tag{15}$$

If the domain D is finite, the homogeneous form of equation (14), viz.

$$\Omega(p)\phi(p) + \int_C \phi(q)\log'|q-p|dq = 0, \quad p \in C, \tag{16}$$

has the obvious non-trivial solution $\phi = 1$, since, by the Cauchy-Riemann equations,

$$\int_C \log'|q-p|dq = -\int_C \frac{d\arg(q-p)}{dq}dq = -\Omega(p). \tag{17}$$

Correspondingly, the transpose homogeneous equation

$$\Omega(p)\lambda(p) + \int_C \lambda(q)\log'|p-q|dq = 0, \quad p \in C, \tag{18}$$

has a non-trivial solution $\lambda(q)$ and, from the Fredholm theory of integral equations, equation (14) has a solution only if its right-hand side g is orthogonal to λ on C, i.e. only if

$$\int_C g(p)\lambda(p)dp = 0. \tag{19}$$

Now we note that the left-hand side of equation (18) represents the inward normal derivative at the point $p \in C$, assuming C to be smooth at this point, of the simple-layer potential

$$V(p) = \int_C \lambda(q)\log|q-p|dq. \tag{20}$$

It follows that V is constant in D, whence, from the definition (15) of g, condition (19) reduces to the usual Gauss condition:

$$\int_C \phi'(q)dq = 0. \tag{21}$$

Only if the prescribed normal derivative satisfies this condition does equation (14), for the interior Neumann problem, have a solution and in this case the solution is not unique since an arbitrary constant may be added to it. A unique solution $\phi(q)$ of equation (14) may be defined by imposing an appropriate auxiliary condition, e.g.

$$\int_{C_0} \phi(q)dq = 0, \qquad (22)$$

where C_0 is the outer boundary of the domain as above. Substitution of this solution, together with the prescribed boundary values $\phi'(q)$, back into Green's third identity (2) yields the solution of the Neumann problem at any point $p \in D + C$.

If the domain D is unbounded, equation (14) always has a unique solution $\phi(q)$ which, together with the prescribed boundary values $\phi'(q)$, generates, via equation (2), a potential ϕ with the asymptotic behaviour given by expressions (9) and (10). In this case, the value of A is determined by the prescribed boundary condition and cannot be forced to zero; an arbitrary constant may be added to ϕ if required, as in equation (13), the same constant being added to g on the right-hand side of equation (14).

THE MIXED PROBLEM

In the mixed boundary-value problem, the function ϕ is prescribed on part C_D of the boundary C and its normal derivative ϕ' is prescribed on the remainder C_N. In this case, Green's boundary formula (3), for which we now introduce the concise notation:

$$I_1[\phi']_C - I_2[\phi]_C = \Omega(p)\phi(p), \quad p \in C, \qquad (23)$$

reduces to a pair of coupled integral equations:

$$\left.\begin{aligned} I_1[\phi']_{C_D} - I_2[\phi]_{C_N} &= I_2[\phi]_{C_D} - I_1[\phi']_{C_N} + \Omega(p)\phi(p), \quad p \in C_D, \\ I_1[\phi']_{C_D} - I_2[\phi]_{C_N} - \Omega(p)\phi(p) &= I_2[\phi]_{C_D} - I_1[\phi']_{C_N}, \quad p \in C_N, \end{aligned}\right\} \qquad (24)$$

for ϕ on C_N and ϕ' on C_D. These equations have a unique solution (Hayes and Kellner, 1972) unless C is a Γ-contour (i.e. $C = \Gamma$), in which case the corresponding homogeneous equations:

$$\left.\begin{aligned} & I_1[\phi']_{\Gamma_D} - I_2[\phi]_{\Gamma_N} = 0, \quad p \in \Gamma_D, \\ & I_1[\phi']_{\Gamma_D} - I_2[\phi]_{\Gamma_N} - \Omega(p)\phi(p) = 0, \quad p \in \Gamma_N, \end{aligned}\right\} \tag{25}$$

have a non-trivial solution:

$$\left.\begin{aligned} & \phi(p) = u(p), \quad p \in \Gamma_N, \\ & \phi'(p) = u'(p) + \lambda(p), \; p \in \Gamma_D, \end{aligned}\right\} \tag{26}$$

where u satisfies Laplace's equation

$$\nabla^2 u(p) = 0, \quad p \in D, \tag{27}$$

with boundary conditions

$$\left.\begin{aligned} & u(p) = 0, \; p \in \Gamma_D, \\ & u'(p) = -\lambda(p), \; p \in \Gamma_N, \end{aligned}\right\} \tag{28}$$

where λ is a non-trivial solution of

$$\int_\Gamma \lambda(q)\log|q-p|dq = 0, \; p \in \Gamma. \tag{29}$$

Proof that formulae (26) give a solution of equations (25) follows by simple substitution, noting that u satisfies equation (23) and boundary conditions (28), while λ satisfies equation (29), i.e. $I_1[\lambda]_\Gamma = 0$. As in the interior Dirichlet problem, this exceptional case may be avoided by making a simple change of scale - a method adopted also by Hayes and Kellner (1972). The solution of equations (24) and the prescribed boundary values together generate, via equation (2), the solution of the mixed problem at any point $p \in D + C$.

THE ROBIN PROBLEM

In the Robin problem for Laplace's equation, the prescribed boundary condition on C takes the form

$$a\phi + b\phi' = c, \tag{30}$$

where a, b and c may be functions of x and y in general and each may be piecewise continuous on C. Provided that a is not identically equal to zero and that the product ab is never positive on C, this problem, of which those above are particular cases, has a unique solution (Kellogg, 1929; Tsuji, 1959).

In this problem, either ϕ or ϕ' may be eliminated from Green's boundary formula (3) at each point $q \in C$ by means of the boundary condition (30). Then, for $p \in C$, equation (3) becomes an integral equation (or a system of coupled integral equations) for those boundary values of ϕ and ϕ' which are not eliminated.

In general, by solving this integral equation and substituting the solution, complemented by the boundary condition (30), back into formula (2), we may obtain the value of ϕ at any point $p \in D + C$.

DISCRETISATION

The methods formulated above may be implemented numerically by dividing the boundary C into N intervals in each of which ϕ and ϕ' are approximated by constants. Denoting these constants by

$$\phi_i \text{ and } \phi_i', \quad i = 1,2,\ldots,N, \tag{31}$$

and applying equation (3) at one "nodal" point q_i in each interval of C, we obtain

$$\sum_{j=1}^{N} \phi_j' \int^{(j)} \log|q-q_i|\,dq - \sum_{j=1}^{N} \phi_j \int^{(j)} \log'|q-q_i|\,dq - \Omega(q_i)\phi_i = 0, \quad i = 1,2,\ldots,N, \tag{32}$$

where $\int^{(j)}$ denotes integration over the jth interval of C. From these equations we eliminate one of the constants (31) in each interval by applying the relevant boundary condition at the corresponding nodal point. In particular, when the Robin boundary condition (30) is applicable, we set

$$\phi = \frac{c}{a} - \frac{b}{a}\phi' \tag{33}$$

when a is greater in magnitude than b at the nodal point and

$$\phi' = \frac{c}{b} - \frac{a}{b}\phi \tag{34}$$

otherwise. We thus obtain a system of N simultaneous linear algebraic equations (in N unknowns) whose solution provides, via equation (2) and the boundary data, the approximation

$$[\sum_{j=1}^{N}(\phi_j'\int^{(j)}\log|q-p|dq - \phi_j\int^{(j)}\log'|q-p|dq)]/\theta(p) \qquad (35)$$

to $\phi(p)$ at any point $p \in D + C$.

In order to evaluate the coefficients in equations (32) and (35), we approximate each interval of C by the two chords which join its end points to the nodal point within it. Then all the integrations can be carried out analytically (Symm and Pitfield, 1974; Jaswon and Symm, 1977). If, for example, AB is one chord of the jth interval of the boundary C, the contribution from AB to the coefficient of ϕ_j' in expression (35) is

$$a \cos \alpha (\log a - \log b) + h (\log b - 1) + a \psi \sin \alpha, \qquad (36)$$

where h denotes the length of AB, a and b are the lengths of pA and pB respectively and α and ψ are the angles BAp and BpA respectively (Figure 1). The corresponding contribution to the coefficient of ϕ_j in expression (35) is simply $-\psi$, from an application of the Cauchy-Riemann equations as in equations (17). It follows that, in equation (32), Ω becomes the "internal" angle, as defined earlier, at the nodal point q_i, of an approximation to the boundary C by a polygon or polygons. (For a multiply-connected domain, whose boundary is given by equation (4), there are m + 1 such polygons.)

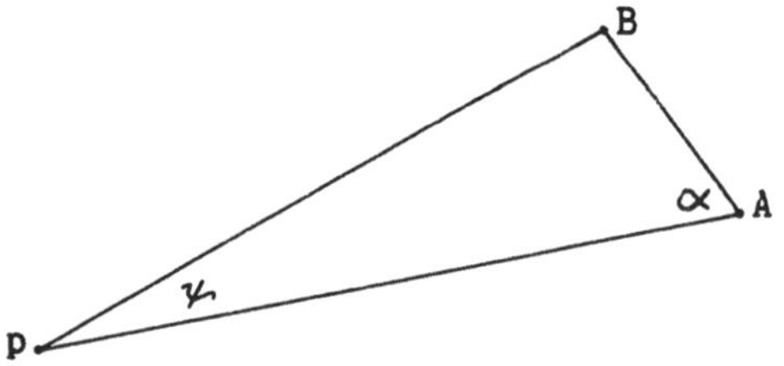

Figure 1. Geometry for analytic integration

In the interior Neumann problem, equations (32) are supplemented by a discrete form of equation (22), viz.

$$\sum_{j=1}^{N_0}\phi_j h_j = 0, \qquad (37)$$

where N_0 is the number of intervals on C_0 and h denotes interval length measured on the corresponding approximating polygon. In this case, the resulting N + 1 equations are solved in the least-squares sense.

In exterior problems, when ϕ is required to be bounded at infinity, equations (32) are modified by the addition of a constant k to the left-hand side, as in equation (13). In the case of a Neumann problem the value of this constant is prescribed. Otherwise k is an extra unknown whose value may be obtained, as part of the boundary solution, by adding an extra equation

$$\sum_{j=1}^{N} \phi'_j h_j = 0, \tag{38}$$

the discrete form of equation (12), to the system. Correspondingly, for the evaluation of $\phi(p)$, the constant k is added to the numerator of expression (35).

In general-purpose computer programs written at NPL (Symm and Pitfield, 1974; Symm, 1980) the solution of the simultaneous linear algebraic equations is obtained by means of library routines which determine the unique solution of the square non-singular system, in the general case, or the unique least squares solution of the rectangular system, in the interior Neumann case, by the method of Peters and Wilkinson (1970). In the Robin problem, the boundary solution is completed by a further application, at each nodal point, of equation (33) or (34) as appropriate. Also the boundary of the domain is scaled down so that its maximum diameter is not greater than unity, this being sufficient to ensure that there is no possibility of non-uniqueness due to C, or C_0, being a Γ-contour.

EXAMPLES

There now follow a number of numerical examples illustrating the methods described above.

Problem 1: An interior Dirichlet problem

To solve Laplace's equation (1) in the L-shaped domain D obtained by removing the square $2 < x,y < 4$ from the square $0 \leqslant x,y \leqslant 4$, given that $\phi = x$ on the boundary C of this domain.

This problem has, of course, the simple analytic solution $\phi = x$ throughout the domain D.

Typical numerical results, for uniform subdivisions of the boundary C with nodal points at the mid-points of the intervals, are presented in Table 1.

Table 1. Results of Problem 1

x	N=16	N=32	N=64
	Values of ϕ at points on y = x		
0.000	- 0.060	- 0.030	- 0.015
1.000	1.006	1.002	1.001
1.500	1.516	1.505	1.502
1.900	1.945	1.917	1.906
2.000	2.017	2.010	2.005

We note that the re-entrant corner of the domain, at the point with coordinates (2.0,2.0), causes no particular difficulty in this example.

Problem 2: An exterior Neumann problem

To solve Laplace's equation (1) in the unbounded domain D outside the square $-1 < x,y < 1$, given that

$$\phi' = \frac{1}{x^2 + y^2} \tag{39}$$

on the square internal boundary C.

If we assume that ϕ has asymptotic behaviour of the form (9) at infinity, this problem has the analytic solution

$$\phi = \log r = \frac{1}{2}\log(x^2 + y^2). \tag{40}$$

Numerical results, for uniform subdivisions of the boundary C, are presented in Table 2, where they may be compared with the analytic solution, which reduces to $\phi = \log x$ on $y = 0$.

Table 2. Results of Problem 2

x	N=16	N=32	N=64	Anal.
	Values of ϕ at points on y = 0			
1.000	0.017	0.003	0.001	0.000
2.000	0.700	0.695	0.694	0.693
4.000	1.396	1.389	1.387	1.386
8.000	2.093	2.083	2.080	2.079
16.000	2.791	2.777	2.774	2.773

These results show typical $O(h^2)$ convergence (Stiller, 1979).

This and the previous example are taken from the report by Symm and Pitfield (1974), which describes an NPL Algorithms Library routine for the solution of Dirichlet, Neumann and mixed problems in arbitrary plane domains.

Problem 3: A Neumann problem in a multiply-connected domain

To solve Laplace's equation (1) in the domain D bounded externally by the circle

$$C_0: \quad x^2 + y^2 = 16, \tag{41}$$

and internally by the circles

$$C_1: \quad x^2 + (y-2)^2 = 1, \qquad C_2: \quad x^2 + (y+2)^2 = 1, \tag{42}$$

given that $\phi' = x'$ everywhere on the boundary C.

This problem has, again, the obvious analytic solution $\phi = x$, satisfying the uniqueness condition (22).

Numerical results, corresponding to subdivisions of each boundary contour into N/3 equal intervals, are shown in Table 3.

Table 3. Results of Problem 3

Values of ϕ at points on y = 0

x	N=12	N=24	N=48	N=96
0.500	0.499	0.481	0.494	0.499
1.000	0.963	0.961	0.989	0.997
1.500	1.397	1.443	1.484	1.496
2.000	1.804	1.924	1.979	1.995
2.500	2.183	2.401	2.474	2.493
3.000	2.530	2.867	2.968	2.992
3.500	2.841	3.314	3.458	3.491
4.000	3.524	3.863	3.964	3.991

It might have been better, in this example, to have divided the outer boundary into more intervals than each of the relatively small inner boundaries. However, the results given here suffice to illustrate the effectiveness of the method for an interior Neumann problem, involving the extra equation (37) and the least-squares solution of the linear system.

Problem 4: An interior mixed problem with a discontinuity

To solve Laplace's equation (1) in the quadrant of the unit disc bounded by the lines

$$C: \quad x = 0,\ y = 0 \text{ and } x^2 + y^2 = 1,\ x > 0,\ y > 0, \tag{43}$$

given the boundary conditions

$$\left.\begin{array}{l} \phi = 1 \text{ on the circular arc,} \\ \phi = 0 \text{ on } y = 0, \\ \phi' = 0 \text{ on } x = 0. \end{array}\right\} \tag{44}$$

This problem has the analytic solution

$$\phi = \frac{2}{\pi} \arctan\{2y/(1 - x^2 - y^2)\} \tag{45}$$

with a boundary discontinuity at the point (1.0,0.0).

Numerical results, for various subdivisions of the boundary, are shown in Table 4, where they may be compared with the analytic solution. For the first four sets of results, the circular arc is divided into N/2 equal intervals with each of the straight sides divided into N/4 equal intervals. The final set of results, marked *, corresponds to a grading of the intervals on the circular arc and on the x-axis so that they decrease in length towards the point of discontinuity of ϕ.

Table 4. Results of Problem 4

		Values of ϕ at selected points					
x	y	N=8	N=16	N=32	N=64	N=64*	Anal.
0.1	0.1	0.1546	0.1266	0.1258	0.1272	0.1277	0.1282
0.5	0.1	0.1372	0.1658	0.1672	0.1676	0.1680	0.1680
0.9	0.1	0.5594	0.5459	0.5323	0.5323	0.5333	0.5335

The slow convergence of the results in this example may be attributed to the presence of the boundary discontinuity. The grading of the intervals, for N = 64, yields a significant improvement over the more uniform boundary subdivision. Details of this grading are included in the report by Symm (1980), which describes a suite of three NPL Algorithms Library routines for the solution of the Robin problem in a simply-connected domain, and from which this example and the next one are taken.

<u>Problem 5</u>: An interior mixed problem with a singularity

To solve Laplace's equation (1) in the rectangular domain bounded by the lines

$$C: \quad x = 7,\ y = 7,\ x = -7 \text{ and } y = 0, \tag{46}$$

given the boundary conditions

$$\left.\begin{aligned} &\phi = 1000 \quad \text{on the side } x = 7, \\ &\phi = 500 \quad \text{for } x < 0 \text{ on } y = 0, \\ &\phi' = 0 \quad \text{on the remainder of C.} \end{aligned}\right\} \tag{47}$$

This problem has no simple explicit solution but the numerical solution of Whiteman and Papamichael (1971), obtained by conformal transformation methods, is essentially analytic. This solution has a boundary singularity, in the form of an infinite discontinuity in the normal derivative ϕ' at the origin.

Numerical results, for various subdivisions of the boundary, are shown in Table 5, where they may be compared with the "analytic" solution. The first three sets of results correspond to uniform subdivisions of the boundary. The results for N = 48 marked * correspond to a slight modification of the uniform subdivision, with the 8 equal intervals on the negative x-axis part of the boundary replaced by 8 graded intervals - decreasing in length towards the singularity at the origin. The final results, marked **, correspond to a similar subdivision with 16 graded intervals on the negative x-axis, 12 equal intervals on the positive x-axis and 4, 8 and 8 equal intervals respectively on the sides x = 7, y = 7 and x = - 7.

Table 5. Results of Problem 5
Values of ϕ at selected points

x	y	N=12	N=24	N=48	N=48*	N=48**	Anal.
-1.0	1.0	581	569	565	563	562	562
0.0	1.0	641	621	612	606	604	604
1.0	1.0	706	685	677	671	670	670
1.0	0.0	691	675	666	657	656	656
2.0	0.0	751	741	735	729	728	728
4.0	0.0	855	850	847	845	844	844

The slow convergence of the first three sets of results in this example may be attributed to the presence of the boundary singularity. Grading the intervals, following Symm (1980), yields a significant improvement very simply. Alternatively, the singular behaviour of the solution at the origin can be incorporated into the integral equation formulation of the problem (Symm, 1973; Papamichael and Symm, 1975); this yields more accurate results than interval grading but involves a little extra effort.

Problem 6: A Robin problem in a multiply-connected domain

To solve Laplace's equation (1) in the domain D bounded externally by the circle

$$C_0: \; x^2 + y^2 = 4, \tag{48}$$

and internally by the circle

$$C_1: \; x^2 + y^2 = 1, \tag{49}$$

given the boundary conditions

$$\phi - 2\phi' = 0 \quad \text{on } C_0 \tag{50}$$

and

$$\phi - \phi' = x \quad \text{on } C_1. \tag{51}$$

This problem has the analytic solution

$$\phi = \frac{x}{2(x^2+y^2)}. \tag{52}$$

This problem is solved numerically by an extension of the NPL Algorithms Library routines for the Robin problem (Symm, 1980) from simply- to multiply-connected domains. Indeed, the first of these routines, which is a boundary discretisation routine, carries over unchanged and makes data input particularly simple in this example, where each boundary is a single circular arc defined by its starting point, (2,0) or (1,0) as the case may be, and its centre (0,0). Each of these arcs is divided into N/2 equal intervals and corresponding numerical results may be compared with the analytic solution in Table 6.

Table 6. Results of Problem 6

		Values of ϕ at selected points				
x	y	N=16	N=32	N=64	N=128	Anal.
1.0	0.0	0.481	0.497	0.499	0.500	0.500
1.2	0.0	0.396	0.412	0.416	0.416	0.417
1.4	0.0	0.341	0.354	0.356	0.357	0.357
1.6	0.0	0.298	0.309	0.312	0.312	0.313
1.8	0.0	0.264	0.275	0.277	0.278	0.278
2.0	0.0	0.251	0.250	0.250	0.250	0.250
0.0	1.0	0.000	0.000	0.000	0.000	0.000
0.2	1.0	0.141	0.095	0.096	0.096	0.096
0.4	1.0	0.171	0.168	0.171	0.172	0.172
0.6	1.0	0.199	0.220	0.220	0.220	0.221
0.8	1.0	0.226	0.241	0.243	0.244	0.244
1.0	1.0	0.239	0.248	0.249	0.250	0.250
1.2	1.0	0.238	0.244	0.245	0.246	0.246
1.4	1.0	0.233	0.235	0.236	0.236	0.236
1.6	1.0	0.224	0.222	0.225	0.225	0.225

Not unnaturally, the worst results occur (near the inner boundary) when N is small. This example was originally run as a test case prior to the solution of a heat conduction problem in a turbine blade with cooling holes.

CONCLUSION

In this paper we have seen how Green's boundary formula may be used to formulate the common boundary-value problems for Laplace's equation in terms of integral equations and how these problems may thereby be solved numerically. We note here that less common boundary-value problems, such as problems in multi-media with interfaces, are equally amenable to this form of solution (Symm, 1977).

Finally, we observe that more accurate numerical results may be obtained from the same formulation by using more accurate representations for ϕ and ϕ' than the piecewise constant approximations (31). In particular, by introducing shape functions as used in finite element methods, we obtain a hierarchy of boundary element methods (Brebbia, 1978; Brebbia and Walker, 1980) based upon the integral equation formulation.

REFERENCES

Brebbia, C. A. (1978) The Boundary Element Method for Engineers. Pentech Press, London.

Brebbia, C. A. and Walker, S. (1980) Boundary Element Techniques in Engineering. Newnes-Butterworths, London.

Christiansen, S. (1975) Integral Equations Without a Unique Solution Can Be Made Useful for Solving Some Plane Harmonic Problems. J. Inst. Maths. Applics., 16, pp. 143-159.

Hayes, J. and Kellner, R. (1972) The Eigenvalue Problem for a Pair of Coupled Integral Equations Arising in the Numerical Solution of Laplace's Equation. SIAM J. Appl. Math., 22, 3, pp. 503-513.

Jaswon, M. A. (1963) Integral Equation Methods in Potential Theory. I. Proc. Roy. Soc. (A), 275, pp. 23-32.

Jaswon, M. A. and Symm, G. T. (1977) Integral Equation Methods in Potential Theory and Elastostatics. Academic Press, London.

Kellogg, O. D. (1929) Foundations of Potential Theory. Springer-Verlag, Berlin.

Papamichael, N. and Symm, G. T. (1975) Numerical techniques for two-dimensional Laplacian problems. Comp. Methods Appl. Mech. Eng., 6, pp. 175-194.

Peters, G. and Wilkinson, J. H. (1970) The Least Squares Problem and Pseudo-Inverses. Computer J., 13, 3, pp. 309-316.

Stiller, S. (1979) Konvergenzverhalten bei der numerischen Lösung von Randwertproblemen mittels eines Doppelschichtpotential-Ansatzes. Computing, 21, pp.233-243.

Symm, G. T. (1973) Treatment of singularities in the solution of Laplace's equation by an integral equation method. NPL Report NAC 31.

Symm, G. T. (1977) Practical applications of an integral equation method for the solution of Laplace's equation. Comput. Electr. Eng., 4, pp. 167-170.

Symm, G. T. (1980) The Robin Problem for Laplace's Equation. NPL Report DNACS 32/80.

Symm, G. T. and Pitfield, R. A. (1974) Solution of Laplace's Equation in Two Dimensions. NPL Report NAC 44.

Tsuji, M. (1959) Potential Theory in Modern Function Theory. Maruzen, Tokyo.

Wendland, W.L. (1980) On Galerkin Collocation Methods for Integral Equations of Elliptic Boundary Value Problems. In "Numerical Treatment of Integral Equations", edited by J.Albrecht and L.Collatz, Birkhäuser Verlag, Basel, pp. 244-275.

Whiteman, J.R. and Papamichael, N. (1971) Numerical Solution of Two Dimensional Harmonic Boundary Problems Containing Singularities by Conformal Transformation Methods. Brunel University Department of Mathematics Report TR/2.

CHEBYSHEV APPROXIMATION ON THE UNIT DISK

Lloyd N. Trefethen

Courant Institute of Mathematical Sciences
New York University

Abstract. We consider the problem of rational Chebyshev approximation of an analytic function on the unit disk, and survey known results related to nearly-circular error curves and to the Carathéodory-Fejér (CF) method for near-best approximation. The real CF (or Chebyshev-CF) method for approximation of a continuous real function on an interval is also described.

1. THE PROBLEM

This paper will present a relatively easygoing account of some recent developments in complex Chebyshev approximation that relate to "the CF method", to "nearly circular error curves", and to "the AAK theory". This should be a good introduction to the second lecture of Gutknecht in this volume [4], where the *CF table* and related matters are studied carefully. Also closely related are the third lecture of Henrici [10], where the ideas described here are applied to analyze the asymptotic behavior of best approximations, and the lectures of Meinguet [12], where the AAK theory is developed in an elegant way at a higher level. Most of the material presented here appears in greater detail in [16] and [17].

Let S be the unit circle $|z| = 1$, D the unit disk $|z| < 1$, and P_m the set of complex polynomials of degree at most m. If f is a function continuous on $\overline{D}$ and analytic in D, it is natural to ask, how well can f be approximated on $\overline{D}$ by a polynomial $p \in P_m$ with respect to the supremum norm? Since

H. Werner et al. (eds.), Computational Aspects of Complex Analysis, 309–323.

f and p are analytic, the maximum modulus principle ensures that it is enough to consider the boundary circle, so we define

$$\|\phi\| = \sup_{z\epsilon S} |\phi(z)| .$$

The *polynomial Chebyshev approximation problem* is this: find a *best approximation (BA)* $p^* \epsilon P_m$ to f such that

$$\|f - p^*\| = \inf_{p\epsilon P_m} \|f - p\| .$$

More generally, we can consider approximation by rational functions of type (m,n) that are constrained to have no poles in $\overline{D}$. Let R_{mn} be the set of such functions. The *rational Chebyshev approximation problem* is then: find a BA $r^* \epsilon R_{mn}$ such that

$$\|f - r^*\| = \inf_{r\epsilon R_{mn}} \|f - r\| .$$

We let E^*, or E^*_{mn}, denote this infimum.

In polynomial approximation, p^* exists and is unique and is characterized by the Kolmogorov criterion, although in practice this characterization is not very useful for computing BAs. In contrast, although r^* exists too, it need *not* be unique for $n > 0$ — a fact first established at this NATO meeting! [8] In the rational case no characterization of BAs is available, and no very satisfactory algorithms for their computation are known. In fact the CF method that we will describe, although in principle it delivers only a near-best approximation, often comes closer to best than can practicably be achieved by other means. For details about the general theory of complex Chebyshev approximation, see the first paper of Gutknecht in this volume [3].

Unlike real Chebyshev approximations, complex BAs on the unit disk are not very important for the construction of function evaluation procedures for computers, perhaps because it is rare that a function defined on all of $\mathbb{C}$ can be reduced to a representation on a disk. However, there are other applications in which the unit circle comes into its own, particularly in the areas of *linear systems theory* [11] and *digital signal processing* [14]. The reason is that both of these problems involve linear processes with constant coefficients that act on a discrete time variable; they are therefore naturally analyzed by Fourier methods, but because the time variable is discrete, its dual variable has a bounded domain, which is conveniently reduced to S through an application of the *z-transform*. The paper of Meinguet in this volume is motivated by systems theory applications [13], while for applications of CF ideas to digital filter design, see [6].

2. THE NEAR-CIRCULARITY PHENOMENON

Our approximation problem has a simple geometric interpretation. Given f and an approximation r, consider the image $(f-r)(S)$, which is called the *error curve* for r. The Chebyshev approximation problem is obviously equivalent to the following: *find* r^* *so that the error curve can be contained in a circle about* 0 *of minimal radius.* See Fig. 1.

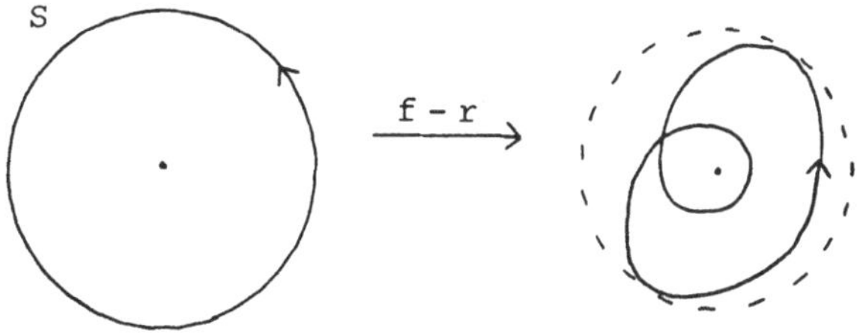

Figure 1. The error curve

A priori, we do not know much about what the error curve for r^* will look like, except that it must touch the boundary circle at at least $m+2$ points. But when best approximations are computed numerically, a remarkable fact emerges: their error curves are often very nearly perfect circles with winding number $m+n+1$. For illustration, consider the plots shown in Fig. 2. In Fig. 2a, the error curve for the type (1,1) Padé approximant to e^z is plotted, and it is evidently a closed loop with winding number $m+n+1 = 3$. Fig. 2b shows on the same scale the error curve for the BA r^*. This curve also has winding number 3, but one can no longer see this, because its modulus varies as z traverses S by less than 1%. In approximation of type (2,1), this figure becomes 0.01%, and it decreases rapidly further as m and n are increased.

The question is, how can this near-circularity phenomenon be accounted for?

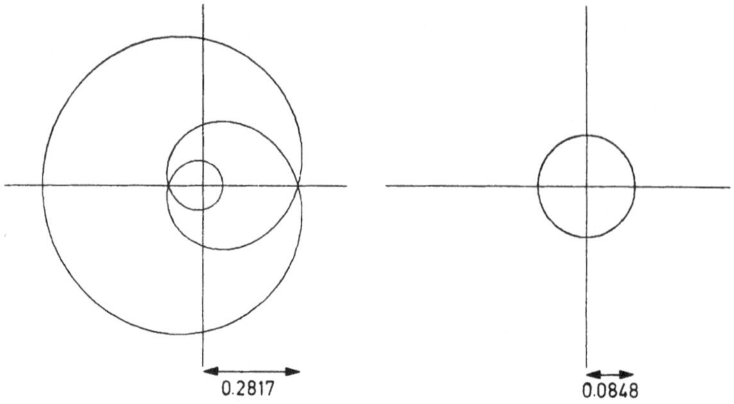

(a) Padé (b) best

Figure 2. Error curves for type (1,1) approximation of e^z .

To begin with, we can observe that if an approximation r happens to have a nearly circular error curve, then it is nearly best. The proof is essentially Rouché's theorem.

THEOREM 1. Suppose the error curve for $r \in R_{mn}$ *has winding number at least* $m+n+1$ *about the origin. Then*

$$\min_{z\in S} |(f-r)(z)| \leq E^* \leq \|f-r\| .$$

In particular, if the error curve is a perfect circle, then $r = r^*$.

Proof. The second inequality is nothing more than the definition of E^* . For the first, suppose to the contrary $\|f-r^*\| < \min_{z\in S} |(f-r)(z)|$. Then simple geometry shows that the funcion $(f-r)-(f-r^*) = r^*-r$ must have the same winding number as $f-r$ on S , which by assumption is at least $m+n+1$. But this is impossible, for r^*-r belongs to $R_{m+n,2n}$, and hence has at most $m+n$ zeros in D . □

Thus "nearly circular implies nearly best", which makes the near-circularity phenomenon at least plausible. Nevertheless, the implication runs in the wrong direction, so we will need additional ideas to see why the phenomenon is in practice so pronounced.

3. THE CARATHÉODORY-FEJÉR METHOD

Another clue to near-circularity can be obtained from the following observation: in many extremal problems in function theory where a supremum norm is involved, solutions occur that involve circles or arcs of circles. For example, a standard proof of the Riemann mapping theorem for a Jordan region Ω containing 0 considers the set of all analytic functions on Ω normalized by $f(0)=0$, $f'(0)=1$. Among these, that function with minimal norm $\|f\|$ is precisely the conformal map from Ω to a disk — which means, it maps the boundary $\partial\Omega$ onto a circle αS . Another well-known example of an extremal problem whose solution involves circles is the *Nevanlinna-Pick* problem of interpolation with minimal norm in the disk ([20], chap. 10).

These problems in effect contain an infinite number of unknown parameters — the Taylor coefficients of an analytic function. In contrast, the Chebyshev approximation problem has a finite number of unknown parameters. This is the essential reason why near-circles rather than perfect circles appear.

Our analysis is based on an extended problem that is closely related to both the above proof of the Riemann mapping theorem and to Nevanlinna-Pick, namely the *Carathéodory-Fejér problem.* Let q be a polynomial of degree d , and consider the set of extensions of q to a Taylor series $\tilde{q}$ that is analytic and bounded in D . Carathéodory and Fejér asked in 1911: what extension $\tilde{q}$, if any, attains minimal norm $\|\tilde{q}\|$? They found the following solution: a minimal extension $\tilde{q}$ exists, it is unique, and it is the Taylor series of a function that is a constant times a *finite Blaschke product* of order $\nu \le d$,

$$\tilde{q}(z) = \sigma \prod_{i=1}^{\nu} \frac{z - z_i}{\bar{z}_i z - 1} \qquad \sigma \in \mathbb{C}, \ z_i \in D .$$

Thus $\tilde{q}$ maps S onto a perfect circle with winding number at most d . Moreover, Schur showed in 1917 that $\tilde{q}$ can be computed analytically by solving a certain matrix eigenvalue problem.

Our original polynomial approximation problem can be viewed as follows: we are given coefficients $a_{m+1}, a_{m+2}, \ldots$, and asked to find $a_0, \ldots, a_m$ so that $\Sigma_0^\infty a_k z^k$ has minimal norm on $\bar{D}$. (Without loss of generality we have assumed that the Taylor series of the given function f begins at degree $m+1$.) The CF problem reverses the prescription: given $a_0, \ldots, a_m$, what infinite set of coefficients $a_{m+1}, a_{m+2}, \ldots$ leads to a series $\Sigma_0^\infty a_k z^k$ of minimal norm? Since the number of unknown parameters is now infinite, circular error curves become possible.

The idea of the *CF method* is perhaps now obvious. First, truncate the given function f at some high order M, so that it takes the form $f(z) = \Sigma_{m+1}^{M} a_k z^k$. Next, apply the CF theorem to construct an extension of f *backwards* to a Laurent series $\tilde{f}(z) = \Sigma_{-\infty}^{M} a_k z^k$ that has minimal norm on S. Finally, with a little luck, the coefficients of negative index in this expansion will be small, so that the coefficients $a_0, \ldots, a_m$ determine a polynomial p^{cf} that is close to p^*.

To describe the CF method more precisely, let us generalize to the rational approximation problem. Given f analytic in D and continuous on $\overline{D}$, we seek a BA $r^* \varepsilon R_{mn}$. Now let $\tilde{R}_{mn}$ denote the set of functions that are meromorphic in $1 \leq |z| \leq \infty$, have $\nu \leq n$ poles in $1 \leq |z| < \infty$ and at most $m - \nu$ poles at ∞ (i.e. a zero at ∞ of order at least $\nu - m$, if $m < \nu$), and are bounded in $1 \leq |z| \leq \infty$ except near these poles. Equivalently, $\tilde{R}_{mn}$ is the set of functions that can be represented in the form

$$\tilde{r}(z) = \sum_{k=-\infty}^{m} d_k z^k \Big/ \sum_{k=0}^{n} e_k z^k ,$$

where the numerator converges in $\mathbb{C} - \overline{D}$ and is bounded there except near ∞. Consider then:

EXTENDED APPROXIMATION PROBLEM. Find $\tilde{r}^* \varepsilon \tilde{R}_{mn}$ *such that*

$$\|f - \tilde{r}^*\| = \inf_{\tilde{r}\varepsilon\tilde{R}_{mn}} \|f - \tilde{r}\| .$$

Like the CF problem, this extended problem has a solution that can be explicitly constructed. For general f of the class considered (in fact for $f \varepsilon L^{\infty}(S)$), the procedure for this is exactly what is taken up in the *AAK theory* published in 1971 by Adamjan, Arov, and Krein [1,13]. Let us however again assume $f \varepsilon P_M$ for some large M, in which case the solution is simpler and can be worked out on the basis of an extension of the Carathéodory-Fejér theorem to rational approximation published by T. Takagi in 1924 [5,15]. (For the intermediate possibility $f \varepsilon R_{MN}$, see the paper [4] by Gutknecht in this volume.)

Let H_{m-n} denote the $(M-m+n) \times (M-m+n)$ Hankel matrix

$$H_{m-n} = \begin{bmatrix} a_{m-n+1} & a_{m-n+2} & \cdots & a_M \\ a_{m-n+2} & & \cdot & \\ \vdots & \cdot & & \\ a_M & & & 0 \end{bmatrix} .$$

Let

$$H_{m-n} = U \Sigma \overline{U}^H$$

be a *singular value decomposition (SVD)* of H_{m-n} — i.e. U is a square unitary matrix ($U^{-1} = U^H$), and Σ is a square diagonal matrix of nonnegative real *singular values* arranged in non-increasing order, $\Sigma = \mathrm{diag}(\sigma_0, \dots, \sigma_{M-m+n-1})$, $\sigma_0 \geq \sigma_1 \geq \dots \geq \sigma_{M-m+n-1}$. In particular, let σ_n be the *nth singular value* of H_{m-n} (starting from the 0th), and let $u^{(n)}$ be the corresponding *nth singular vector,* namely the nth column of U , which we write in the form

$$u^{(n)} = (u_0, u_1, \dots, u_{M-m+n-1})^T .$$

(If the coefficients a_k are real, the SVD reduces to an eigenvalue decomposition.) Then the following is the basic theorem, due to Carathéodory and Fejér, Schur, Takagi, and Trefethen, that the rational CF method is based on:

THEOREM 2 [17]. *The function* f *has a unique BA* $\tilde{r}^*$ *in* $\tilde{R}_{mn}$ *, and it is given by*

$$(f - \tilde{r}^*)(z) = \sigma_n z^{m-n+1} \frac{u_0 + u_1 z + \dots + u_{M-m+n-1} z^{M-m+n-1}}{\overline{u}_0 + \overline{u}_1 z^{-1} + \dots + \overline{u}_{M-m+n-1} z^{-M+m-n+1}} .$$

The error curve $(f - \tilde{r}^*)(S)$ *is a perfect circle about the origin with radius* $\|f - \tilde{r}^*\| = \sigma_n$ *, and if* σ_n *is a simple singular value of* H_{m-n} *, it has winding number exactly* $m+n+1$. □

The *rational CF method* now consists of constructing $\tilde{r}^*$ and then truncating it to obtain an approximation $r^{cf} \in R_{mn}$ that is hopefully near-best. Here is the recipe. In practice, many of these computations are best performed numerically with the Fast Fourier Transform, as indicated (see [9]).

Step 1. Given f analytic in $\overline{D}$, compute its Taylor coefficients $a_0, \dots, a_M$ for some large M (FFT).

Step 2. Set up the Hankel matrix H_{m-n} and compute its nth singular value and vector.

Step 3. Factor the denominator of the Blaschke product in Thm. 2 to determine its n (or fewer) zeros outside $\overline{D}$ (FFT). The polynomial with these zeros is the denominator q^{cf} of r^{cf} .

Step 4. Multiply by q^{cf} to obtain the numerator $\tilde{p}^*$ in the representation

$$\tilde{r}^*(z) = \sum_{k=-\infty}^{m} d_k z^k \Big/ \sum_{k=0}^{n} e_k z^k .$$

Compute coefficients $d_0, \ldots d_m$ of this function (FFT).

Step 5. Truncate terms of negative degree in the numerator of $\tilde{r}^*$ to obtain the *CF approximant*

$$r^{cf}(z) = \sum_{k=0}^{m} d_k z^k \Big/ \sum_{k=0}^{n} e_k z^k .$$

The truncation $\tilde{r}^* \to r^{cf}$ described in Step 5 is only one ("Type 1") of several reasonable methods for obtaining an approximation in $\tilde{R}_{mn}$ from $\tilde{r}^*$. Various others are mentioned in [5], and it is not yet clear which is best.

4. ACCURACY OF THE CF METHOD

The CF method construction leads immediately to a beautiful result:

THEOREM 3 [17]. $\sigma_n(H_{m-n}) \le E^*_{mn}$.

Proof. Theorem 2 implies $\sigma_n = \|f - \tilde{r}^*\|$, and since $R_{mn} \subseteq \tilde{R}_{mn}$, one has also $\|f - \tilde{r}^*\| \le \|f-r^*\| = E^*$. □

This theorem gives some algebraic insight into best approximations that is far from trivial.

Now for all we know a priori, the bound of Thm. 3 could be very crude. But the following table of results for $f(z) = e^z$ shows that in practice, just the opposite can be the case:

(m,n)	$\sigma_n(H_{m-n})$	E^*_{mn}
(0,0)	1.258	1.344
(1,0)	.5575	.5586
(2,0)	.177374	.177377
(3,0)	.043368927	.0433689...

TABLE 1

Obviously the inequality in Thm. 3 is sometimes virtually an equality. Indeed the ellipsis in the last line of Table 1 reflects the fact that we have not succeeded in computing the error E^*_{30}

for e^z more accurately than by applying Thm. 1 to r^{cf}. Although the examples in this paper involve only e^z, similar results hold for many other functions.

For more insight let us look in detail at the problem $f(z) = e^z$, $(m,n) = (1,1)$. Here the SVD construction leads to the following extended BA in $\tilde{R}_{11}$:

$$\tilde{r}^*(z) = \frac{\cdots + 0.00000983z^{-2} + 0.00024668z^{-1} + 0.99613054 + 0.58955195z}{1 - 0.43416584z}$$

Obviously the terms of negative degree in the numerator are very small. Truncating them gives

$$r^{cf}(z) = \frac{0.99613054 + 0.58955195\,z}{1 - 0.43416584z},$$

and since $\tilde{r}^* - r^{cf}$ is small, r^{cf} must have a nearly circular error curve, which by Thm. 1 means it is near best. In fact, numerical computation gives the following representation for r^*, which is obviously almost the same as r^{cf}:

$$r^*(z) = \frac{0.99625 + 0.58952\,z}{1 - 0.43414\,z}.$$

Ideally, a general theory would be available that would show exactly why the CF method is so accurate for a function like e^z, and delineate just what functions the method performs well for. Unfortunately, Thm. 3 is the only fully general theorem regarding CF approximation that we have. However, the following *asymptotic* results have been obtained in [17] and [18], and show that at least in a limiting sense, the CF method is highly accurate, and error curves are nearly circular.

We will need a normality condition.

ASSUMPTION A. The Padé approximant r^p *to* f *of order* (m,n) *has* n *finite poles, and its Taylor series agrees with* f *exactly through the term of degree* $m+n$.

Essentially this assumption requires that the determinant of a certain Hankel matrix section of H_{m-n} be nonzero, which is a standard hypothesis in Padé approximation; see the lectures in this volume by Brezinski [2].

Now for any $\varepsilon > 0$, consider approximation of $f(\varepsilon z)$ on $\overline{D}$. (Or equivalently, consider approximation of $f(z)$ on $|z| \le \varepsilon$.) By a fairly lengthy but elementary argument, one derives the following estimates for the accuracy of the CF approximant:

THEOREM 4 [17,18]. *As* $\varepsilon \to 0$,

$$\|f - r^*\| - \sigma_n = O(\varepsilon^{2m+2n+3})$$

and

$$\|r^{cf} - r^*\| = O(\varepsilon^{2m+2n+3})$$

uniformly for all BAs r^* , *and*

$$\|f - r^{cf}\| - \|f - r^*\| = O(\varepsilon^{2m+2n+4}) . \square$$

These orders of accuracy are remarkably high, when one considers that $\|f - r^*\|$ has order ε^{m+n+1} as $\varepsilon \to 0$. Thus the CF method has a *relative* accuracy of $O(\varepsilon^{m+n+2})$. Padé approximation, by contrast, has relative accuracy $O(\varepsilon)$.

Furthermore, the CF method also gives information about the BA (or BAs) r^* . Since the error curve of $\tilde{r}^*$ is perfectly circular, Thm. 4 together with the associated bound $\|\tilde{r}^* - r^{cf}\| = O(\varepsilon^{2m+2n+3})$ implies that the error curve of r^* must be nearly circular:

THEOREM 5 [17,18]. *As* $\varepsilon \to 0$,

$$\|f - r^*\| - \min_{z \in S} |(f - r^*)(z)| = O(\varepsilon^{2m+2n+3}) .$$

Moreover, this estimate is sharp in that there exist functions for which the left hand side has magnitude at least const. $\varepsilon^{2m+2n+3}$ *for some fixed constant.* $\square$

Naturally one would also like estimates for the limit $m, n \to \infty$ with a fixed value of ε . A theorem in this direction for the case $n = 0$ is given in [16], but the general rational approximation problem has not been treated.

5. REAL CF APPROXIMATION

The CF method and the phenomenon of near-circularity can be transplanted to a more general domain Ω by means of a conformal map of the exterior of $\overline{\Omega}$ onto the exterior of $\overline{D}$. The resulting procedure makes use of a Hankel matrix of Faber series coefficients, and is called the *Faber-CF method*. For details see the second lecture by Gutknecht in this volume [4]. Gutknecht has also developed *Laurent-CF* and *Fourier-CF* methods for related problems, which extend the CF idea in analogy to the work of Gragg, et al. on Laurent-Padé and Fourier-Padé approximation [4,5].

Here we will discuss the particularly interesting case of transplantation to a real interval, where a *real CF* or *Chebyshev-CF* method is obtained that turns out to be even more powerful than on the disk. (Related ideas were developed earlier by S. Darlington, D. Elliott, B. Lam, and others; see [19] for a discussion and references.) The real CF method is presented at length in [7] and [19].

Let $F(x)$ be a continuous real function on $I = [-1,1]$, which for simplicity as before we will assume has a finite expansion in Chebyshev polynomials,

$$F(x) = \sum_{k=0}^{M} a_k T_k(x) , \quad a_k \in \mathbb{R} .$$

If $x \in I$ and $z \in S$ are related by $x = \operatorname{Re} z = \frac{1}{2}(z + z^{-1})$, then a widely known but underutilized formula expresses $T_k(x)$ in terms of z :

$$T_k(x) = \operatorname{Re} z^k = \frac{1}{2}(z^k + z^{-k}) .$$

To apply a CF method to F, we consider the analytic function f defined by

$$f(z) = \sum_{k=0}^{M} a_k z^k ,$$

so that one has

$$F(x) = \operatorname{Re} f(z) = \frac{1}{2}(f(z) + f(z^{-1})) .$$

If the CF/Takagi construction is carried out for f — so that we are working with the singular value (or eigenvalue) decomposition of a Hankel matrix of Chebyshev coefficients of F — we obtain the optimal approximation $\tilde{r}^* \in \tilde{R}_{mn}$ to f on S,

$$\tilde{r}^*(z) = \sum_{k=-\infty}^{m} d_k z^k \Big/ \sum_{k=0}^{n} e_k z^k .$$

By cross-multiplication, the real part of $\tilde{r}^*$ can be written in the form

$$\operatorname{Re} \tilde{r}^*(z) = \frac{1}{2}(\tilde{r}^*(z) + \tilde{r}^*(z^{-1}))$$

$$= \frac{\sum_{k=0}^{\infty} \hat{d}_k (z^k + z^{-k})}{\sum_{k=0}^{n} \hat{e}_k (z^k + z^{-k})} = \frac{\sum_{k=0}^{\infty} \hat{d}_k T_k(x)}{\sum_{k=0}^{n} \hat{e}_k T_k(x)}$$

for some new sets of coefficients $\{\hat{d}_k\}$, $\{\hat{e}_k\}$. Assuming $m \geq n$, we now truncate this expression in the obvious way to obtain the *Chebyshev-CF approximant*

$$R^{cf}(x) = \frac{\sum_{k=0}^{m} \hat{d}_k T_k(x)}{\sum_{k=0}^{n} \hat{e}_k T_k(x)} .$$

(If $m < n$, the definition of R^{cf} is more complicated. See [19].)

Now by construction, R^{cf} belongs to R_{mn} and has real coefficients. But beyond this, why should we expect it to be close to the real BA R^* to F ? To see the answer, observe that if $R^{cf}(x) \approx \operatorname{Re} \tilde{r}^*(z)$, then

$$(F - R^{cf})(x) \approx \operatorname{Re} (f - \tilde{r}^*)(z) .$$

By Thm. 2, $(f - \tilde{r}^*)(S)$ is a perfect circle with winding number $m+n+1$ (assuming $\sigma_n(H_{m-n})$ is simple), and therefore $\operatorname{Re}(f-\tilde{r}^*)(z)$ equioscillates $m+n+2$ times between values $\pm\sigma_n$ as z traverses the upper half of S. Therefore R^{cf}, correspondingly, must have a nearly-equioscillating error curve on I, which by the well-known theorem of de la Vallée Poussin [12], implies that R^{cf} is near-best.

Thus the justification of the real CF construction is a matter of showing $R^{cf}(x) \approx \operatorname{Re} \tilde{r}^*(z)$, that is, the truncated terms $\hat{d}_{m+1}, \hat{d}_{m+2}, \ldots$ are small. Again, it would be nice to have a non-asymptotic theory for this, but none is available; in fact even Thm. 3 fails to extend (except with the loss of a factor of two) to Chebyshev-CF approximation. In analogy to Thm. 4, on the other hand, one can derive a very satisfactory asymptotic result. The following theorem relates to approximation of $F(\varepsilon x)$ on $[-1,1]$, and requires again that F satisfy Assumption A. Note that the powers of ε here are even higher than in Thm. 4, a result of the fact that fewer terms are truncated here in going from $\tilde{r}^*$ to R^{cf} than in the complex CF method.

THEOREM 6 [19]. *As* $\varepsilon \to 0$,

$$\|F - R^*\|_I - \sigma_n = O(\varepsilon^{3m+2n+3})$$

and

$$\|R^{cf} - R^*\|_I = O(\varepsilon^{3m+2n+3}) .$$

The second estimate can be replaced by $O(\varepsilon^{3m+2n+4})$ *in the special case* $n=0$. □

As in the complex case, the asymptotic accuracy of the real CF method is overwhelmingly apparent in many computational examples where ε is not at all small. Table 2 shows results for polynomial approximation of $F(x) = e^x$. A comparison with the numbers of Table 1 confirms that, as the asymptotic results suggest, the CF method is even more powerful on I than on $\overline{D}$. (From a computational point of view, however, the CF method for $\overline{D}$ remains probably the more important, because for approximation on I rapidly convergent alternatives such as the Remes algorithm are available [12].)

(m,n)	$\sigma_n(H_{m-n})$	E^*_{mn}
(0,0)	1.196	1.175
(1,0)	.2787994	.27880...
(2,0)	.0450173878	.04501738...
(3,0)	.005528370108712	.0055283701087...

TABLE 2

Since Thm. 6 is the analog for real CF approximation of Thm. 4 for the complex case, one naturally wonders, what about a real analog of Thm. 5? Sure enough, it is easy to derive by CF methods the following result: as $\varepsilon \to 0$, best approximation error curves equioscillate up to $O(\varepsilon^{3m+2n+3})$. This would be very interesting, were it not of course well known that error curves in real Chebyshev approximation equioscillate exactly! In other words the sharpness statement in Thm. 5 is very significant, and does not extend to real approximation.

Nevertheless, it seems that the high accuracy of the real CF method must reveal something about the structure of BAs in real Chebyshev approximation. One can, for example, state a somewhat awkward theorem to the effect that as $\varepsilon \to 0$, real BA error functions approach the real parts of Blaschke products up to $O(\varepsilon^{3m+2n+3})$ [7,19]. It remains to be seen whether this fact can be recast in some way that makes its significance for approximation on the interval as clear as that of the near-circularity results for the unit disk.

REFERENCES

[1] Adamjan, V., Arov, D. and Krein, M.: *Analytic properties of Schmidt pairs for a Hankel operator and the generalized Schur-Takagi problem,* Math. USSR Sbornik 15 (1971), pp. 31-73.

[2] Brezinski, C.: *Outlines of Padé approximation,* this Proceedings, 1982.

[3] Gutknecht, M.: *On complex rational approximation I: The characterization problem,* this Proceedings, 1982.

[4] Gutknecht, M.: *On complex rational approximation II: The Carathéodory-Fejér method,* this Proceedings, 1982.

[5] Gutknecht, M.: *Rational Carathéodory-Fejér approximation on a disk, a circle, and an interval,* to appear.

[6] Gutknecht, M., Smith, J.O. and Trefethen, L.: *The Carathéodory-Fejér (CF) method for recursive digital filter design,* subm. to IEEE Trans. Acoustics Speech and Signal Processing.

[7] Gutknecht, M. and Trefethen, L.: *Real polynomial Chebyshev approximation by the Carathéodory-Fejér method,* SIAM J. Numer. Anal. 19 (1982), pp. 358-71.

[8] Gutknecht, M. and Trefethen, L.: *Nonuniqueness of complex rational Chebyshev approximations on the unit disk,* to appear.

[9] Henrici, P.: *Fast Fourier methods in computational complex analysis,* SIAM Review 21 (1979), pp. 481-527.

[10] Henrici, P.: *Topics in computational complex analysis III: The asymptotic behavior of best approximations to analytic functions on the unit disk,* this Proceedings, 1982.

[11] Kailath, T.: *Linear Systems,* 1981.

[12] Meinardus, G.: *Approximation of functions: theory and numerical methods,* Springer, Berlin-Heidelberg-New York, 1967.

[13] Meinguet, J.: *A simplified presentation of the Adamjan-Arov-Krein approximation theory,* this Proceedings, 1982.

[14] Rabiner, L. and Gold, B.: *Theory and Application of Digital Signal Processing,* Prentice-Hall, Englewood Cliffs, NJ, 1975.

[15] Takagi, T.: *On an algebraic problem related to an analytic theorem of Carathéodory and Fejér,* Japan J. Math. 1 (1924), pp. 83-93, and 2 (1925), pp. 13-17.

[16] Trefethen, L.: *Near-circularity of the error curve in complex Chebyshev approximation,* J. Approx. Theory 31 (1981), pp. 344-67.

[17] Trefethen, L.: *Rational Chebyshev approximation on the unit disk,* Numer. Math. 37 (1981), pp. 297-320.

[18] Trefethen, L.: *Circularity of the error curve and sharpness of the CF method in complex Chebyshev approximation,* subm. to SIAM J. Numer. Anal.

[19] Trefethen, L. and Gutknecht, M.: *The Carathéodory-Fejér method for real rational approximation,* SIAM J. Numer. Anal., to appear.

[20] Walsh, J.: *Interpolation and Approximation by Rational Functions in the Complex Domain,* Amer. Math. Soc., Providence, RI, 1935 (5th ed., 1969).

CALCULATIONS OF SINGULARITIES FOR SOLUTIONS OF ALGEBRAIC DIFFERENTIAL EQUATIONS

Helmut Werner

University of Bonn
Federal Republic of Germany

Abstract: In this introduction we first define nonlinear splines and describe their application to the solution of initial value problems of ordinary differential equations. Stability and convergence results for polynomial splines and their generalization to nonlinear splines are given. The Painlevé theory is referred to for the motivation of the selection of nonlinear spline families in order to integrate the differential equations in the neigborhood of movable singularities of their solution. For two special cases the asymptotic behaviour of the estimation of the location of the singularity is analysed.

INTRODUCTION

In this paper we deal with the numerical solution of ordinary differential equations as described by the problem

$$\mathbb{P}: \quad y' = f(x,y) \quad \text{and} \quad y(x_0) = y_0 .$$

We denote the solution of $\mathbb{P}$ by y and assume it be defined in I. Our concern is numerical accuracy and efficiency and not the care for weakest assumptions under which our statements can be proven. We therefore make only the assumption that a function be "smooth" or "sufficiently smooth" to indicate that it is continuous and possesses all the derivatives needed for our investigations.

H. Werner et al. (eds.), Computational Aspects of Complex Analysis, 325–360.

In a systematic treatment of numerical methods for differential equations one may distinguish between discrete and continuous methods. A discrete method gives approximations of y only at a set of discrete points while continuous methods give functions approximating y. Usually these functions depend on a certain number of parameters. Practically the difference of both methods is not as big as one may think. Of course continuous methods can be interpreted as discrete ones if only the resulting approximations at the grid points are considered. On the other hand discrete approximating values u_j at x_j may be interpolated, usually taking into account that the value of the derivative is associated with (x_j,u_j) by the differential equation. Methodically some of the proofs for the results for continuous solutions reduce to standard results on one step methods as they are found in the classical book of Henrici on "discrete variable methods".

Some numerical analysts nowadays say that it is not worthwhile to invent new methods for solving "well behaved" initial value problems. What is needed are techniques for cases with certain complications, e.g. stiffness. The results presented here lend themselves easily to the situation that a solution y becomes singular and therefore causes difficulties for standard routines for initial value problems. We will show the strength of this method by a number of examples in the second part of this paper.

For the continuous approximation of y we propose non-linear splines. Roughly speaking these are functions of a certain differentiability class in all of I whose restrictions to subintervals associated with a fixed subdivision of I are of a preselected type that depends on a fixed number of parameters and is easy to handle numerically, e.g. polynomials, rationals, exponentials, algebraic functions. So non-linearity says that the dependence on the parameters used need not be linear. On the other hand the non-linear splines allow many problems to be treated with equidistant knots, i.e. a constant step size in our situation, while linear techniques will require a refined step size control.

For theoretical purposes it is useful to choose the parametrisation in each subinterval in a canonical way, i.e. by the coefficients of the Taylor expansion of the spline about a knot. Then the non linearity will only appear in the higher order derivatives. In this lecture we will indicate that this simplifies some proofs as compared to earlier publications. One may argue that the use of non-linear splines has the following merits.

- The family of splines used can be adjusted to a priori known properties of the solutions of the differential equation at hand, e.g. to the type of singularities of those solutions.
- We gain high accuracy with comparatively small numerical work, get a stable numerical procedure, comparatively small round off errors and thrifty use of calculation time. Asymptotic expansions make it feasible to apply extrapolation techniques.
- We will not apply it here but remark only that use of splines is advisable in cases in which one has to use the solution at intermediate points, and not only at the knots. Such applications arise in connection with difference differential equations when the delay is not fixed but variable.

The complex analysis enters the picture if one looks for a judicious choice of the spline functions. To simulate the behaviour of the solution y it is useful to know its analytic character. It is a result of the last century by Painlevé that scalar differential equations with $f(x,y)$ polynomial have solutions that possess movable algebraic singularities. For systems and higher order differential equations the discussion of the function theoretical properties of the solution is a large field of research, compare the contribution of Farwig to this meeting.

If we know the analytic and asymptotic behaviour of y in the neigborhood of its singularities, we can choose families of functions for the construction of the spline that show a similar character. Then only the parameters are to be tuned properly. The results of these trials are striking. There are also good partial theoretical results to understand the numerically observed phenomena, but a complete theory is still lacking.

Here we survey material scattered in a number of papers and hope to give a unified treatment. It would not be difficult to generalize most of the results of polynominal splines as applied to differential equations to non-linear splines, but we feel that the situation analysed here is numerically the most useful one.

Higher order matching of the differential equation or use of several points in one subinterval heavily increases the work for programming and implementation for each application. We feel that methods with simple structure but tailormade trial functions for generating the spline are superior and more efficient.

DEFINITION OF NON-LINEAR SPLINES

First we generalize the well known concept of a spline in a way that is suitable for our purpose. Given "a degree" n and "the order of smoothness" q, $q < n$, one may think of a spline as a function that is globally of class C^q, i.e. q times continuously differentiable, and depends locally on $n + 1$ parameters. Usually this local function is a polynomial of degree n, this is, however, by no means necessary. We see examples below. The number $d := n - q$ is usually called the defect. To give a formal definition, assume that I is the interval $[x_o, x^+]$ and denote by $X := \{x_o, x_1, \ldots, x_m\}$ a set of knots contained in I such that $x_{j-1} < x_j$ holds for $j = 1, \ldots, m$. We write $I_j := [x_{j-1}, x_j]$. Given classes of functions $t_j(z, c_o, \ldots, c_n)$ for $o \le z \le x_j - x_{j-1}$ and a domain $D \subset \mathbb{R}^{n+1}$ for $c_o, \ldots, c_n$ we call $u(x)$ a nonlinear spline (with respect to these classes), if

1) $u(x) = t_j(x - x_{j-1}, c_o, \ldots, c_n)$ for $x \in I_j$,

2) $u(x) \in C^q$.

Usually we require that t_j satisfies certain differentiability assumptions, for instance the existence of $n + 2$ times continuous derivatives with respect to x and differentiability with respect to every parameter c_i.

For simplicity we may assume in the following considerations that the classes t_j are the same for all intervals I_j.

The spline may be parametrized in several ways. It is convenient for theoretical considerations, if its derivatives are used. This motivates the following definition:

The representation of the spline $u(x)$ has canonical form if there is $S(z, x, c_o, \ldots, c_n)$ such that

$$t_j(x - x_{j-1}, c_o, \ldots, c_n) = c_o + c_1 \frac{z}{1!} + \ldots + c_n \frac{z^n}{n!} + \frac{z^{n+1}}{(n+1)!} S(z, x_{j-1}, c_o, \ldots, c_n)$$

with $z = x - x_{j-1}$,
in other words we write down the Taylor expansion of t_j about the left hand endpoint of I_{j-1} and $S(0, x_{j-1}, c_0, \ldots, c_n)$ is the value of the $(n+1)$st derivative at this point.

Remark.
It is obvious that the function O is a candidate for a function S - it reproduces the polynomial splines. Therefore every statement in the following holds for polynomial splines in particular. A consequence is that no result in this general setting can be better than the special result for polynomial splines.

A second example is given by the so called special rational splines which in former publications is written and given by

$$t_j(z,\ldots) = c_0 + c_1 z + \frac{1}{2} c_2 \cdot \frac{z^2}{1+dz} .$$

It is obvious that

$$t_j(z,\ldots) = c_0 + c_1 z + c_2 \frac{z^2}{2} + \frac{1}{2} c_2(-d)\, z^3 + + \frac{1}{2} c_2 \cdot \frac{d^2 z^4}{1+dz} .$$

The connection between c_3 and d is established by

$$c_3 = -3c_2 \cdot d \quad \text{and} \quad S = 12 c_2 d^2 \cdot \frac{1}{1+dz} .$$

We observe that in this case the domain D and the region of continuity for z are coupled. They depend upon the parameters, since $1 + dz > 0$ is necessary for continuity and differentiability of S .

NUMERICAL TREATMENT OF ORDINARY DIFFERENTIAL EQUATIONS BY MEANS OF SPLINES

Consider the following problem

$$\mathbb{P}: \quad y' = f(x,y)$$
$$y(x_0) = y_0 ,$$

here we assume that $f : G \to \mathbb{R}$ is given in an appropriate domain and satisfies suitable differentiability properties. Furthermore the point (x_0, y_0) should be contained in the domain G . We look for a function $y : I \to \mathbb{R}$ which is continuously differentiable, solves the differential equation and attains the prescribed initial value.

Numerical methods to solve differential equations can be split into discrete ones and continuous ones. Here we are concerned with the continuous methods that means we do not only calculate the approximation at some fixed points of the interval but we construct functions which are approximations of y . These functions shall be non-linear splines as defined above where the functions t_j respectively S are specified later. In each subinterval they will depend on $n+1$ parameters. For reasons soon to be apparent we will restrict ourselves to $n \leq 3$ and $d = 1$.

Let h be fixed in a proper way, i.e. by some step size control. We write

$$x_j = x_o + j \cdot h \quad \text{for} \quad j = 1,\dots,m$$

with

$$h = \frac{x^* - x_o}{m} \quad .$$

We propose to use the following <u>algorithm</u>.

<u>Start</u>: Define

$$u(x_o,h) = y_o$$

$$D^i u(x_o,h) = y^i(x_o) = \frac{d^{i-1}}{dx^{i-1}} f(x,y(x))\big|_{x=x_o}$$

$$i = 1,2,\dots,q \; .$$

These initial data are determined by f and the initial condition directly.

<u>Iterative definition</u> of u in the interval I_{j+1} . Assume $q+1$ values of the function and its derivatives are prescribed by the initial conditions or the values brought forward from the adjacent left-hand interval I_j :

$$u_j^{(k)} = u^{(k)}(x_j - 0,h) \quad \text{for} \quad k = 0,\dots,q \; .$$

Then d equations for derivatives at the right-hand endpoint of the interval are used to determine the remaining parameters

$$u^{(k)}(x_{j+1}-0,h) = D^{k-1}f(x,u(x,h))|_{x\to x_{j+1}}$$

for $k = 1,\ldots,d$.

The resulting spline segment $u(x,h)$ in I_{j+1} then defines $u_{j+1}, u'_{j+1},\ldots,u^q_{j+1}$ so that the construction can be continued in I_{j+2} unless the stopping condition ist satisfied:
Either I is covered or the graph $(x,u(x,h))$ is not in the domain G of f for every $x \in I_{j+1}$.

LINEAR FAMILIES

Let us first review the classical case that t is of the class of cubic polynomials. The theory was worked out by Loscalzo and Talbot [3]. Generalizations were given by Schoenberg and Varga 1969 [7]. We summarize the results on linear splines as they are found in a definitive form by Mülthei, 1979 - 1980 [4].

Consider polynomial splines that are made up piecewise of polynomials of degree n and that are globally of the smoothness class C^q where $q = n-d$, and let d denote the defect with respect to differentiability of the spline.

Under appropriate regularity conditions concerning f and for sufficiently small h the $n + 1$ relations uniquely determine a polynomial in I_{j+1} . With above notations we have the following result, comp. Mülthei [4].

i) The above sketched algorithm is divergent for $h \to 0$ if

$$n \geq 2d + 2 .$$

For $d = 1$ this was already observed by Loscalzo and Talbot. Mülthei critizises the previous proofs and shows simply by a counterexample that the spline method can diverge under the said relation between n and d .

ii) Case $n \leq 2d + 1$.

1) If $n \leq 2d$, therefore $q \leq d$, then one has some kind of a convergent implicit one step method. In this case

$$u_{j+1} - u_j$$

is a linear combination of the $q + 1$ data at the left and the d values $f(x,y(x))$, $\frac{d}{dx}f(\ldots),\ldots$ at the right endpoint of the interval I_{j+1}.
The order of consistency is given at the knots by

$$y^{(i)}(x_j) - u^{(i)}(x_j,h) = O(h^n)$$

$$\text{for } i = 0,\ldots,d \quad \text{and} \quad \forall j ,$$

and for the whole interval by

$$y^{(i)}(x) - u^{(i)}(x,h) = O(h^{\min(n,n+1-i)}) ,$$

$$\text{for } i = 0,\ldots,n \quad \text{and} \quad \forall x \in I .$$

2) If $n = 2d + 1$, d odd, then one has an implicit two step method which expresses $u_{j+1} - u_{j-1}$ as a combination as described before with evaluations at x_{j-1}, x_j, x_{j+1}. This time the order of consistency at the knots is

$$y^{(i)}(x_j) - u^{(i)}(x_j,h) = O(h^{n+1})$$

$$\text{for } i = 0,\ldots,d \quad \text{and} \quad \forall j ,$$

for the whole interval

$$y^{(i)}(x) - u^{(i)}(x,h) = O(h^{n+1-i})$$

$$\text{for } i = 0,\ldots,n \quad \text{and} \quad x \in I .$$

These methods may be looked upon as generalizations of the Milne-Simpson method.
If d is even one has to use special arguments that are also developed by Mülthei.

For applications values of the defect d that are greater than 1 seem not very useful even in the polynomial case. This is due to the fact that the derivatives of f have to be available for every argument, i.e. subroutines for f_x, f_y, ... are needed in the numerical calculations. This is certainly an undesirable feature. Mülthei nowadays proposes to use several interior points of I_{j+1} with $d = 1$ (one equation at each point, $u' = f(x,u(x,h))$ that is) if one wishes to increase the order of accuracy.

CONSISTENCY AND CONVERGENCE FOR NON-LINEAR SPLINES

For the non-linear splines we consider the simplest of the cases quoted before. We take only one point in I_j (i.e. the endpoint x_j) in addition to x_{j-1} and we choose $d = 1$ and $n \leq 3$ to ensure stability - at least for the polynomial case.

We consider the situation that a piece of a solution, say $y(x)$ with $x \in I := [x_0,x^+]$, is approximated and $y(x)$ is continuous and several times continuously differentiable. Under appropriate technical assumptions we will sketch how to generalize the previously stated results for polynomial splines to nonlinear ones.

Lemma:
Let $y(x)$ solve $y' = f(x,y)$ and $y(x^*) = y^*$ for given $(x^*,y^*) \in G$ and f is $n + 1$ times continuously differentiable.
$S(z,x^*,c_0,\dots,c_n)$ be defined for $o \leq z \leq h$, $(c_0,\dots,c_n) \in D$, and it be sufficiently smooth.
Assume $(y,y',\dots,y^{(n)})|_{x=x^*} \in D$.

Then there is a spline segment

$$u(x) = u(x,h) = c_0 + c_1 z + \dots + c_n \cdot \frac{z^n}{n!} + \frac{z^{n+1}}{(n+1)!} \cdot S(z,x^*,c_0,\dots,c_n)$$

with $z = x - x^*$,
such that

$$c_j = y^{(j)}(x^*) \quad \text{for} \quad j = o,\dots,n-1$$

and the defining equation

$$u'(x^*+h) = f(x^*+h,u(x^*+h,h))$$

holds.

Furthermore the n th order derivative satisfies

$$u^{(n)}(x^*,h) = c_n(h) =$$

$$= y^{(n)}(x^*) + h\,\frac{y^{n+1}(x^*) - S(o,x^*,y^*,\dots,y^{(n)}(x^*))}{n} + O(h^2)$$

for $h \le h_o$ sufficiently small.

<u>Proof:</u> We insert the derivative of the canonical expression of u into the defining relation to find that $c := c_n(h)$ satisfies

$$c\cdot\frac{h^{n-1}}{(n-1)!} = f(x^{**},u(x^{**},h)) - c_1 - \dots -$$

$$- c_{n-1}\frac{h^{n-2}}{(n-2)!} - \frac{h^n}{n!}\cdot S(h,x^*,c_o,\dots,c) -$$

$$- \frac{h^{n+1}}{(n+1)!}\cdot S'(\dots)$$

$$=: \frac{h^{n-1}}{(n-1)!}\,\Phi(c,h)$$

where $x^{**} = x^* + h$.

We observe that on the right hand side the parameter c is hidden in u , S and S' .
The value of the parameter c which determines the spline in $[x^*,x^{**}]$ is a fixed point of Φ .
We differentiate Φ with respect to c and find

$$\frac{\partial\Phi}{\partial c} = \frac{(n-1)!}{h^{n-1}}\cdot[f_y(\dots)\cdot(\frac{h^n}{n!} + \frac{h^{n+1}}{(n+1)!}\cdot S_c) -$$

$$- \frac{h^n}{n!}S_c - \frac{h^{n+1}}{(n+1)!}\cdot S_{xc}]$$

$$= \frac{h}{n}(f_y - S_c) + \frac{h^2}{n(n+1)}(f_y\cdot S_c - S_{xc}) = O(h) .$$

This shows that for h sufficiently small Φ is a contracting operator.

It is a matter of technicalities to see that starting with $c^{(o)} = y^{(n)}(x^*)$ an iteration $c^{(j+1)} = \Phi(c^{(j)},h)$ will converge to c^* for h sufficiently small. Then $c_o,\dots,c_{n-1},c^*$ give a vector of parameters in D .

To prove the differentiability of $c_n(h)$ with respect to h we will differentiate the fixed point equation

$$c = \Phi(c,h) .$$

Therefore we have to study $\frac{\partial \Phi}{\partial h}$.
We may use

$$\tilde{f}(x^*,h) = (n+1) \int_0^1 t^n \cdot y^{(n+1)}(x^*+h-th)dt$$

and the Taylor expansion

$$y(x^{**}) = y(x^*+h) = y(x^*) + y'(x^*) \cdot h + \dots +$$
$$+ y^{(n)} \cdot \frac{h^n}{n!} + \frac{h^{n+1}}{(n+1)!} \cdot \tilde{f}(x^*,h)$$

to write

$$w := u(x^{**},h) - y(x^{**}) = [c_n(h) - y^{(n)}(x^*)] \cdot$$
$$\cdot \frac{h^n}{n!} + [S(h,\dots,c_n) - \tilde{f}(x^*,h)] \cdot \frac{h^{n+1}}{(n+1)!}$$

and therefore

$$f(x^{**},u(x^{**},h)) = f(x^{**},y(x^{**})) +$$
$$+ \int_0^1 f_y(x^{**},y(x^{**}) + t \cdot w) \cdot w\, dt$$
$$= y'(x^{**}) + \int_0^1 f_y(\dots)\, w\, dt .$$

This is introduced into Φ to obtain

$$\Phi(c,h) = \frac{(n-1)!}{h^{n-1}} \cdot [y'(x^{**}) + \int_0^1 f_y(\dots) \cdot w\, dt -$$
$$- c_1 - \dots - c_{n-1} \frac{h^{n-2}}{(n-2)!} -$$
$$- \frac{h^n}{n!} \cdot S(h,\dots) - \frac{h^{n+1}}{(n+1)!} \cdot S'(h,\dots)] =$$
$$= y^{(n)}(x^*) + \frac{h}{n} \{[\tilde{f}(x^*,0) - S(0,\dots)] +$$
$$+ f_y(x^*,y(x^*)) \cdot [c_n(h) - y^{(n)}]\} + O(h^2) ,$$

and we remark that $O(h^2)$ contains terms which are actually differentiable w.r.t. h to give an expression of the order $O(h)$.
Furthermore

$$\Phi(c,o) = y^n(x^*)$$

and

$$\left.\frac{\partial\Phi}{\partial h}\right|_{h=o} = \frac{1}{n}\,[\underbrace{\tilde{f}(x^*,o)}_{y^{n+1}(x^*)} - S(0,\ldots) + f_y(x^*,y^*)\cdot$$

$$\cdot \underbrace{(c_n(0) - y^n(x^*))}_{=o}]$$

$$= \frac{1}{n}\,[y^{n+1}(x^*) - S(0,\ldots)]\ .$$

Hence

$$\left.\frac{\partial c}{\partial h}\right|_{h=o} = \frac{\partial\Phi}{\partial h} + \underbrace{\frac{\partial\Phi}{\partial c}}_{o \text{ at } h=o}\cdot\frac{\partial c}{\partial h} = \frac{\partial\Phi}{\partial h}\ .$$

We could also calculate the higher order derivatives if the formal differentiability requirements are met by $f(x,y)$ and $S(z,x,\ldots,c)$ - which we here assume. This shows the last ascertion of our lemma.

We summarize the result for the <u>local error</u>:

$$w(z) := w(z,x^*,h) := u(x,h) - y(x)$$

$$\text{with } z = x - x^* \le h$$

$$= \frac{(y^{n+1}(x^*)-S(0,\ldots))}{n}\,h\cdot\frac{z^n}{n!} + S(0,\ldots)\cdot$$

$$\cdot\frac{z^{n+1}}{(n+1)!} - \tilde{f}(x^*,z)\cdot\frac{z^{n+1}}{(n+1)!} + O(h^{n+2})\ .$$

Since $\tilde{f}(x^*,o) = y^{n+1}(x^*)$, we finally have

$$w(z,x^*,h) = \frac{y^{n+1}(x^*)-S(o,\dots)}{n!}\left[\frac{h\cdot z^n}{n} - \frac{z^{n+1}}{n+1} + \right.$$

$$\left. + O(h^{n+2})\right]$$

and because of the possibility of expressing remainder terms by means of integral representations the terms collected in $O(h^{n+2})$ may be differentiated and it is seen that every differentiation reduces the power of h by 1.

In particular

$$w'(z) = \frac{y^{n+1}(x^*)-S(o,x^*,c_o,\dots,c_n(o))}{n!}(h\cdot z^{n-1}-z^n) +$$

$$+ O(h^{n+1}) .$$

We see that $w'(h) = O(h^{n+1})$ which verifies our construction.
In general we have

$$\frac{d^j w(z)}{dz^j} = O(h^{n+1-j}) \quad \text{for} \quad j = o,\dots,n+1 .$$

This shows that the starting values for the spline u at x^{**} namely the derivatives $u^{(j)}(x^*,h)$ inherit errors of the order $O(h^{n+1-j})$.

Therefore it would be necessary to repeat the considerations condensed in the above Lemma with initial data

$$c_j = y^j(x^{**}) + w^{(j)}(o,x^{**},h) , \quad j = o,\dots,n-1$$

with smooth functions $w^{(j)}(o,x^{**},h)$ of h and of orders $O(h^{n+1-j})$ to find u and the error for the intervall $[x^{**},x^{**}+h]$.

Again the fixed point theorem of Banach and the implicit function theorem may be applied to show that for h sufficiently small $c_n(h)$ becomes a smooth function of h. It is left to the reader to show that the following corollary holds.

Corollary.

Let $w^j(o,x^*,h)$ denote smooth functions of h for which

$$w^j(o,x^*,h) = e_j(x^*) \cdot h^{n+1-j} + O(h^{n+2-j}), \quad j=1,\dots,n$$

holds and denote

$$e_{n+1}(x^*) := S(o,x^*,y(x^*),\dots,y^n(x^*)) - y^{n+1}(x^*) \quad .$$

Then $e_n(x^*,h)$ is of the form

$$\frac{e_n(x^*,h)}{(n-1)!} := \frac{c_n(h)-y^n(x^*)}{(n-1)!} =$$

$$= - \sum_{\substack{v=1 \\ v \neq n}}^{n+1} \frac{1}{(v-1)!} \cdot e_v \cdot h + O(h^2) \quad .$$

The quantities e_j form a vector that gives the asymptotic error coefficient of the spline and its derivatives in comparison with the local solution $y(x)$ with the initial date $y(x^*) = y^*$.

In standard fashion one may prove the Lipschitz property of the method - $f(x,y)$ being sufficiently smooth - and use arguments of one step methods to show the convergence of the method, i.e.

$$u(x,h) \to y(x) \quad \text{in all of} \quad I := [x_o, x^+]$$

where now $y(x)$ again denotes the solution of the original initial value problem.

Furthermore the global order of convergence is n which is equal to the local order of consistency minus 1 .

ORDER OF CONVERGENCE IN CASE $n = 3$

In the previous section we stated that the global order of convergence is by 1 smaller than the local order of consistence. If n is odd, however, this is not the case. We will not give all details but look at the growth of the leading terms of the expansion of the error and its derivatives if we perform two successiv steps of the algorithm.

We take the formulas of the previous section and replace x^* by x_o, x^{**} by x_1. If we start with the values e_j exactly equal to 0 just as in the lemma then we have for $x^* = \{x_o, x_1\}$ the error expansion described in the corollary.
Denote in particular

$$e(x_k) := \begin{pmatrix} e_o(x_k) \\ . \\ . \\ . \\ e_{n+1}(x_k) \end{pmatrix} \qquad e(x_o) := \begin{pmatrix} 0 \\ \vdots \\ 0 \\ e_n(x_o,h) \\ e_{n+1}(x_o) \end{pmatrix} .$$

If one uses again the Taylor expansion to express

$$w^j(z,x,h) = u^j(x,h) - y^j(x) \quad ,$$

it is seen that the values

$$\lim_{z\to h} h^j w^j(z,x_o,h) = \sum_{l=j}^{n+1} \frac{e_l(x_o)}{(l-j)!} h^{n+1} + O(h^{n+2})$$

are obtained. $e_n(x_o)$ was found to be a linear combination of the other components e_j. The relations are easily expressed by means of matrices.

For $n = 3$ these read

$$e(x_1-0) =$$

$$= \begin{pmatrix} 1 & 1 & 1/2 & 1/6 & 1/24 \\ 0 & 1 & 1 & 1/2 & 1/6 \\ 0 & 0 & 1 & 1 & 1/2 \\ 0 & 0 & 0 & 1 & 1 \\ 0 & 0 & 0 & 0 & 1 \end{pmatrix} \cdot \begin{pmatrix} 1 & 0 & 0 & 0 & 0 \\ 0 & 1 & 0 & 0 & 0 \\ 0 & 0 & 1 & 0 & 0 \\ 0 & -2 & -2 & 0 & -1/3 \\ 0 & 0 & 0 & 0 & 1 \end{pmatrix} \cdot e(x_o) + O(h)$$

$$= \begin{pmatrix} 1 & 2/3 & 1/6 & 0 & -1/72 \\ 0 & 0 & 0 & 0 & 0 \\ 0 & -2 & -1 & 0 & 1/6 \\ 0 & -2 & -2 & 0 & 2/3 \\ 0 & 0 & 0 & 0 & 0 \end{pmatrix} \cdot e(x_o) + O(h) \quad .$$

The component e_3 is eliminated.
The last matrix also shows that the leading coefficient of the error of the first derivative is damped out. Hence it suffices to consider e_0, e_2, e_4 .

The components e_j for $j = 0,1,2$ are continuous since y and u are twice continuously differentiable. e_4 changes just like its arguments only by $O(h)$, i.e. $e_4(x_1+0) = e_4(x_0+0) + O(h)$, since it depends on the quantities $\tilde{f}(\tilde{z},x_1^0)$ and $S(z,x_1,c_0,\ldots,c_n)$ and c_j for $j = 0,\ldots,n$ change only of the order 1 in h .
The above equations now read

$$\begin{pmatrix} e_0 \\ e_2 \\ e_4 \end{pmatrix}(x_1) = \begin{pmatrix} 1 & 1/6 & -1/72 \\ 0 & -1 & 1/6 \\ 0 & 0 & 1 \end{pmatrix} \cdot \begin{pmatrix} e_0 \\ e_2 \\ e_4 \end{pmatrix}(x_0) + O(h) .$$

We repeat this process to find the vector for the next knot

$$\begin{pmatrix} e_0 \\ e_2 \\ e_4 \end{pmatrix}(x_2) = \begin{pmatrix} 1 & 1/6 & -1/72 \\ 0 & -1 & 1/6 \\ 0 & 0 & 1 \end{pmatrix}^2 \cdot \begin{pmatrix} e_0 \\ e_2 \\ e_4 \end{pmatrix}(x_0) + O(h) =$$

$$= \begin{pmatrix} 1 & 0 & 0 \\ 0 & 1 & 0 \\ 0 & 0 & 1 \end{pmatrix} \cdot \begin{pmatrix} e_0 \\ e_2 \\ e_4 \end{pmatrix}(x_0) + O(h) ,$$

because the square of the matrix is easily seen to be the unit matrix.

We learn from this formula that the leading coefficients of the error vector grow only by $O(h)$ with every step of $2h$. For $h \to 0$ the number of steps increases proportional to $1/h$, therefore the error vector remains bounded in all of I . This is a rough sketch for the proof of the fact that we get $O(h^{n+1})$ for the error globally. More precisely by the one step technics one could get exponential estimates for it.

It is also interesting to investigate the <u>discontinuity of e_3</u> at the knots.

If we assume $e_1(x_o) = 0$ from the outset, we see from the Taylor expansion

$$e_3(x_1-0) = -2e_2(x_o) + \frac{2}{3} e_4(x_o) .$$

If we use the corollary of the last section to calculate the parameters for the interval $I_2 = [x_1,x_2]$ we get

$$e_3(x_1+0) = -2e_2(x_1) - \frac{1}{3} e_4(x_1) + O(h)$$

$$= -2(-e_2(x_o) + \frac{1}{6} e_4(x_o)) - \frac{1}{3} e_4(x_o) + O(h)$$

$$= 2e_2(x_o) - \frac{2}{3} e_4(x_o) + O(h) .$$

This shows the symmetry of the jump of e_3, namely

$$-e_3(x_j+0) = e_3(x_j-0) + O(h) .$$

By the already stated results on polynomial splines it makes no sense to use higher order splines, because every formula with $d = 1$, n odd, $n > 3$ would only lead to an instable method.

We now concentrate on the question of proper selection of the generating functions of the splines.

APPLICATION TO POLYNOMIAL DIFFERENTIAL EQUATIONS

To give examples of the foregoing theory we have to impose additional assumptions on to $f(x,y)$ the right-hand side of the differential equations.
Let

$$f(x,y) = p_m(x) \cdot y^m + p_{m-1}(x)y^{m-1} + \ldots + p_o(x) , \qquad m > 1$$

and $p_j(x)$ polynomials. These differential equations have been investigated in the complex domain and they are known to have solutions with algebraic singularities. According to the theory of Painlevé a solution having a singularity at x^* may be expanded with respect to the "local variable"

$$t = x - x^*$$

into a series

$$y(t) = c \cdot t^{\mu} \cdot (1+c_1 t^{\gamma}+c_2 t^{2\gamma}+ \ldots) ,$$

where

$$\mu = \frac{1}{1-m}$$

and

$$c^{m-1} = \frac{\mu}{p_m(x^*)} \quad \text{if} \quad p_m(x^*) \neq 0 .$$

This may be found heuristically by inserting the formal series $y(t)$ into the differential equation. In the same way one establishes that γ is a multiple of $|\mu|$. For the formal proofs the interested reader is referred to books on differential equations in the complex domain, e.g. E. Hille [2].

The form of the above expansion suggests a family of splines that contains terms of the form $(z + \text{const})^{\alpha}$, where α may be chosen in accordance with the right-hand side of the differential equation or may be treated as one of the spline parameters. In either case the spline may become singular in the interval I.

1st CASE: SPLINES WITH FIXED EXPONENT α

To simulate the first term of the foregoing expansion and in accordance with the canonical form we construct the spline $\tilde{u}$ from pieces of the form

$$t(z,u,u',u'',u''') :=$$

$$:= u+u'\cdot z+ \frac{u''\cdot b^2}{\alpha(\alpha-1)}\cdot[(1+\frac{z}{b})^{\alpha}- \frac{\alpha\cdot z}{b} -1] \quad \text{for } \alpha \neq 0,1,2 ,$$

$$= u+u'\cdot z+u''\cdot b^2[\frac{z}{b} - \ln(1+\frac{z}{b})] \quad \text{for } \alpha = 0 ,$$

$$= u+u'\cdot z+u''\, b^2[\frac{z}{b} - (1+\frac{z}{b})\cdot \ln(1+\frac{z}{b})] \quad \text{for } \alpha = 1 ,$$

$$= u+u'\cdot z+u''\, \frac{z^2}{2} + u'''\cdot \frac{z^3}{3!} \quad \text{for } \alpha = 2 .$$

Here z is the local coordinate.
It is easy to see that u , u' , and u'' are the corresponding value resp. derivatives of $\tilde{u}$ at $z = 0$.

Furthermore it is convenient for our further purposes to introduce b instead of u''', which may be expressed by b and u''.

The intrinsic meaning of the parameter b is clear - it determines the location $\tilde{x}$ of the singularity of $\tilde{u}$, namely

$$\tilde{z} = -b \quad \text{resp.} \quad \tilde{x} = x_j + \tilde{z} = x_j - b ,$$

if the function $\tilde{u}$ is already determined

in $[x_j, x_{j+1}]$.

These splines prove useful for estimating the locations where the solution of the initial value problem becomes singular. To estimate the singularity x^* of y we use $\tilde{x}_j = x_j - b_j$, the place where the restriction of $u(x,h)$ to I_{j+1} would become singular. But which j should we take?

In performing the recursion we stop if a singularity $\tilde{x}_j$ of the spline is found that lies in the interval I_{j+2} next to I_{j+1}.

ESTIMATION OF MOVABLE SINGULARITIES OF THE SOLUTION $y(x)$

From the above functions (for $\alpha \neq 1,2$) it is easily seen that

$$u'(x,h) = u_j' + \frac{u_j''\cdot b}{\alpha-1}\left[\left(1+\frac{z}{b_j}\right)^{\alpha-1}-1\right] \quad \text{for } x \in I_{j+1} .$$

The parameter b_j is chosen such that the equation

$$u'(x_{j+1},h) = u_j' + \frac{u_j''\cdot b_j}{\alpha-1}\left[\left(1+\frac{h}{b_j}\right)^{\alpha-1}-1\right] =$$

$$= f(x_{j+1},u_{j+1})$$

holds. This equation transcribes to

$$\frac{1}{\alpha-1}\cdot\frac{\left(1+\frac{h}{b_j}\right)^{\alpha-1}-1}{h/b_j} = \frac{f(x_{j+1},u_{j+1})-u_j'}{h\cdot u_j''}$$

and this equation is actually solved, if possible, in each subinterval I_j .

We want to exploit this equation to get an appraisal of b_j in comparison to $x_j - x^*$, the distance of x_j from the singularity x^* of y .

To illustrate the quality of these formulas we treat an example in which the precise location of the singularity is known, hence the error can be seen explicitly.

Example:

$$y' = 1 + y^2 + y^4$$

$$y(0) = 1 .$$

An elementary but lengthy calculation shows

$$x^* = \frac{\pi\sqrt{3}}{12} - \frac{\ln 3}{4} = 0.178\ 796\ 769 \ldots .$$

With $t = x - x^*$ the solution y has the expansion

$$y(t) = -(3t)^{-1/3} + \frac{1}{5}(3t)^{1/3} + \frac{3}{25}\cdot(3t) + \ldots$$

that is $\mu = -1/3$ and $\gamma = 2/3$.

If spline approximation with fixed exponent α are calculated with $h = 0.015\ 625 = 2^{-6}$ up to $x_j =$ 0.15625 and then $\tilde{x}_j = x_j - b_j$ is used to estimate x^* we get the following results for different values of α .

α	-1	-1/2	-1/3	-1/5	-1/10
$\tilde{x}_j$	.18234	.17970	.17896	.17815	.17763

It is not too surprising that $\alpha = \mu$ gives the best approximation. What is remarkable, however, is the monotonicity of $\tilde{x}_j$ in dependence on the exponent α .

We set out to give an explanation of this phenomenon.

It was stated before that fourth order convergence of $u(x,h)$ to $y(x)$ will take place in any compact interval in which $y(x)$ is regular. In fact it is even seen that $u'(x,h)$ converges to $y'(x)$ with fourth order at every knot x_j, if convergence takes place at all.

This suggests that we replace $u(x,h)$ by $y(x)$ in the previous equation under the assumption that the systematic error causes a difference between b_j and $x_j - x^*$ that is mainly controlled by another effect. After this substitution b_j appears only on the left-hand side of the preceding equation.

Let

$$V(x,h) := \frac{f(x+h,y(x+h))-y'(x)}{h\cdot y''(x)} \quad \text{for fixed } x ,$$

then

$$V(x,h) = \frac{y'(x+h)-y'(x)}{h\cdot y''(x)}$$

$$= 1 + \frac{h}{2}\frac{y'''}{y''} + \frac{h^2}{6}\cdot\frac{y^{IV}}{y''} + \ldots .$$

This series converges for $|h| < |x - x^*|$, assuming that there is no other singularity of y close by.

With $v := \frac{h}{b}$ we derive an expansion of the left-hand side of the above equation from

$$G(v) := \frac{1}{\alpha-1}\cdot\frac{(1+v)^{\alpha-1}-1}{v}$$

$$= 1 + \frac{\alpha-2}{2}\cdot v + \frac{\alpha-2}{2}\cdot\frac{\alpha-3}{3}\cdot v^2 + \ldots .$$

The equation

$$G(\tfrac{h}{b}) = V(x,h)$$

gives

$$\frac{\alpha-2}{2}\cdot\frac{1}{b} + \frac{\alpha-2}{2}\cdot\frac{\alpha-3}{3}\cdot\frac{h}{b^2} + \ldots =$$

$$= \frac{1}{2}\cdot\frac{y'''}{y''} + \frac{h}{6}\cdot\frac{y^{IV}}{y''} + \ldots .$$

We summarize these findings:

For fixed x as $h \to 0$ we get

$$b = (\alpha-2) \cdot \frac{y''}{y'''} \qquad (\alpha \neq 2,1)$$

and the convergence is linear.

If $t = x - x^*$ denotes the distance of x from the singularity x^* of y then

$$b = \frac{2-\alpha}{2-\mu} \cdot t(1+c^* \cdot t^\gamma + \ldots) \ .$$

The last relation is found by the formal differentiation of the above expansion of y about x^* and substitution into

$$\frac{y''(x)}{y'''(x)} = \frac{c \cdot \mu \cdot (\mu-1) \cdot t^{\mu-2} + c_1 (\mu+\gamma)(\mu+\gamma-1) \cdot t^{\mu+\gamma-2} + \ldots}{c \cdot \mu \cdot (\mu-1)(\mu-2) t^{\mu-3} + c_1 (\mu+\gamma)(\mu+\gamma-1)(\mu+\gamma-2) t^{\mu+\gamma-3} + \ldots} \ .$$

Furthermore we may compare

$$x^* = x - t \quad \text{and} \quad \tilde{x} = x - b \ .$$

By the above formula

$$\tilde{x} = x - \frac{2-\alpha}{2-\mu} \cdot t + O(t^{1+\gamma}) \cong x - t + \frac{\alpha-\mu}{2-\mu} \cdot t + O(t^{1+\gamma})$$

$$\cong x^* + \frac{\alpha-\mu}{2-\mu} \cdot t \ .$$

For small negative t, i.e. if we integrate the initial value problem with $h > 0$ by the described spline method and if x approaches a singularity then we have (asymptotically) the <u>monotonicity relation</u>:

If $\mu < 2$ and $\alpha < \mu$ then $\tilde{x} > x^*$,

$\alpha > \mu$ then $\tilde{x} < x^*$.

This explains the behaviour of the numbers seen in the previous example.

On the other hand one may use the obtained information to subject $\tilde{x}$ to a correction

$$x^* \cong \tilde{x} - \frac{\alpha-\mu}{2-\mu} \cdot t \cong \tilde{x} - \frac{\alpha-\mu}{2-\alpha} b =: \hat{x} .$$

In the above example, where $x^* = 0.178796...$ this correction yields the following table of results:

	α	-1	-1/2	-1/3	-1/5	-1/10
$h = 2^{-6}$	$\tilde{x}_j$	.18234	.17970	.17896	.17815	.17763
$x_j = 10 \cdot h$	$\hat{x}_j$	.17654	.17814	———	.17948	.18001
$h = 2^{-7}$	$\tilde{x}_j$	.18180	.17956	.178825	.17824	.17780
$x_j = 21 \cdot h$	$\hat{x}_j$	.17786	.17853	———	.17910	.17933

It is seen that the estimation of x^* by $\hat{x}$ is again monotonic, which is understandable again from the above expansions. It should, however, be kept in mind that we neglected the errors of the numerical integration, therefore caution in using this procedure is advisable.

The correction does not work if $\alpha = \mu$, because $\frac{\alpha-\mu}{2-\mu} b \equiv 0$.

In the next section we propose an extrapolation scheme that is applicable in this case.

THE EXTRAPOLATION TECHNIQUE FOR THE SINGULARITY IN CASE $\alpha = \mu$

If $\alpha = \mu$ the expansion of b is simplified to be

$$b = t + c_1^* \cdot t^{1+\gamma} + \dots .$$

This is the parameter value b corresponding to a spline osculating y at the point $x = x^* + t$ that is for the limit case $h = 0$. We do not know, however, and do not wish to compute c_1^* by cumbersome calculations that would have to be repeated for every equation.

The expansion of b may be inverted to give

$$t = b + d_1(-b)^{1+\gamma} + \dots$$

(remember $b < 0$ if we integrate with $h > 0$ and approach a singularity from the left).

To get more information we may consider b_j associated with x_j for different values of j. Though t_j is unknown, we can infer that

$$t_j - t_i = (j-i)h .$$

This makes it feasible to eliminate the leading higher order terms with the unknown coefficients $d_1, \dots$. If we take two terms, for example,

$$t_j = b_j + d_1(-b_j)^{1+\gamma}$$
$$t_{j+1} = b_{j+1} + d_1(-b_{j+1})^{1+\gamma}$$

then

$$h = b_{j+1} - b_j + d_1[(-b_{j+1})^{1+\gamma} - (-b_j)^{1+\gamma}] ,$$

so that d_1 can be evaluated and may be used to calculate the extrapolated value

$$\tilde{\tilde{x}}_{j+1} = \tilde{x}_{j+1} - d_1(-b_{j+1})^{1+\gamma} .$$

It is clear how to generalize this method to more terms.

One may systematize the elimination procedure by rewriting

$$x^* \cong x_j + t_j = \tilde{x}_j + d_1(-b_j)^{1+\gamma} + d_2(-b_j)^{1+\gamma+|\mu|} + \dots$$

into the form of a system of homogeneous linear equations for the "unknowns" $(1, d_1, d_2, \dots, d_n)$, i.e.

$$0 = 1\cdot(\tilde{x}_j - \tilde{x}^*) + d_1(-b_j)^{1+\gamma} + d_2(-b_j)^{1+\gamma+|\mu|} + \dots$$
$$0 = 1\cdot(\tilde{x}_{j+1} - \tilde{x}^*) + d_1(-b_{j+1})^{1+\gamma} + d_2(-b_{j+1})^{1+\gamma+|\mu|} + \dots$$

$$\dots .$$

Here $\tilde{x}^*$ is the yet unknown extrapolated value that approximates x^* .

Since there is a nontrivial solution of this system its determinant must vanish.

$$\det \begin{pmatrix} \tilde{x}_j - \tilde{x}^* & (-b_j)^{1+\gamma} & (-b_j)^{1+\gamma+|\mu|} \ldots \\ \tilde{x}_{j+1} - \tilde{x}^* & (-b_{j+1})^{1+\gamma} & (-b_{j+1})^{1+\gamma+|\mu|} \ldots \end{pmatrix} = 0.$$

Expanding with respect to the first column we see that

$$\tilde{x}^* \cdot \sum_{i=0}^{n} A_{i1} = \sum_{i=0}^{n} \tilde{x}_{j+1} \cdot A_i \ ,$$

where the A_{i1} are the obvious cofactors of the expansion of the determinant with respect to the first column. It is beyond the scope of this introduction to give a detailed analysis and comparison of the different sources of error or to provide exhaustive proofs.

We conclude by giving the result of the extrapolation in case $\alpha = \mu = -\frac{1}{3}$, and $n = 1$ for the previously given example.

From

x_j	$\tilde{x}_j$	$\lvert b_j \rvert$
0.1640625	0.178871	0.014808
0.171875	0.178825	0.006950

we find $\tilde{x}^* = 0.178807\ldots$ which compares excellently with the exact value $x^* = 0.178796\ldots$.

2nd CASE: SPLINES CONTAINING TERMS WITH VARIABLE EXPONENT α

The above form of a solution of the differential equation in the neighborhood of a singularity x^* suggests also the following alternative choice of functions t for the generation of the splines:

$$u(x_j+z,h) = t(z;x_j,u,\ldots,u''') \quad \text{for} \quad 0 \leq z = x-x_j \leq h$$

with

$$t(z;x,u,u',u'',u''') = u + u' \cdot \frac{b}{\alpha} [(1+\frac{z}{b})^{\alpha}-1] , \quad \alpha \neq 0,1.$$

The parameters α and β are used instead of u'' and u''' , but it is easy to establish the connection. From

$$t'(z;\dots) = u' \cdot (1 + \frac{z}{b})^{\alpha-1}$$

$$t^{(i)}(z;\dots) = \frac{\alpha-i+1}{b} \cdot t^{(i-1)}(0;\dots) \cdot [1 + \frac{z}{b}]^{\alpha-i} \quad \text{for } i > 1$$

we obtain

$$u''/u' = \frac{\alpha-1}{b}$$

$$\text{hence } \frac{1}{b} = u''/u' - u'''/u''$$

$$u'''/u'' = \frac{\alpha-2}{b} \qquad = \frac{(u'')^2-u' \cdot u'''}{(u')^2} \cdot \frac{u'}{u''} ,$$

which transforms into

$$\frac{1}{b} = -[\ln(\frac{u''(z;\dots)}{u'(z;\dots)})]\Big|'_{z=0} ,$$

$$\frac{1}{\alpha-1} = \frac{u'(0;\dots)}{u''(0;\dots)} \cdot \frac{1}{b} .$$

If $\alpha < 0$ the above function becomes singular if $1 + \frac{z}{b}$ tends to zero, i.e. at the point

$$\underline{x}(x) = x - b(x) .$$

This fact will again later be used to estimate the location of the singularity of the approximated function y . If we use y,y',y'',y''' at x to replace u,u',u'',u''' we approximate y by an osculating function of the class t .

If y satisfies the differential equation $y' = f(x,y)$ then

$$\frac{y''}{y'} = f_y(x,y) + f_x(x,y)/f(x,y)$$

and similarly the logarithmic derivative can be expressed by f and its derivatives. Remarkably, if we take the trouble of evaluating the derivatives of f analytically then we get the quantities b, α with the same accuracy as y , i.e. if numerical integration is performed,

$$y(x) \sim u(x) + O(h^4) .$$

This means very high accuracy for the derivatives and associated terms. A warning - the coefficient of the O-Relation may become large if x approaches the singularity x^* . It would be an interesting research topic to study this behaviour in more detail.

We will see by numerical evidence that the error of numerical integration influences the error of the singularity estimate less than the error inherent to our procedure. (This may be called the systematic error.) This is the justification that we replace the numerical solution $u(x,h)$ by the "theoretical" solution $y(x)$ of the differential equation in our analysis.

In a recent paper [11] it was shown, how the osculating function could be used to get asymptotically an estimate of an interval that includes the singularity x^* of the solution to an initial value problem $\mathbb{P}$.

Suppose that $\underline{x}(x) > x$ as defined above is a monotonically increasing concave function of x in some neighborhood $x_0 \leq x < x^*$ of the singularity. Furthermore let x and $x - h$, $h > 0$ belong to this interval. Letting $\Delta_t(x,x-h)\underline{x}(t)$ be the first difference quotient of $\underline{x}$ we set

$$\overline{x}(x) = x - \frac{b(x)}{1-\Delta_t(x,x-h)\underline{x}(t)} .$$

(Assume the denominator to be positive.)

Then

$$\underline{x}(x) < x^* < \overline{x}(x) .$$

Remark: The key fact in this inclusion theorem is that $\underline{x}(x)$ has some analytical property from which the existence of such a neighborhood of monotonicity can be deduced. For example the existence of a series expansion of

$\underline{x}(x)$ with leading coefficients of known sign could provide such an information.

The proof of this inclusion is almost immediate. We have to estimate the zero x^* of the function $z(x) := \underline{x}(x) - x$. This function $z(x)$ is concave just as $\underline{x}(x)$. Hence the secant through the points $(x-h, z(x-h))$ and $(x, z(x))$ will majorize $z(t)$ for $x < t < x^*$. This secant cuts the x-axis at

$$\bar{x}(x) = x + \frac{z(x)}{-\Delta(x-h,x)z(\cdot)} = x - \frac{b(x)}{1-\Delta(x-h,x)x(\cdot)} .$$

Hence

$$x^* \le \bar{x}(x) .$$

To apply this inclusion one could numerically integrate the initial value problem, calculate $b(x)$ and see whether $-b(x) = z(x)$ is a concave function. Finally from the values of b at adjacent grid points the estimate for x^* may be obtained.

But we will in this case just as before give a look at asymptotic expansions and **their use in locating a** singularity.

SINGULARITY ESTIMATIONS OF THE SOLUTIONS OF INITIAL VALUE PROBLEMS

We may ask how the solution y of an initial value problem can be continued up to a singular point. Assume that the right hand side of the differential equation is an m-th degree polynomial in y . Therefore we use splines from the family defined in the foregoing section. The approximating spline may be iteratively continued until we reach an interval $I_{j+1} := [x_j, x_{j+1}]$ such that the analytic continuation of $u|_{I_{j+1}}$ will become singular within I_{j+2} , i.e. b found in I_{j+1} satisfies $-2h \le b_j < -h$. Then

$$x = x_j - b_j$$

provides an estimate for the singularity of y .

If we repeat the calculations with a different step size, say $h/2$, we may get a slightly different b at the same point x_j . Therefore we should write

$$b = b(x_j, h) \ .$$

It can be shown that

$$b(x_j, h) \to b(x_j) = [\ln(\frac{y''}{y'})]\Big|'_{x=x_j} \quad \text{for} \quad h \to 0$$

and that this convergence is linear in h for fixed x_j. (Compare with [11].) Hence we may apply the extrapolation procedure:

$$b(x) \doteq b(x,0) = 2b(x,h/2) - b(x,h) \ .$$

We search for an x such that $b(x,0) = 0$. As before we use the expansion in powers of t^γ of $y(t)$ to see that there is an appropriate expansion

$$b(t) = t\cdot(1 + a_1 t^\gamma + a_2 t^{\gamma+|\mu|} + \ldots) \ .$$

Unfortunately the value t_j that corresponds to x_j is unknown, in fact, it is exactly the quantity we want to calculate. Inverting this expansion we find

$$t = b + d_1 \cdot b^{1+\gamma} + d_2 \cdot b^{1+\gamma+|\mu|} + \ldots$$

with unknown coefficients d_j.

For $b := b(x_j,0), b(x_{j+1},0),\ldots$ the corresponding values t_j, t_{j+1} satisfy $t_j - t_{j+1} = x_j - x_{j+1}$ which is equal to the step size h. If we use the expansion of t_j for several values of j we may eliminate the quantities t_j by forming differences and calculate the coefficients $d_1, d_2, \ldots$ from the resulting equations we may then use the foregoing formula to calculate t_j.

Practically, the elimination of the coefficients d_j can be achieved in the following way. If x_j is added to both sides of the last equation we find

$$x^* = x_j + t_j = \tilde{x}(x_j) + d_1 \cdot b(x_j, 0)^{1+\gamma} + \ldots ,$$

$$\text{with} \quad \tilde{x}(x_j) := x_j + b(x_j, 0) \ .$$

Subtracting x^* on both sides, we get the homogeneous linear equation

$$A_n \cdot \begin{pmatrix} 1 \\ d_1 \\ d_2 \\ \cdot \\ \cdot \\ \cdot \end{pmatrix} = \begin{pmatrix} 0 \\ \cdot \\ \cdot \\ \cdot \\ 0 \end{pmatrix}$$ for the quantities $1, d_1, d_2, \ldots, d_n$

where A_n denotes an $(n+1) \times (n+1)$ matrix

$$\begin{pmatrix} \tilde{x}(x_j)-x^* & b(x_j,0)^{1+\gamma} & \ldots & b(x_j,0)^{1+\gamma+|\mu|(n-1)} \\ \cdot & \cdot & & \cdot \\ \cdot & \cdot & & \cdot \\ \cdot & \cdot & & \cdot \\ \tilde{x}(x_{j+n})-x^* & b(x_{j+n},0)^{1+\gamma} & \ldots & b(x_{j+n},0)^{1+\gamma+|\mu|(n-1)} \end{pmatrix}.$$

Since there is a solution of the homogeneous system the first component of which is equal to 1 , we have det $A_n = 0$.

Expanding this determinant with respect to the first column (the quantities A_{j1} being the cofactors of the determinantal expansion) one finds,

$$\tilde{x}^* \cdot \sum_{i=0}^{n} A_{j1} = \sum_{i=0}^{n} \tilde{x}(x_{j+i}) \cdot A_{j1} .$$

Before giving an example we remark once more that we have made the assumption that the error of the x^* estimation is primarily due to the systematic error and that the calculation of the values $u(x,h)$ by the spline technique is sufficiently accurate to interchange y and u .

As an example we choose again the initial value problem

$$y' = 1 + y^2 + y^4$$

$$y(0) = 1$$

the solution of which can be found by elementary integration. The solution has a singularity at $x^* = \frac{\pi}{12} \cdot \sqrt{3} - \frac{\ell n\ 3}{4} \doteq .178\ 796\ 769$. In [11] inclusion techniques as described before were applied to this example. We will show here how well the extrapolation performs in this regard.

For $h = .015\ 625$ we find

j	x_j	$-b(x_j,h)$	$-b(x_j,h/2)$	$-b(x_j,0)$	$\tilde{x}(x_j)$
8	.125	.051 534	.051 146	.050 758	.175 758
9	.140 625	.037 296	.036 944	.036 592	.177 217
10	.156 25	.022 364	.022 190	.022 016	.178 266

Remarks: It is convenient to work with $-b$ since b is always negative.

With the above expansion of the determinant A_n we obtain for

$$n = 1:\ (j=9,10) \quad \tilde{x}^* \doteq .179\ 053 \quad \tilde{x}^*-x^* \doteq .257\cdot 10^{-3}\ ,$$

$$n = 2:\ (j=8,9,10) \quad \tilde{x}^* \doteq .178\ 951 \quad \tilde{x}^*-x^* \doteq .155\cdot 10^{-3}\ .$$

If we take

$h = 0.003\ 906\ 25$ and $x_j = h\cdot j$ then the values are

j	$-b(x_j,h)$ $\cdot 10^8$	$-b(x_j,h/2)$ $\cdot 10^8$	$-b(x_j,0)$ $\cdot 10^8$	$\tilde{x}(x_j)$ $\cdot 10^8$
42	145 6408	145 3738	145 1068	178 573 18
43	107 4629	107 2505	107 0381	178 672 56
44	68 9570	68 8251	68 6932	178 744 32

For $n = 1$:

$$(j = 43,44) \quad \tilde{x}^* \doteq .178\ 8099 \quad \tilde{x}^*-x^* \doteq 1.31\cdot 10^{-5}\ ,$$

For $n = 2$:

$$(j = 42,43,44) \quad \tilde{x}^* \doteq .178\ 8006 \quad \tilde{x}^*-x^* \doteq .39\cdot 10^{-5}\ .$$

We numerically confirm the observation of [12] that

the first coefficient in the expansion of $t(b)$ is relatively small,

$$t(b) = b(1 + d_1 b^{2/3} + d_2 b^{4/3} + \dots)$$

$$d_1 \doteq 0.170\,238 \quad , \quad d_2 \doteq 1.567\,376 \ .$$

DEPENDENCE OF SINGULARITY ON INITIAL DATA

Consider $y' = f(x,y)$ in $(0,1)$. Determine (if possible) y_0 such that

$y(x)$ with $y(0) = y_0$ satisfies

$$\lim_{x \to 1} |y(x)| = \infty \ .$$

For fixed $N \in \mathbb{N}$ let $h = 1/N$.

A <u>shooting technique</u> is applied to select $C = C(h)$ such that the spline solution $u(x,h)$ with

$u(0,h) = C(h)$ satisfies

$$\lim_{x \to 1} |u(x,h)| = \infty \ .$$

We take $C(h)$ as approximation to y_0 and ask for the asymptotic behaviour of $C(h) - y_o^o$.

Examples have shown a sometimes unexpected high order of convergence with respect to h, see [12].

Refering to the previous considerations, we concentrate on the systematic error of the spline method.

1) Let $y_h(x)$ denote that solution of the differential equation which satisfies

$$y_h(0) = C(h) \ .$$

The singularity will probably not occur exactly at $x = 1$ but in a neighborhood, say at

$$x^* = 1 + \sigma(h) \ .$$

2) We assume that there is a differentiable relation between $y_h(0)$ and σ as independent variable such that

$$\frac{\partial y_h(0)}{\partial \sigma} \neq 0 .$$

3) Finally we explicitly assume that the shift $\sigma(h)$ of the singularity primarily depends upon the systematic error, not the round off and the propagated error.

   Practically $y_h(x)$ is used instead of $u(x,h)$ in the following considerations.

A consequence of these assumptions is the fact that we may study the dependence between h and σ instead of h and $y_h(0)$.

The parameter u'' or its equivalent b as pointed out before is determined from

$$u'(jh+h;h) = f(jh+h,u(jh+h,h)) .$$

The equation reduces to

$$\frac{u'_{j+1}}{u'_j} = \left(1 + \frac{h}{b}\right)^{\alpha-1} \quad \text{with } u'_j := u'(jh,h) .$$

Again elimination of α by means of b furnishes

$$\left(\frac{u'(jh+h,h)}{u'(jh,h)}\right)^{\frac{u'_j}{h\cdot u''_j}} = \left(1 + \frac{h}{b}\right)^{\frac{b}{h}} .$$

Specialization to $j = N - 2$ and the condition $b = -2h$ yield the relation for the dependence of σ on h . Now we replace u by y_h , taking into account that

$$u'_j = f(x_j,u_j) \doteq f(x_j,y_h(x_j)) = y'_h(x_j) .$$

We thus find the defining equation

$$F(\sigma,h) := \left(\frac{y'_h(x_{N-1})}{y'_h(x_{N-2})}\right)^{\frac{y'_h(x_{N-2})}{hy''_h(x_{N-2})}} = 4 .$$

We now introduce

$$t = x - x_h^* = x - (1 + \sigma(h)) \quad \text{and}$$

$$y_h(t) = c \cdot t^{\mu}(1+c_1 t^{|\mu|}+c_2 t^{2|\mu|}+c_3 t^{3|\mu|}+ \ldots) ,$$

$$y_h'(t) = c\mu t^{\mu-1}+|\mu| \cdot c_2 \cdot ct^{|\mu|-1}+2|\mu| c_3 ct^{2|\mu|-1}+ \ldots .$$

It is possible that some of the c_j might vanish.

Observation: 1. Since the defining equation contains only first and second order derivatives the second term of the expansion of $y_h(t)$ the constant term of the expansion, i.e. $c \cdot c_1$, does not influence $F(\sigma,h)$.
Therefore in all expansions of F the difference between the first and second exponent of t is (at least) $2|\mu|$.

2. The coefficients c, c_j may be dependent upon σ. Since x_{N-2}, x_{N-1} correspond to $t_{N-2} = -2h-\sigma$, $t_{N-1} = -h-\sigma$ one may expand F to find that

$$\sigma = h^{1+\gamma} \cdot \text{const} + o(h^{1+\gamma}) .$$

Here γ denotes the difference of the exponents of the first and second term of the expansion of y_h' about $x^* = 1$ for which the coefficient are different from zero.

The details are given in [12]. It is also pointed out there that it might seem that the numerical convergences of σ and hence $y_h(0)$ is of higher order than $1 + \gamma$. This may be due to the fact that the coefficient of $h^{1+\gamma}$ in the expansion of σ is relatively small as compared to that of $h^{1+\gamma+|\mu|}$ (or any higher one). In any case Richardson's extrapolation is a feasable way to improve the convergence of

$$y_h(0) = y_0 + d_1 \cdot h^{1+\gamma}+ d_2 h^{1+\gamma+|\mu|} + \ldots ,$$

with $h = h_0, h_0/2, h_0/4, \ldots$. (For numerical examples see [12].)

REFERENCES

[1] Arndt, H., Lösung von gewöhnlichen Differentialgleichungen mit nichtlinearen Splines, Num. Math. 33, 323-333 (1979).

[2] Hille, E., Ordinary differential equations in the complex domain, J. Wiley & Sons, NewYork-London-Sydney-Toronto, (1976).

[3] Loscalzo, F.R., and Talbot, T.D., Spline function approximations for solutions of ordinary differential equation, SIAM J. Numer. Anal. 4, 433-445 (1967).

[4] Mülthei, H.N., Splineapproximationen von beliebigem Defekt zur numerischen Lösung gewöhnlicher Differentialgleichungen, Teil I, II, III, Numer. Math. 32, 146-157, 343-358 (1979), Numer. Math. 34, 143-154 (1980).

[5] Rooij van, P.L.J., and Schurer, F., A bibliography on spline functions II, TH Report 73-WSK-01, Technological University Eindhoven, Netherlands.

[6] Runge, R., Lösung von Anfangswertproblemen mit Hilfe nichtlinearer Klassen von Spline-Funktionen, Dissertation Münster, 1972.

[7] Schoenberg, I.J. (ed.), Approximations with special emphasis on spline functions, Academic Press, NewYork-London, (1969).

[8] Varga, R., Error bounds for Spline Interpolation, p. 367-388 in [7], (1969).

[9] Werner, H., Neuere Entwicklungen auf dem Gebiete der nichtlinearen Splinefunktionen, ZAMM 58, 86-95 (1978).

[10] Werner, H., An introduction to non-linear splines, in Polynomial and Spline Interpolation, B.N. Sahney (ed.), Reidel, Dordrecht, Boston-London, 247-304 (1979).

[11] Werner, H., Extrapolationsmethoden zur Bestimmung der beweglichen Singularitäten von Lösungen gewöhnlicher Differentialgleichungen, in: Numer. Math., R. Ansorge, K. Glashoff, B. Werner (ed.), ISNM 49, p. 159-176 (1979).

[12] Werner, H., Spline Functions and the Numerical Solution of Differential Equations, in: Special Topics of Applied Mathematics, ed. by. J. Frehse, D. Pallaschke, U. Trottenberg, North-Holland Publishing Company, Amsterdam-NewYork-Oxford, 173-192 (1980).

[13] Werner, H., and Wuytack, L., Nonlinear Quadrature Rules in the Presence of a Singularity, Comp. & Maths. with Appls., Vol. 4, p. 237-245, Pergamon Press Ltd. (1978).

[14] Werner, H., and Zwick, D., Algorithms for numerical integration with regular splines, Schriftenreihe des Rechenzentrums der Universität Münster, Nr. 27 (1977).

ON THE USE OF A CLASS OF ALGEBRAIC MAPPINGS TECHNIQUES FOR SOME PROBLEMS IN COMPLEX ANALYSIS

David Y. Y. Yun

IBM Research, Math Sciences Department
Box 218, Yorktown Heights, N. Y. 10598

Abstract: In computer algebra, there are two predominant classes of constructive techniques useful for a large variety of algebraic problems, particularly in the multivariate case. Both rely on a projection mapping technique to cast the problem into a simpler domain, by computing residues (or modular images). The difference between the two techniques lies mainly in their method of lifting the solution from the simpler domain back up to the original domain of concern. The class based on the Chinese Remainder Theorem utilizes multiple images for reconstruction, while the one based on the Hensel Lemma relies on a single modulus. In this paper, we concentrate on the former technique. We give a general description of the technique and offer a perspective that unifies two other well known (algebraic) techniques in complex numerical analysis -- fast Fourier transform and polynomial evaluation and interpolation. The essential supporting algebraic operations, such as polynomial multiplication, division with remainder, and the extended Euclidean computation, are also discussed to give them the proper algebraic setting. This leads to another unification to two important problem areas of numerical analysis -- rational approximation and solving Toeplitz equations. These problems are given an asymtotically fast yet practically realizable algorithmic solution. Finally, this class of techniques is further demonstrated by applying them to derive algebraic identities that underly even faster algorithms, such as FFT and the multiplication of complex numbers.

I. Residue Mapping Techniques

In computer algebra many computational operations must deal with multivariate expressions. These variables interact and the expressions often grow in size. That is why simplification is the essence of algebraic computation. Yet simplification must

H. Werner et al. (eds.), Computational Aspects of Complex Analysis, 361–377.

often be intermixed with many other operations. For example, after each polynomial multiplication there should be a simplification to gather all terms with the same degrees. After most operations there is a need to keep the resulting expression as simple as possible. It is not really cost effective if simplification is only used for the clean up job. The most effective method suitable for the purpose of both performing some operations and keeping the resulting expressions simple is by ***projection***. In its simplest form, a projection might be viewed as evaluating a variable at a particular point. In general, it takes on the following schematic form:

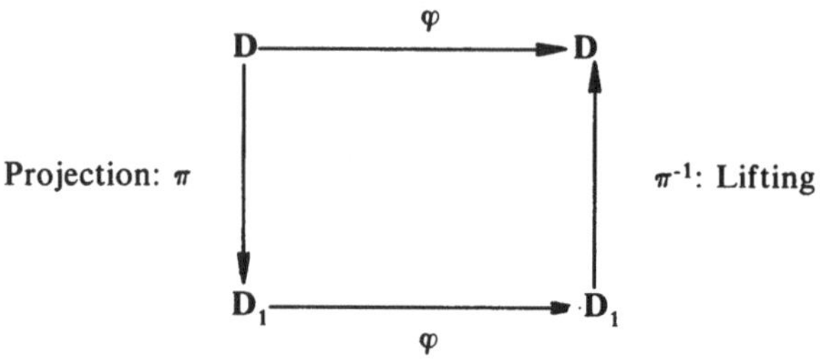

Here, **D** is the domain of computation and φ is the operation to be performed on elements of **D**. For example, **D** could be the domain of polynomials, rational functions, or power series in one or more variables; and φ could be the operation of multiplication, greatest common divisors (GCD), factorization, reversion, or solving equations. The schematic diagram depicts a method of performing the operation φ by projection and lifting in the operator sense of

$$\varphi(\mathbf{D}) = \pi^{-1}\ \varphi_{\pi}\ \pi(\mathbf{D}).$$

Namely, to perform φ on **D** the operation is carried out by first projecting the given elements of **D** via a mapping π to a sub-domain $\mathbf{D}_1$ of **D**; then performing the corresponding operation φ_π within the sub-domain; and finally lifting the results from $\mathbf{D}_1$ up to the original domain of concern **D** via the inverse mapping π^{-1}.

Over the years of effort to improve the capabilities of computer algebra, two predominant classes of mapping techniques emerged -- Chinese Remaindering and P-adic Construction. Both use ***residue computation*** as the projection mapping π, which we will discuss shortly. The lifting techniques (π^{-1}) differ, but are based on two well known algebraic constructive techniques -- the Chinese Remainder Theorem [KNUT72] and the Hensel Lemma [YUN76]. Both methods have permeated essentially all of computer algebra by underlying numerous constructive algorithms.

In this paper, we will only concentrate on the first class of techniques, their specializations, interrelations, and applications. Included in this class are pairs of invertible mappings, π and π^{-1}: residue(s) computation (also known as computing the modular representation) and the Chinese Remainder Algorithm (CRA); polynomial evaluation and interpolation; the forward Fast Fourier Transform and the Inverse FFT (denoted forthwith by FFT and IFFT respectively). The second class of p-adic constructions can not be covered here due to lack of space. Interested readers can consult the following references [ZASS69], [YUN76], [WANG79], [YUN80], [KNUT81], and [LIPS81]. This class includes the p-adic construction based on the Hensel lemma and the more familiar Newtonian iteration on algebraic domains [LIPS76]. The major distinction between these two classes of mapping techniques is that the first class uses multiple moduli (to compute multiple residue images of the result) and the second uses a single modulus (or single residue image).

II. Residue Computation and Chinese Remainder Algorithm

The type of algebraic objects we consider for the remainder of this paper will be generally a Euclidean domain. The ***modular representation*** of an element A of a Euclidean domain with respect to a given set of k **moduli**, M_1, M_2, ..., M_k, is a set of ***residues*** or remainders, R_1, R_2, ..., R_k, which result from dividing A by M_i for $i=1,2,...,k$. The computation of residues R_i w.r.t. the given moduli M_i represents a projection mapping π. The inverse mapping is provided by the constructive proof of the Chinese Remainder Theorem (circa 100BC, [KNUT81]). We give an algorithmicized version of the theorem for the Euclidean domain:

Given residues $(R_1, R_2,...,R_k)$ corresponding to
moduli $(M_1, M_2,..., M_k)$, which are pairwise relatively prime,
there exists an (effectively computable) element R of the same domain such that $R \equiv R_i \pmod{M_i}$, for $i = 1,2,...,k$.
And R is uniquely determined if $\text{size}(R) < \text{size}(\prod_{i=1}^{k} M_i)$.

The computational algorithm, often known as the Chinese Remainder Algorithm (CRA), can be expressed in one line, as a generalized ***Lagrange Interpolation Formula*** (cf. [YUN79]):

$$\sum_{i=1}^{k} R_i C_i D_i \text{ modulo } M, \quad \text{where } C_i = M/M_i \text{ and } D_i = C_i^{-1} \text{ modulo } M_i.$$

We will discuss several special cases of residue computation and CRA:

Consider the Euclidean domain of polynomials in x over the rationals $\mathbf{Q}[x]$. The given moduli $M_i = P_i(x)$, $i=1,...,k$, are pairwise relatively prime polynomials in $\mathbf{Q}[x]$. Let $P(x)=P_1P_2...P_k$. If $A(x)$ is a polynomials satisfying $\deg(A)<\deg(P)$, then A is uniquely represented by the set of residues R_i w.r.t. P_i Also, given R_i and P_i the unique A with degree less than $\deg(P)$ can be computed by CRA or the generalized Lagrange Interpolation Formula. This is the concept of ***modular representation*** and ***Chinese remainder algorithm*** for polynomials.

Another case is the well known operation of polynomial ***evaluation and interpolation***. Here M_i is atill in Q[x] but further restricted to linear polynomials x-b_i, i=1,...,k (i.e. $P_i(x)$ = x-b_i). For a given polynomial $A(x)$, the residue computation R_i = A mod x-b_i becomes evaluation of $A(x)$ at b_i = $A(b_i)$. Since the b_i's are distinct, the P_i's are pairwise relatively prime. The inverse mapping is precisely the process of polynomial interpolation, specifically here Lagrange interpolation process. The computation process can be made more efficient in this special case as shown in [AH&U74].

The last special case is when the b_i's are not arbitrary points but roots of unity ω^i. In this case the pair of invertible mappings is the ***Discrete Fourier Transform*** and the ***inverse Discrete Fourier Transform***. Given a vector (c_0, c_1, ..., c_k), we associate a polynomial $C(x)$ of degree k with the c_i's as its coefficients. Residue computation is evaluation of this polynomial at ω^i, $C(\omega^i)$ = d_i, resulting in a vector of d_i's known as the DFT of the vector of c_i's. The inverse mapping corresponds to interpolation w.r.t. the d_i's. The importance of this pair of mappings stems from the popularity of FFT, where its algebraic properties are somewhat neglected since the domain of concern for FFT is usually the complex numbers.

The computational efficiency of FFT, O(n log n) for a n-point transform, is the cause of its popularity [C&T65]. Borodin and Munroe [B&M75] give the algorithm and analysis for the evaluation and interpolation mappings in a general setting, with a cost of O(n $\log^2$ n) where n is the degree of the polynomial. This particular mapping technique (sometimes known as the modular homomorphism technique) is a popular, practical method of computations with polynomials in computer algebra systems. Its importance in algebraic computation was established in the early 1970's through the works of Brown [BROW71] and Collins [COLL71] on polynomial GCD and related problems. However, the use there is almost exclusively restricted to moduli which are linear polynomials of the form x-b. The more general form of the technique, residue computation and CRA, is needed in the simplest of domains -- the integers. Its use for (large) integer calculations has a long history in number theory. The utility and efficiency considerations in the more complex domains of multivariate polynomials and algebraic functions motivated the work by Yun [YUN79], which gives an algorithm and cost analysis of CRA in the most general setting and still maintain the efficiency of O(n $\log^2$ n).

The interrelationship of these algebraic mappings reaches much deeper than what we already discussed. In fact, as we consider some of the necessary basic operations supporting the class of residue computation and Chinese remaindering, many additional relations come to the surface. First we consider the most fundamental operation for polynomials -- multiplication:

Given polynomials $F(x) = \sum_{i=0}^{n} f_i x^i$, $G(x) = \sum_{i=0}^{n} g_i x^i$, the product polynomial

$$H(x) = \sum_{k=0}^{n+m} h_k x^k = F(x)G(x)$$

can be computed by the usual formula

$$h_k = \sum_{j=0}^{k} f_j \, g_{k-j} \, .$$

But the formula suggests immediately convolution and FFT. Indeed, it is well known (c.f. [AH&U74]) that this polynomial can be carried out via FFT, and thereby achieving O(n log n) polynomial multiplication. However, FFT computation requires roots of unity, hence numerical approximation. Although the evaluation and interpolation mappings cost O(n $\log^2 n$), they allow the flexibility of choosing a set of arbitrary (distinct) rational or even integral (hence exact) points. The method is simply to choose $m+n+1$ distinct integral values $x_0,\ldots,x_{m+n}$. Evaluate F and G at these points. Computed the point-wise product $H(x_i) = F(x_i)G(x_i)$, $i = 0,\ldots,m+n$. And finally interpolate to the polynomial H from these point-wise values. In operator notation with F denoting FFT, we have the convolution of the vector of coefficients of $F(x)$ and $G(x)$, $\vec{f} \circledast \vec{g} = \mathrm{IFFT}(\mathrm{FFT}\,\vec{f} \cdot \mathrm{FFT}\,\vec{g})$. The use of FFT for multiplication also points out the connection of viewing a polynomial as a vector of coefficients. From this point of view, another form of multiplication will prove to be useful later. Let $\vec{h}$ denote the vector of coefficients for $H(x)$.

$$\vec{h} = \vec{f} \circledast \vec{g} = \begin{bmatrix} f_0 & 0 & \cdot & 0 \\ f_1 & \cdot & \cdot & \cdot \\ \cdot & \cdot & \cdot & 0 \\ f_n & \cdot & f_1 & f_0 \\ 0 & f_n & \cdot & f_1 \\ \cdot & \cdot & \cdot & \cdot \\ 0 & \cdot & 0 & f_n \end{bmatrix} \begin{bmatrix} g_0 \\ \cdot \\ \cdot \\ g_m \end{bmatrix} = \begin{bmatrix} g_m & 0 & \cdot & 0 \\ \cdot & \cdot & \cdot & \cdot \\ g_1 & \cdot & g_m & 0 \\ g_0 & g_1 & \cdot & g_m \\ 0 & \cdot & \cdot & \cdot \\ \cdot & \cdot & \cdot & g_1 \\ 0 & \cdot & 0 & g_0 \end{bmatrix} \begin{bmatrix} f_0 \\ \cdot \\ \cdot \\ f_n \end{bmatrix}$$

That is the product polynomial $H(x)$ can also be obtained as a vector $\vec{h}$ by the multiplication of a Toeplitz matrix of coefficients of F and the vector of coefficients of G or by the multiplication of a Toeplitz matrix of coefficients of G and the vector of coefficients of F. A further relation to the FFT is revealed when we reorder the elements of a (5-point) FFT matrix:

$$\begin{bmatrix} A_0 \\ A_1 \\ A_2 \\ A_4 \\ A_3 \end{bmatrix} = \begin{bmatrix} \omega^0 & \omega^0 & \omega^0 & \omega^0 & \omega^0 \\ \omega^0 & \omega^1 & \omega^2 & \omega^4 & \omega^3 \\ \omega^0 & \omega^2 & \omega^4 & \omega^3 & \omega^1 \\ \omega^0 & \omega^4 & \omega^3 & \omega^1 & \omega^2 \\ \omega^0 & \omega^3 & \omega^1 & \omega^2 & \omega^4 \end{bmatrix} \begin{bmatrix} a_0 \\ a_1 \\ a_2 \\ a_4 \\ a_3 \end{bmatrix} .$$

This shows the embedding of a Hankel circulant (the lower right principal minor) within the FFT matrix, i.e. the reason for FFT to be a special case of Hankel or Toeplitz matrix operations.

Next, we consider another fundamental polynomial operation -- division with remainder, which by definition is to find quotient $Q(x)$ and remainder $R(x)$ given $F(x)$ and $G(x)$ such that

$$F(x) = G(x)Q(x) + R(x), \text{ size } R < \text{size } G$$

Define the ***reverse*** of a degree-n polynomial $P(x)$ to be $P^r(x) = x^n P(\frac{1}{x})$. We can derive

$$x^n F(\frac{1}{x}) = [x^m G(\frac{1}{x})][x^{n-m} Q(\frac{1}{x})] + x^{n-m+1}[x^{m-1} R(\frac{1}{x})]$$

where deg $F = n$ and deg G $= m$. In terms of reverse polynomials, that means

$$F^r(x) = G^r(x)Q^r(x) + x^{n-m+1} R^r(x)$$

The significance of this equation is further revealed by

$$Q^r(x) = \frac{F^r}{G^r} \mod x^{n-m+1}$$

Since deg$(Q)=n$-m, it is now clear that the quotient polynomial can be completely determined via a truncated power series division (without remainders). This domain contains series of at most n-m+1 terms, so that the computation effort is reduced. The essence of this derivation is that the computation of the quotient of division with remainder is only proportional to the difference of the degrees of F and G (n-m) and not to n or m, which could be much greater.

Another important operation we must consider is the ***Extended Euclidean Algorithm*** (EEA). The use of Euclidean algorithm to compute GCD's is one of the oldest known algorithms [KNUT72]. It needs no further reiteration. But, its slight generalization, EEA, yields such useful by-products that deserves some discussion here. More specifically, recall that the generalized Lagrange interpolation formula requires the computation of C_i^{-1} mod M_i. EEA provides a solution to this problem. Given A and B, EEA finds solutions V and W to the equation

$$AV + BW = \text{GCD}(A,B).$$

When dealing with finite field, algebraic number, and/or algebraic function arithmetic where GCD$(A, B)=1$, this equation provides the often indispensible operation of computing inverse of A with respect to the modulus B or vice versa, i.e. $V = A^{-1}(\text{mod } B)$ or $W = B^{-1}(\text{mod } A)$. The algorithm EEA can be stated simply:

With $U_0 = A$ and $U_1 = B$;
Initialize $V_0 \leftarrow 1$; $V_1 \leftarrow 0$; $W_0 \leftarrow 0$; $W_1 \leftarrow 1$;
Iterate for $1 \leq i \leq k$

$U_{i+1} \leftarrow U_{i-1} - Q_i\, U_i$ by division, together with the iterations
$V_{i+1} \leftarrow V_{i-1} - Q_i\, V_i$
$W_{i+1} \leftarrow W_{i-1} - Q_i\, W_i$
Then $V = V_k$ and $W = W_k$ satisfy $AV + BW = \mathrm{GCD}(A, B)$.

In fact, a similar relation holds for each i:

$$U_0\, V_i + U_1\, W_i = U_i\ (= A\, V_i + B\, W_i),$$

which provides the essential relation for rational function approximation to be discussed later and the discovery of faster versions of the extended Euclidean computation discussed below.

Several asymptotically fast extended Euclidean algorithms, achieving $O(n \log^2 n)$, have been developed. We briefly mention three here since they achieve the efficiency while maintaining the practicality.

- The most general fast algorithm is PRSDC (for *P*olynomial *R*emainder *S*equence by *D*ivide and *C*onquer) developed by Gustavson and Yun [G&Y79]. Given U_0, U_1, and a positive rational number $r < n = \mathrm{size}(U_0)$, algorithm PRSDC computes the 2 by 3 matrix

$$M_j = \begin{bmatrix} U_j & W_j & V_j \\ U_{j+1} & W_{j+1} & V_{j+1} \end{bmatrix}, \text{ where } \begin{bmatrix} U_j \\ U_{j+1} \end{bmatrix} = \begin{bmatrix} W_j & V_j \\ W_{j+1} & V_{j+1} \end{bmatrix} \begin{bmatrix} U_0 \\ U_1 \end{bmatrix}$$

  for such a j that $\mathrm{size}(U_j) \geq r$ and $\mathrm{size}(U_{j+1}) < r$. This algorithm is useful for computing arbitrary (s, t) Padé approximants asymptotically fast.
- A more specialized version of PRSDC is called EMGCD (for *E*xtended *M*iddle *GCD*) [BGY80]. It also computes the matrix M_j with the same properties except r is specifically $n/2$. The reason for this particular valueof r is that it is the only one needed for solving a n by n Toeplitz or Hankel system of equations.
- The algorithm HGCD given in [AH&U74], which is a rendition of an algorithm originally proposed by Moenck [MOEN73], was developed primarily to compute polynomial GCD's fast. Applied to U_0 and U_1, it yields

$$R = \begin{bmatrix} W_j & V_j \\ W_{j+1} & V_{j+1} \end{bmatrix}, \text{ such that } R \begin{bmatrix} U_0 \\ U_1 \end{bmatrix} = \begin{bmatrix} U_j \\ U_{j+1} \end{bmatrix}.$$

  where $\deg(U_j) > n+1/2$ and $\deg(U_{j+l}) \leq n+1/2$. The difference between HGCD and EMGCD or PRSDC is minor, but evident from the degree conditions. It is more convenient to apply the condition of PRSDC, however. Some minor errors in the statements of algorithm HGCD have also been corrected. It is important to accredit HGCD for containing the essence of the divide and conquer idea that made the asymptotically fast algorithm possible.

III. Rational Approximation and Toeplitz Equations

Two areas of complex numerical analysis where techniques of computer algebra (some discussed above) have found significant relations and utility are ***rational Hermite approximation*** and the solution of linear systems of ***Toeplitz equations***. Given an analytic function $f(x)$ and a bounded sequence of points (some of which may be repeated) $x_0, x_1, x_2, \ldots, x_{m+n}$, the rational function $R_{mn}(x) = U(x)/V(x)$ with $\deg(U) < m$ and $\deg(V) < n$ is called the (m, n)th ***Rational Hermite Interpolant*** of f if

$$f(x) - R_{mn}(x) = g(x) \prod_{i=0}^{m+n} (x - x_i) \tag{1}$$

for some analytic function $g(x)$.

There are three extreme cases of the Hermite interpolation problem:

- If any point x_i is repeated m_i times then the interpolant coincides with $f(x_i)$ and it first m_i derivatives at x_i. Hermite solved the case $(m,n) = (N,0)$, then the interpolant is called the ***Hermite Interpolating Polynomial***.
- When the sequence of points are distinct it is called ***Cauchy Interpolation*** and
- When all the points are the same it is called ***Padé Approximation***.

In the algebraic formulation, a rational function $R_{mn}(x) = U(x)/V(x)$ is said to solve the ***Modified Hermite Interpolation Problem*** if

$$U(x) \equiv f(x)V(x) \mod \prod_{i=0}^{N} (x - x_i) \tag{2}$$

If $R_{mn}(x)$ solves equation (1) then equation (2) is automatically satisfied. For some choices of m and n equation (1) may have no solutions, and in that case there is a parameterized family of solutions to equation (2). However, each solution $(U(x), V(x))$ to equation (2) then yields the same rational function. This unique function is called the (m,n)th ***Rational Interpolant*** for $f(x)$. Thus the set of rational interpolants for $f(x)$, which is called the ***Rational Interpolation Table*** for $f(x)$, contains all solutions to the problem of rational Hermite interpolation.

In 1974, Wuytack [WUYT74] showed that the modified rational Hermite interpolation problem always has a solution. Warner [WARN76] later showed that all solutions to the problem can be computed by Kronecker's algorithm [KRON81]. McEliece and Shearer [M&S78] and Gustavson and Yun [G&Y79] independently discovered that Padé approximants can be computed by the Extended Euclidean Algorithm. It is shown in [G&Y79] that Kronecker's algorithm and EEA are essentially the same, but can be actually carried out asymptotically fast. Briefly, let

$$U_0 = \prod_{i=0}^{m+n} (x - x_i) \quad \text{and} \quad U_1(x) = \sum_{i=0}^{m+n} a_i x^i,$$

which is a power series representation of the function $f(x)$. The extended Euclidean algorithm in the form of algorithm PRSDC applied to U_0 and U_1 computes to a particular index j, according to the given (m, n), to yield U_j, V_j, and W_j such that

$$U_j(x) = f(x)V_j(x) + \prod_{i=0}^{m+n} (x-x_i) \; W_j(x). \tag{3}$$

Thus U_j/V_j is the (m, n)th rational Hermite interpolant (or Padé approximant) of $f(x)$. Assuming the usual definition for the rational interpolation table (or Padé Table, depending on whether the points are distinct or not), the following results can be established (respectively):

- Each step of the extended Euclidean computation gives rise to a unique entry (in lowest terms) of the rational interpolation table (Padé Table).
- The rational function U_j/V_j obtainable via the extended Euclidean computation yields $\deg(Q_j)$ equal entries of the rational interpolation table (Padé Table) along the $(m+n)$th anti-diagonal.
- All entries along the $(m+n)$th anti-diagonal of the rational interpolation table (Padé Table) for a power series are computed uniquely by the extended Euclidean computation.

For the case $m = n$ and all the points are the same (Padé case), equating coefficients of x^n, x^{n+1}, ..., x^{2n} in the relation (corresponding to eqn. (3)) for the (n, n)th Padé approximant, we get a Toeplitz system:

$$\begin{bmatrix} a_n & & & a_0 \\ & \cdot & & \\ & & \cdot & \\ & & & \cdot \\ a_{2n} & & & a_n \end{bmatrix} \begin{bmatrix} v_0 \\ \cdot \\ \cdot \\ \cdot \\ v_n \end{bmatrix} = \begin{bmatrix} u_n \\ 0 \\ \cdot \\ \cdot \\ \cdot \\ 0 \end{bmatrix}$$

where the n by n matrix, denoted by T, is Toeplitz. The vectors $u = (u_0,\ldots,u_n)^T$ and $v = (v_0,\ldots,v_n)^T$ are the coefficients of the (n, n)th Padé approximant $(U_j(x),V_j(x))$. This observation and the above results indicate that Euclid's algorithm can be adapted to solve Toeplitz systems of equations. Namely, this equation provides the solution to a special case, $Tv = u_n e_0$, of the system of Toeplitz equation of interest, $Tz = b$:

$$\begin{bmatrix} a_n & \cdots & a_1 & a_0 \\ a_{n+1} & & & a_1 \\ \cdot & & & \cdot \\ \cdot & & & \cdot \\ \cdot & & & \cdot \\ a_{2n} & \cdots & a_{n+1} & a_n \end{bmatrix} \begin{bmatrix} z_0 \\ z_1 \\ \cdot \\ \cdot \\ \cdot \\ z_n \end{bmatrix} = \begin{bmatrix} b_0 \\ b_1 \\ \cdot \\ \cdot \\ \cdot \\ b_n \end{bmatrix} \qquad \begin{matrix} T_{ij} = a_{i-j+n} \\ \\ i,j = 0,\ 1,\ldots,\ n \end{matrix}$$

Several observations are important to solving Toeplitz equations: T is $(n+1)$ by $(n+1)$ but only contains $2n+1$ distinct elements; T^{-1} is not Toeplitz and may have $(n+1)^2/2$ distinct elements; T is persymmetric (symmetric about the anti-diagonal) and so is T^{-1}. A theorem by Trench [TREN64] indicates that the first step of solving a Toeplitz system of equations is to solve for x and y of $Tx = e_0$ and $T^t y = e_0$. The solution z can then be expressed in terms of x and y. His suggestion was used by Zohar [ZOHA69] and given a convolutional setting by Kailath, Viera, and Morf [KVM78]. A theorem synthesized by Gustavson and Yun [G&Y79], which is a compaction of two theorems due to Gohberg and Semencul [G&S74], provides the most general and appropriate setting to solve Toeplitz equations along the line of Trench's idea.

Theorem : Let the Toeplitz matrix

$$\tilde{T} = \begin{bmatrix} a_n & \bullet & \bullet & a_0 & a_{-1} \\ \bullet & \bullet & & \bullet & \bullet \\ \bullet & & \bullet & \bullet & \bullet \\ a_{2n} & & & a_n & \bullet \\ a_{2n+1} \bullet & & \bullet & \bullet & a_n \end{bmatrix}$$

be a bordering of the Toeplitz matrix T with one additional row and column consisting of all the same elements except two at the corners. Suppose $x = (x_0,\ldots,x_{n+1})^T$ and $y^R = (y_{n+1},\ldots,y_0)^T$ are solutions of $\tilde{T}x = e_0$ and $\tilde{T}y^R = e_{n+1}$ and suppose $x_0 = y_0 \neq 0$. Then T is invertible and it's inverse S is formed according to the formula (4):

$$S = \frac{1}{x_0}\left\{ \begin{bmatrix} x_0 & 0 & \bullet & 0 \\ x_1 & \bullet & \bullet & \bullet \\ \bullet & \bullet & \bullet & 0 \\ x_n & \bullet & x_1 & x_0 \end{bmatrix} \begin{bmatrix} y_0 & y_1 & \bullet & y_n \\ 0 & \bullet & \bullet & \bullet \\ \bullet & \bullet & \bullet & y_1 \\ 0 & \bullet & 0 & y_0 \end{bmatrix} - \begin{bmatrix} y_{n+1} & 0 & \bullet & 0 \\ y_n & \bullet & \bullet & \bullet \\ \bullet & \bullet & \bullet & 0 \\ y_1 & \bullet & y_n & y_{n+1} \end{bmatrix} \begin{bmatrix} x_{n+1} & x_n & \bullet & x_1 \\ 0 & \bullet & \bullet & \bullet \\ \bullet & \bullet & \bullet & x_n \\ 0 & \bullet & 0 & x_{n+1} \end{bmatrix} \right\} \quad (4)$$

Furthermore, suppose x and y^R solve $Tx = e_0$ and $Ty^R = e_n$ and $x_0 = y_0 \neq 0$. Then $T^{-1} = S$ is given by formula (4) with x_{n+1} and y_{n+1} set equal to zero (which expresses the original Trench formula).

This theorem provides the basis to solve a system of Toeplitz equations, where it is certainly assumed $\mathrm{Det}(T) \neq 0$. However, if $x_0 = 0$ then formula (4) can no longer be used. This is precisely the reason for needing this general theorem involving $\tilde{T}$ which offers two free parameters as additional degrees of freedom. We can choose values for a_{-1} and a_{2n+1} so that $\mathrm{Det}(\tilde{T}) \neq 0$. Then $x_0 = \tilde{T}_{11}^{-1} = \mathrm{Det}(T)/\mathrm{Det}(\tilde{T}) \neq 0$. Thus formula (4) can again be used and we have a stronger statement:

for solving $Tz = b$ it is always possible to find x and y of formula (4) such that $x_0 = y_0 \neq 0$.

Formula (4) expresses the inverse S of a Toeplitz matrix T as a difference of two products of triangular Toeplitz matrices. But the matrix multiplications, which will be too costly to carry out, can be avoided. To solve $Tz = b$ we can perform four matrix-vector multiplications to obtain the solution $z = Sb$. We now recall the observation earlier that the multiplication of Toeplitz matrices and the vector b given by

$$\begin{bmatrix} x_0 & 0 & \cdot & 0 \\ x_1 & \cdot & \cdot & \cdot \\ \cdot & \cdot & \cdot & 0 \\ x_n & \cdot & x_1 & x_0 \\ x_{n+1} & x_n & \cdot & x_1 \\ 0 & \cdot & \cdot & \cdot \\ \cdot & \cdot & \cdot & x_n \\ 0 & \cdot & 0 & x_{n+1} \end{bmatrix} \begin{bmatrix} b_0 \\ \cdot \\ \cdot \\ b_n \end{bmatrix} \quad \text{and} \quad \begin{bmatrix} y_{n+1} & 0 & \cdot & 0 \\ y_n & \cdot & \cdot & \cdot \\ \cdot & \cdot & \cdot & 0 \\ y_1 & \cdot & y_n & y_{n+1} \\ y_0 & y_1 & \cdot & y_n \\ 0 & \cdot & \cdot & \cdot \\ \cdot & \cdot & \cdot & y_1 \\ 0 & \cdot & 0 & y_0 \end{bmatrix} \begin{bmatrix} b_0 \\ \cdot \\ \cdot \\ b_n \end{bmatrix}$$

are precisely the concatenations of the four matrices in formula (4) and clearly correspond to polynomial multiplications. In fact, the multiplications are truncated and can be carried out via FFT with the appropriate ordering of the coefficients x_i, y_i, and b_i. Specifically and assuming we have the simpler case where $x_0 \neq 0$, let the vectors $(x_0,\dots,x_n)^T$ and $(y_0,\dots,y_n)^T$ be normalized by x_0 into f and g respectively. Now correspond to each vector, we consider the polynomials $F(x)$, $G(x)$, and $B(x)$. Then the multiplications in terms of polynomials are

$$P^r(x) \leftarrow F(x)B^r(x) \pmod{x^{n+1}}$$
$$Q^r(x) \leftarrow G^r(x)B^r(x) \pmod{x^{n+1}}$$
$$z \leftarrow x_0[G(x)P(x) - F^r(x)Q(x)] \pmod{x^{n+1}}$$

Thus, it is clear that the process of solving a Toeplitz equation $Tz = b$ is dominated by solving for x and y such that $Tx = e_0$ and $Ty^r = e_n$ (or if $x_0 = 0$ then $\tilde{T}x = e_0$ and $\tilde{T}y^R = e_{n+1}$), which is $O(n \log^2 n)$, rather than the cost of actually getting the solution vector z, which is $O(n \log n)$ since only polynomial multiplications are needed.

IV. Derivation of Fast Algorithms by Mapping Techniques

Now we will give a few examples, relevant to complex numerical computation, to illustrate some other utility of the algebraic mapping techniques that were just described. This set of examples are all algebraic identities, which appear somewhat magical to someone who had seen them for the first time. We will derive the identities underlying the "fast" algorithms for the 7-point FFT and multiplying complex numbers through the use of techniques based on residue computation and Chinese remaindering. The aim is to show that an n-point FFT can be carried out in $O(n)$ scalar multiplications and that multiplying complex numbers requires only three or even two scalar multiplies depending on the underlying field.

First, some necessary theoretical results further strengthening the power of this mapping technique, in terms of its computational efficiency. It has long been known that multiplying polynomials

$$X(t) = \sum_{i=0}^{m} x_i\, t^i \quad \text{and} \quad Y(t) = \sum_{i=0}^{n} y_i\, t^i$$

over a field F requires a munimum of $m+n+1$ field element multiplications. To computationally achieve this minimum, the evaluation and interpolation techniques can be used in two schemes as follows:

choose $m+n+1$ distinct elements $\alpha_0,\alpha_1,\ldots,\alpha_{m+n}$ of F, then

$$X(t)*Y(t) = \text{Interp}\{X(\alpha_i)*Y(\alpha_i), i = 0,\ldots, m+n\} \bmod \prod_{i=0}^{m+n} (t - \alpha_i) \qquad (S1)$$

or choose only $m+n$ distinct elements $\beta_1,\beta_2,\ldots,\beta_{m+n}$, of F, then

$$X(t)*Y(t) = \text{Interp}\{X(\beta_i)*Y(\beta_i), i = 1,\ldots, m+n\} \bmod \prod_{i=1}^{m+n} (t - \beta_i)$$

$$+\, x_m y_n \prod_{i=1}^{m+n} (t - \beta_i) \qquad (S2)$$

In fact, any algorithm for multiplying $X(t)$ and $Y(t)$ in $m+n+1$ multiplies is equivalent to either scheme (S1) or (S2) (up to scaling by an element of F) [WINO78]. As we have already pointed out, the evaluation and interpolation mapping techniques can be generalized if we replace

$(t - \alpha_i)$by $\quad M_i(t)$

$X(\alpha_i)$by $\quad X(t) \bmod M_i(t)$

$Y(\alpha_i)$by $\quad Y(t) \bmod M_i(t)$

and Interpolation by CRA. We will, in fact, use this general form in the following examples.

Earlier at this meeting, Prof. Merz refered to the fast algorithm (of Winograd) for the 7-point discrete Fourier transform, without proof or derivation. We are now well equiped to present the derivation here, using the same notation of Merz:

$$\begin{bmatrix}\psi_0\\ \psi_1\end{bmatrix} = \begin{bmatrix}\Omega_0 & \Omega_1\\ \Omega_1 & \Omega_0\end{bmatrix}\begin{bmatrix}\phi_0\\ \phi_1\end{bmatrix} = \begin{bmatrix}\Omega_0 & 0\\ \Omega_1 & \Omega_1\\ 0 & \Omega_0\end{bmatrix}\begin{bmatrix}\phi_0\\ \phi_1\end{bmatrix}$$

$$= \Omega(t)\phi(t) \mod t^2 - 1$$

where $\Omega(t) = \Omega_0 + \Omega_1 t$, $\phi(t) = \phi_0 + \phi_1 t$.
$t^2 - 1 = (t-1)(t+1)$, so let $M_1(t) = t - 1$, $M_2(t) = t + 1$, i.e. $\alpha_1 = 1, \alpha_2 = -1$.
Now evaluate: $\Omega(S1) = \Omega_0 + \Omega_1$, $\phi(S1) = \phi_0 + \phi_1$.
So $m_1 = (\Omega_0 + \Omega_1)(\phi_0 + \phi_1) = \Omega(S1)\phi(S1) = \Omega(t)\phi(t) \mod t - 1$
$\Omega(-1) = \Omega_0 - \Omega_1$, $\phi(-1) = \phi_0 - \phi_1$.
So $m_2 = (\Omega_0 - \Omega_1)(\phi_0 - \phi_1) = \Omega(-1)\phi(-1) = \Omega(t)\phi(t) \mod t + 1$
By definition $\psi(t) = \psi_0 + \psi_1 t = \psi(t) \mod t^2 - 1 = \Omega(t)\phi(t) \mod t^2 - 1$.
But the residue images of $\psi(t)$ have already been obtained $m_1 = \psi(t) \mod t - 1$, $m_2 = \psi(t) \mod t + 1$. Since the extended Euclidean algorithm yields $\frac{-1}{2}(t-1) + \frac{1}{2}(t+1) = 1$,

$$m_1(t+1)\frac{-1}{2} + m_2(t-1)\frac{1}{2} = \psi(t) \mod t^2 - 1,$$

by the CRA. That is

$$\frac{1}{2}(m_1 + m_2) + \frac{1}{2}(m_1 - m_2)t = \psi_0 + \psi_1 t$$

Hence $\psi_0 = \frac{1}{2}(m_1 + m_2)$, $\psi_1 = \frac{1}{2}(m_1 - m_2)$.

The second part of the derivation is to compute

$$\begin{bmatrix}c_0\\ c_1\\ c_2\end{bmatrix} = \begin{bmatrix}a_0 & a_2 & a_1\\ a_1 & a_0 & a_2\\ a_2 & a_1 & a_0\end{bmatrix}\begin{bmatrix}b_0\\ b_1\\ b_2\end{bmatrix} \quad \text{i.e.} \quad \begin{aligned} c_0 &= a_0 b_0 + a_1 b_2 + a_2 b_1\\ c_1 &= a_1 b_0 + a_0 b_1 + a_2 b_2\\ c_2 &= a_0 b_2 + a_2 b_0 + a_1 b_1\end{aligned}$$

Let
$$\begin{aligned} A(t) &= a_0 + a_1 t + a_2 t^2\\ B(t) &= b_0 + b_1 t + b_2 t^2\\ C(t) &= c_0 + c_1 t + c_2 t^2\end{aligned} \quad \text{so that } C(t) = A(t)B(t) \mod t^3 - 1.$$

$$t^3 - 1 = (t-1)(t^2 + t + 1) \text{ and } \frac{(-t-2)}{3}(t-1) + \frac{1}{3}(t^2 + t + 1) = 1 \text{ by EEA.}$$

So $p_0(t^2 + t + 1)\frac{1}{3} + p_1(t-1)(\frac{-t-2}{3})\ C(t) \mod t^3 - 1$.

$p_0 = m_0 = A(t)B(t) \mod t - 1 = (a_0 + a_1 + a_2)(b_0 + b_1 + b_2)$.
$p_1 = A(t)B(t) \mod t^2 + t + 1$
$= ((a_0 - a_2) + (a_1 - a_2)t) \times ((b_0 - b_2) + (b_1 - b_2)t) \mod t^2 + t + 1$.

This multiplication $\times$ will be carried out by scheme (S2). We need a modulus with degree = 2, but we have the freedom of choosing it. We choose a very simple one, $t(t+1)$.

mod t: $m_1 = (a_0-a_2)(b_0-b_2)$

mod $t+1$: $m_2 = (a_0-a_1)(b_0-b_1)$

Since $(-1)t + 1(t+1) = 1$, $m_1(t+1)-m_2t = \times \mod t^2+t$

Thus, $m_1(t+1)-m_2t + m_3(t^2+t) = m_1 + (m_1 - m_2 + m_3)t + m_3t^2$

where $m_3 = (a_1-a_2)(b_1-b_2)$.

$p_1 = \times \mod t^2+t+1 = (m_1-m_3) + (m_1-m_2)t$

We finally have

$$m_0(t^2+t+1)\frac{1}{3} + [(m_1-m_3) + (m_1-m_2)t](t-1)(\frac{-t-2}{3}) \mod t^3 - 1$$

$$= \frac{m_0}{3}(t^2+t+1) + \frac{m_3-m_1}{3}(t^2+t-2) + \frac{m_2-m_1}{3}(t^2-2t+1) = C(t)$$

i.e.
$$c_0 = (m_0 + m_1 + m_2 - 2m_3)/3$$
$$c_1 = (m_0 + m_1 - 2m_2 + m_3)/3$$
$$c_2 = (m_0 - 2m_1 + m_2 + m_3)/3$$

In the above derivations, only the m's are the actual field multiplications that needs to be carried out.

The next example regards multiplying complex numbers:

$$z_0 + iz_1 = (x_0 + ix_1)(y_0 + iy_1)$$

We first consider the case where the field of constants is the rationals **Q**. As before we consider the corresponding problem of polynomial multiplication:

$$Z(t) = X(t)Y(t) \mod t^2 + 1$$

Carrying out the calculation following scheme (S2) with the modulus chosen to be $t(t-1)$, we get

$$m_1 = x_0y_0,\ m_2 = (x_0 + x_1)(y_0 + y_1),\ m_3 = x_1y_1.$$

and $z_0 = m_1-m_3$, $z_1 = -m_1 + m_2-m_3$,

i.e.
$$x_0y_0-x_1y_1 = x_0y_0-x_1y_1.$$
$$x_1y_0-x_0y_1 = -x_0y_0 + (x_0 + x_1)(y_0 + y_1)-x_1y_1.$$

Multiplying the first equation by z_0 and the second one by z_1 and add, we can then gather terms with respect to

$$x_0:\ y_0z_0 + y_1z_1 = y_0(z_0 - z_1) + (y_0 + y_1)z_1$$
$$x_1:\ -y_1z_0 + y_0z_1 = (y_0 + y_1)z_1-y_1(z_0 - z_1)$$

corresponding to a multiplication of complex number with just some renaming of the variables,

$$x_1 + ix_0 = (y_0 + iy_1)(z_1 + iz_0)$$

i.e. accomplishing the same objective with a different set of three multiplications in the field:

renaming z_1 by x_0 and z_0 by x_1,
$m_1 = x_0(y_0 + y_1)$, $m_2 = y_0(-x_0 + x_1)$, $m_3 = y_1(x_0 + x_1)$,
and $z_0 = m_1 - m_3$, $z_1 = m_1 + m_2$.

In fact, it has been proved that when the field of constants is **Q**, three multiplications is minimum [WINO80]. However, if the field is **Q**(i), the extension field of **Q** by i, then since a factorization is now possible $t^2 + 1 = (t + i)(t - i)$, a different derivation (using evaluation and interpolation) leads to:

$$m_1 = (\frac{1}{2}x_0 + \frac{i}{2}x_1)(y_0 + iy_1),$$

$$m_2 = (\frac{1}{2}x_0 - \frac{i}{2}x_1)(y_0 - iy_1),$$

and $z_0 = m_1 + m_2$, $z_1 = -im_1 + im_2$. The result is that only two field multiplications are needed, when the field is extended to allow factorization of the critical modulus, in this case $t^2 + 1$.

References

[AH&U74] Aho, A. V., Hopcroft, J. E., and Ullman, J. D. *The Design and Analysis of Computer Algorithms*, Book, Addison Wesley, Reading, Mass., 1974

[B&M75] Borodin, A., and Munro, I. (1975) *The Computational Complexity of Algebraic and Numeric Problems*, Elsevier, New York.

[BROW71] W. S. Brown, "On Euclid's Algorithm and the Computation of Polynomial Greatest Common Divisors", *JACM*, Vol. 18, No. 4, Oct. 1971, pp 478-504.

[BGY80] R.P. Brent, F.G. Gustavson, and D. Y. Y. Yun, Fast Computation of Pade Approximants and the Solution of Toeplitz Systems of Equations, *The Journal of Algorithms, Vol. 1,* Academic Press, June 1980, pp. 259-295.

[COLL71] Collins, G. E., The calculation of multivariate polynomial resultants, *JACM*, Vol. 18, No. 4, Oct. 1971, pp. 515-532.

[C&T65] Cooley, J. W. and Tukey, J. W., An algorithm for the machine calculation of complex Fourier series, *Math of Computation 19,* 1965, pp. 297-301.

[G&S72] Gohberg, I. C., and Semencul, A. A., "On the Inversion of Finite Toeplitz Matrices and their Continuous Analogs", Mat. Issled., 2

(1972), pp. 201-233. (In Russian) Also see Gohberg, I. C. and Feldman, I. A., "*Convolution equations and projection methods for their solutions*", Translation of Math. Monographs, Vol. 41, Amer. Math. Soc., 1974.

[GRAG72] Gragg, W. B., "The Padé Table and its Relation to Certain Algorithms of Numerical Analysis", SIAM Rev., Vol. 14, No. 1, Jan. 1972, pp. 1-62.

[G&Y79] Gustavson, F. G. and Yun, D. Y. Y., Fast Computation of Padé Approximants and Toeplitz Systems of Equations via the Extended Euclidean Algorithm, IBM Research Report RC 7551, March 1979.

[KVM78] Kailath, T., Viera, A., and Morf, M., "Inverses of Toeplitz Operators, Innovations, and Orthogonal Polynomials", SIAM Rev., Vol. 20, No. 1, Jan. 1978, pp. 106-119.

[KNUT81] Knuth, D. E., *The Art of Computer Programming, Vol. 2 -- Seminumerical Algorithms,* Addison-Wesley, second edition, 1981.

[KRON81] Kronecker, L., "Zur Theorie der Elimination einer Variabelen aus zwei algebraisthen Gleichungen", Monatsber. Konigl. Preuss. Akad. Wiss. Berlin, 1881, pp. 535-600.

[LIPS76] Lipson, J. D., Newton's method: a great algebraic algorithm, *Proc. 1976 ACM Symposium on Symbolic and Algebraic Computation,* Aug. 1976, 260-270.

[LIPS81] Lipson, J. D., *Elements of Algebra and Algebraic Computing,* Addison-Wesley, 1981.

[M&S78] McEliece, R. J. and Shearer, J. B., "A Property of Euclid's Algorithm and an Application to Padé Approximation", SIAM J. Appl. Math., Vol. 34, No. 4, June 1978, pp. 611-615.

[MOEN73] R. Moenck, Fast Computation of GCD's, *Proc. of 5th Annual ACM Symposium on Theory of Computing,* 1973, 142-151.

[M&K77] Morf, M. and Kailath, T., Recent Results in Least-Square Estimation Theory, Annals of Economic and Social Measurements, June 1977, pp. 261-274.

[TREN64] Trench, W. F., "An algorithm for the inversion of finite Toeplitz Matrices", J. SIAM 12, 3, Sept. 1964, pp. 515-522.

[WANG79] P. S. Wang and L. P. Rothschild, Factoring Multivariate Polynomials Over the Integers, *Math of Computation,* Vol. 29, No. 131, July 1975, pp. 935-950.

[WARN74] Warner, D. D., *Hermite Interpolation with Rational Functions*, Thesis, Univ. of Calif. at San Diego, June 11, 1974.

[WARN78] Warner, D. D., "Kronecker's Algorithm for Hermite Interpolation with an Application to Sphere Drag in a Fluid-filled Tube". Proc. of the Workshop on Padé Approx., Feb. 1976, Eds. D. Bessis, J. Gliewicz, P. Mery, CNRS Marseilles, pp. 48-51.

[WINS78] Winograd, S., "On Computing the Discrete Fourier Transform", *Mathematics of Computation,* Vol. 32, No. 141, Jan. 1978, 175-199.

[WUYT74] Wuytack, L., On some aspects of the rational interpolation problem, *SIAM J. on Numerical Analysis*, Vol. 11, No. 1, March 1974, pp. 52-60.

[YUN76] Yun, D. Y. Y., Algebraic algorithms using p-adic consturctions, *Proc. 1976 ACM Symposium on Symbolic and Algebraic Computation,* Aug. 1976, 248-259.

[YUN79] Yun, D. Y. Y., "Uniform Bounds for a Class of Algebraic Mappings Related to Residue Computation and Chinese Remaindering", IBM R. C. 7185, Yorktown Heights, N. Y., June 16, 1978, *SIAM J. Computing,* Vol. 8, No. 3, Aug. 1979, pp. 348-356.

[Y&G79] Yun, D. Y. Y. and Gustuvson, F. G., Fast Computation of the Rational Hermite Interpolant and Solving Toeplitz Systems of Equations via the Extended Euclidean Algorithm, *Lecture Notes in Computer Science, No. 72,* edited by G. Goos and J. Hartmanis, Springer-Verlag, New York, June 1979

[YUN80] Yun, D. Y. Y., *The Hensel Lemma in Algebraic Manipulation* one of 28 in "Outstanding Dissertations in the Computer Sciences Series", Garland Publishing, Inc., 1980.

[ZASS69] Zassenhaus, H., On Hensel Factorization I, *Journal of Number Theory,* Vol. 1, 1969, pp. 291-311.

[ZOHA69] Zohar, S., "Toeplitz Matrix Inversion: The Algorithm of W. F. Trench", J. ACM 16, 4, Oct. 1969, pp. 592-601.

COMPUTER ALGEBRA AND COMPLEX ANALYSIS

David Y. Y. Yun

IBM Research, Math Sciences Department
Box 218, Yorktown Heights, N. Y. 10598

Abstract: Taking complex analysis to mean complex numerical analysis, I perceive my mission here to be that of disseminating the algebraic approach taken by computer algebraists to many mathematical problems, which arise from and are important to complex analysis. In turn, complex numerical analysis can be, and have been, providing essential theoretical and computational results for computer algebra. The cross fertilization should and must continue in order that computational mathematics progress with the joint aid of both tools, rather than branching into orthogonal pursuits with disparate approaches. First, we discuss the different issues and principal concerns of computer algebra. Then, the algebraic approach to a long standing problem in calculus or complex analysis, indefinite integration in closed form, will be motivated and derived through examples. Algorithmic solution to the basic, thought provoking, problem of rational function integration as well as theoretical foundation underlying the algorithm for elementary function integration will be discussed. Further issues and approaches will be illustrated through another central (implicitly essential) problem of computer algebra, that is simplification of symbolic and algebraic expressions. We conclude by showing a set of computer executed problems in integration to reveal some of the new capabilities added to the arsenal of a mathematician through the efforts of computer algebra.

I. Introduction

The underlying principles for computer algebra are substantially different from those of numerical analysis that it is important to state and clarify here:

(a) The most essential principle of computer algebra is the use of ***exact arithmetic***. Since the issue is as basic as determining whether a term $x/3$ in an expression

H. Werner et al. (eds.), Computational Aspects of Complex Analysis, 379–393.

cancels with another, 0.333333*x, arithmetic must be carried out with high precision. Computer algebra programs often utilize "indefinite precision" rational (integral) arithmetic to preserve accuracy and determine ***exact cancellation***.

(b) The second principle of computer algebra is to ***defer numerical approximation*** until the latest possible step of any problem solving process. A typical example lies in the process of finding zeros of a polynomial. Real or complex coefficients of the given polynomial must be presented to a digital computer in an approximated format, most likely in floating point. This initial approximation of the input is equivalent to a rational approximation. Trusting the accuracy of the given data means relying on the correctness of the rational coefficients, which is our implicit assumption (as does a numerical analyst with the floating point numbers). We would then view the given polynomial with rational coefficients as an exact representation of the input data. Such a polynomial can be first factored into "square-free" factors (hence determining all multiplicities of zeros), and then into irreducible factors (over the rationals) without further introducing any numerical error. Based on this principle and only when necessary, would computation of real (floating point) quadratic factors or complex zeros be carried out numerically. In fact, for many problems it suffices to work with a symbolic representation (say, α where $P(\alpha) = 0$ and $P(x)$ is one of the factors over the rationals), or a rational isolating interval (rectangle for complex zeros).

(c) Numerical errors aside, a binary operation on two numbers (in floating point) yields another number, which in fact takes less storage. Such a simplicity of data structures in numerical analysis is not enjoyed by computer algebra programs. A case in point is $(a + b)*(c + d)$ yields $ac + bc + ad + bd$ requiring eight as opposed to four literals for representation. Both of these forms must be represented differently in some data structure (where an intermediate form might be $a(c + d) + b(c + d)$). When other algebraic objects, such as Gaussian rationals, polynomials, power series, rational functions, algebraic and transcendental functions, as well as matrices and tensors thereof, are included, the ***variety*** and combination ***of data structures*** often become confusing for even the most experienced computer algebraists.

(d) The complexity of objects could be managable on the abstract level. However, computer processing of these algebraic objects requires careful choice and implementation of the machine representation for each form of the object. In addition, some simplification must be carried out by machine lest expressions become unmanagable or even meaningless. The problem is further aggravated by the fact that ***machine representation and simplification*** can often be an issue of personal preference. Any of the three forms of $(a + b)*(c + d)$ could be the preferred form of input or output. Thus, the conversion among various representations becomes a requirement for not only user interface but also interprogram communication.

Most of these computational issues are natural considerations to mathematicians, in complex analysis or computer algebra, as well as to most scientists and engineers performing computations on machines. It is natural for us to carry out exact algebraic derivations before specific numerical computations. The field of computer algebra, in its attempt to provide a computational tool for symbolic and algebraic manipulation, strives to provide an algebraic preprocessor for numerical computation. Numerical examples can also form the basis for algebraic generalizations. Even though certain mathematical problems can be solved by either purely algebraic or numerical means, it is a frequent experience in scientific computing to find the combination of algebraic and numerical computations more appropriate and profitable. Our objective is one of encouraging interaction between algebraic and numerical computations, in theory, algorithms, and program usage.

In the next sections, we develop the algebraic algorithmic approach to the classic problem in calculus - indefinite integration of elementary functions in close form. Using this typically analytic problem as the motivating vehicle, several principles of symbolic and algebraic computation will be demonstrated. The simple but important problem of rational function integration will be solved in Section 2. This workhorse algorithm also motivates the techniques used in integrating the more general class of elementary functions. The algorithm analysis reveals some of the fundamental units of computational cost for computer algebra.

II. Rational Function Integration

The computational problem of integrating rational functions, as viewed by complex analysts (c.f. [HENR74]), can be stated as follows:

Given a rational function $R(x)$, find b_i, c_i in $\mathbf{C}$ for $1 \leq i \leq k$ and a rational function $S(x)$ such that

$$\int R(x)dx = S(x) + \sum_{i=1}^{k} c_i \log(x-b_i) \tag{1}$$

where b_i are the distinct poles of $R(x)$.

Usually and without loss of generality, it is assumed that $R(x)$ and $S(x)$ are "proper" rational functions, i.e., if $R(x) = A(x)/B(x)$, then $A(x)$ and $B(x)$ are relatively prime and $\deg(A) < \deg(B)$. It is easy to see that every rational function can be uniquely represented as the sum of a polynomial and a proper rational function. Thus, the polynomial part of a rational function can be separated and integrated beforehand. Computationally, the poles b_i are found by repeated approximations and c_i are the residues of $R(x)$ at b_i (depending on the multiplicity of b_i as roots of the denominator of $R(x)$). The point to emphasize is that all these quantities, b_i and c_i, are only approximations over $\mathbf{C}$ in current computers.

This problem (eq. (1)) in complex analysis or calculus is quite easy. In fact, it is considered a "well-worn subject" by Henrici (pp. 562). However, as we retreat from the idealized analytical domain of the complex numbers, **C**, (or even the continuous space of the real numbers) to dealing with exact computation using the rational number field, **Q**, as the ground domain, the same problem takes on added intrigue. Exact results are given as long as possible by the computer, thus minimizing the chance for error in approximation and increasing the possibility of obtaining symbolic solutions (with unspecified parameters). The problem of rational function integration from the perspective of computer algebra, which stress exact results and algebraic approaches, is stated as follows:

Given a proper rational function $A(x)/B(x)$ where both $A(x)$ and $B(x)$ are in $\mathbf{Q}[x]$, compute $C(x)$, $D(x)$, $U(x)$, and $V(x)$ in $\mathbf{Q}[x]$ such that $V(x)$ has no multiple root and

$$\int \frac{A(x)}{B(x)}dx = \frac{C(x)}{U(x)} + \int \frac{D(x)}{V(x)}dx. \tag{2}$$

Obviously, if $V(x)$ can be completely factored into linear factors in $\mathbf{Q}[x]$, i.e., $V(x) = (x-b_1)(x-b_2)...(x-b_k)$, then

$$\int \frac{D(x)}{V(x)}dx = \int \frac{c_1}{x-b_1}dx + \ldots + \int \frac{c_k}{x-b_k}dx$$

where c_i are constants in **Q** which can easily be computed as residues of $D(x)/V(x)$ at $x=b_i$ or simply $D(b_i)/V'(b_i)$
(with $'$ denoting derivative with respect to x or d/dx). As a result,

$$\frac{A}{B} = \frac{C}{U} + \sum_{i=1}^{k} c_i \log(x-b_i),$$

which is the same solution as that in the complex domain. But in this case C, U, c_i, and b_i can all contain other numerically unspecified parameters.

Unfortunately, such a complete factorization of $V(x)$ can not always be done even in $\mathbf{Q}[x]$, much less for a symbolic expression with other parameters. In fact, it can rarely be done. The fundamental reason is that complete algebraic specification of all distinct roots of a polynomial in $\mathbf{Q}[x]$ requires finding an algebraic extension of the field **Q** (written as $\mathbf{Q}(\alpha)$ where α satisfies some irreducible minimum polynomial $M(x)=0$ in $\mathbf{Q}[x]$) sufficiently large so that all distinct roots can now be expressed as distinct polynomials in α. This problem is essentially that of finding the splitting field of polynomials which is an exponential computational process [TRAG76]. In addition, as eq. (1) indicates, the space of rational functions is not closed under the operation of integration. For these reasons, it is often necessary to separate the rational function integration problem into two distinct phases: first being the determination of precisely those polynomials in eq. (2) and the second being the much more difficult problem of finding the splitting field of $V(x)$ so that roots and constants, b_i and c_i, can be computed in $\mathbf{Q}(\alpha)$ and solution as eq. (1) can be obtained in $\mathbf{Q}(\alpha)[x]$. Horowitz [HORO69] shows that no sum of the form $\sum_{i=1}^{k} c_i \log(x-b_i)$, can be non-trivally equal to a rational function. Therefore, we can distinguish the result of integration of rational functions in eq. (2) as a **rational part** (the rational function $C(x)/U(x)$) plus a **transcendental part** (an integral of a rational function, $D(x)/V(x)$, whose denominator has no root of multiplicity greater than 1). Trager [TRAG76] gives an exponentially bounded algorithm for solving phase two — the transcendental part — of the problem. These algebraic algorithms can be used and generalized to preserve the exactness and unspecified parameters as long as possible. Essentially, only when the exponential growth of the transcendental part of the integration problem becomes too overwhelming, would there be a need to depart from exact solution methods and resort to numerical approximation. As we mentioned in the last section, this is an excellent way of keeping numerical approximation errors to a minimum.

Yun [YUN77] gave a detailed description of the algorithms available to solve this problem and derived an upper bound of $O(\log(n)M(n))$ for the improved version of Hermite's algorithm [HERM12], where $n = \deg(B)$ and $M(n)$ is the cost of multiplying two polynomials of degree n. When $M(n)$ can be carried out in $O(n \log n)$, say by FFT methods, then the cost of integrating rational functions have a realizable algorithm of $O(n{**}2 \log(n))$. In fact, Yun showed the stronger result of $O(\log(n)M(n))$ bound for each of the four steps in the algorithm description bellow:

Input: A proper rational function $A(x)/B(x)$, $A(x)$ and $B(x)$ primitive in $D[x]$

(1) Square-free deomposition of $B(x)$:

Find $B_i(x)$ in $D[x]$ such that $\gcd(B_i,B_i') = 1$ for all $i \le k$,

$\gcd(B_i,B_j) = 1$ for all $i \neq j \le k$, and $B(x) = \prod_{i=1}^{k} B_i^{\,i}(x)$.

Let $\deg(B_i) = n_i$ and $\deg(B) = n$, then $n = \sum_{i=1}^{k} i\, n_i$.

(2) Square-free partial fraction decomposition:

Find $A_i(x)$, $i = 1,\ldots,k$

such that $\deg(A_i) < \deg(B_i^{\,i})$ and

$$\frac{A(x)}{B(x)} = \sum_{i=1}^{k} \frac{A_i(x)}{B_i^{\,i}(x)} \tag{3}$$

(3) Complete square-free partial fraction decomposition:

Find $A_{ij}(x)$ for $1 \le j \le i$, $1 \le i \le k$,

such that $\deg(A_{ij}) < \deg(B_i)$ for all $j \le i$ and

$$\frac{A(x)}{B(x)} = \sum_{i=1}^{k} \sum_{j=1}^{i} \frac{A_{ij}(x)}{B_i^{\,j}(x)} \tag{4}$$

(4) Repeated integration by parts:

Find $C_{ij}(x)$ and $D_i(x)$ such that $\deg(C_{ij}) < \deg(B_i)$, $\deg(D_i) < \deg(B_i)$, and

$$\int \frac{A(x)}{B(x)} dx = \sum_{i=2}^{k} \sum_{j=1}^{i-1} \frac{C_{ij}(x)}{B_i^j(x)} + \sum_{i=1}^{k} \int \frac{D_i(x)}{B_i(x)} dx. \tag{5}$$

Output: as required by eq. (2) can be given immediately

$$\frac{C(x)}{U(x)} = \sum_{i=2}^{k} \sum_{j=1}^{i-1} \frac{C_{ij}(x)}{B_i^j(x)} \text{ and}$$

$$\frac{D(x)}{V(x)} = \sum_{i=1}^{k} \frac{D_i(x)}{B_i(x)},$$

hence $U(x) = B_2^1 B_3^2 \ldots B_k^{k-1}$ and $V(X) = B_1 B_2 \ldots B_k$ so that $UV = B$.

III. Integration by Educated Guess

Armed with an efficient algorithm for rational function integration and motivated by the techniques underlying it, we now consider an example in a larger class of functions: $\int \log z\, dz$. Fresh from dealing with polynomials and rational function integrands, we may have the urge of treating $\log z$ as an **independent variable** θ over the ground field of functions $\mathbf{Q}(z)$. The problem then takes a familiar polynomial form $\int \theta\, dz$. We further make a bold **conjecture** (as if we are still dealing with rational functions) that the integral of a linear polynomial in θ will be one of second degree, which has the form of $B_2\theta^2 + B_1\theta + B_0$, where B_i naturally lies in the ground field $\mathbf{Q}(z)$. Now we have the integration problem transformed into an equation

$$\int \theta dz = B_2\theta^2 + B_1\theta + B_0.$$

Differentiate with respect to z yields

$$\theta = B_2'\theta^2 + (\tfrac{2}{z}B_2 + B_1')\theta + (\tfrac{1}{z}B_1 + B_0'),$$

where $'$ denotes derivative w.r.t. z. Since the first assumption is to treat θ as an independent variable, we can now simply equate the appropriate powers of θ and obtain three linear ordinary differential equations: $0 = B_2'$, $1 = \frac{2}{z}B_2 + B_1'$, $0 = \frac{1}{z}B_1 + B_0'$. Each of the equations can be solved by rational function integration. The steps are carried out as follows:

$$B_2 = \text{constant} = b_2,$$

$$z + b_1 = 2b_2\theta + B_1 \rightarrow b_2 = 0 \text{ and } B_1 = z + b_1,$$

$$-1 = \tfrac{1}{z}b_1 + B_0' \rightarrow -z + b_0 = b_1\theta + B_0 \rightarrow b_1 = 0 \text{ and } B_0 = -z + b_0.$$

The constant of integration from one equation is determined by the subsequent step, leaving only one undetermined constant, which is the expected constant for the original integration problem. And the result, which can easily be verified, is

$$\int \theta \, dz = 0\theta^2 + z\theta + (-z + c_0) = z \log z - z + \text{constant}.$$

The case involving exponentials expressions can be treated similarly. $\int (ze^z + e^z)dz = \int \theta(z + 1)dz$, where $\theta = e^z$. The same assumptions and equation solving yields the solution ze^z.

The more interesting case is $\int e^{z^2}$, which is the "error function" and known to be not integrable as an elementary function. Blindly following the steps of the previous examples, let $\theta = e^{z^2}$, and conjecture the integral to have the form $B_1\theta + B_0$. (B_2 will be determined to be 0 anyway, so it is not included.) Then the result of differentiating both side of the equation is $\theta = (B_1' + 2zB_1)\theta + B_0'$ Solving the first linear differential equation over $\mathbf{Q}(z)$, $1 = (B_1' + 2zB_1)\theta + B_0'$, we find that the solution must be a polynomial in z. In fact, it is sufficient to just consider the linear form $B_1 = az + b$, where a and b are in $\mathbf{Q}$. Now the equation becomes $1 = a + 2z(az + b)$, which leads to a set of contradictory set of linear equations: $2a = 0$, $2b = 0$, $a = 1$. The conclusion is naturally that the integral does not exist. However, the desire to make such a conclusion points out the need for a more rigorous foundation for the assumptions made so far.

If the assumptions for integration by educated guess were carefully examined, we find several influences from rational function integration. First, there is the desire to treat functions in z (the variable of integration) as independent variables. And the expression representing the integrand will be treated as mutivariate rational functions in those variables. Thus, the true algebraic independence of these functions must be established. Second, in order to conjecture the form of the integral, it is essential that a theory be established so as to ascertain the form of the integral for a given class of functions. In fact, as the previous example shows, the contradiction will not necessarily lead to the conclusion of non-integrability unless the theory states that there is only one form possible for the integral. We will define the class of elementary functions and state the necessary theoretical results in the next section.

IV. Functional Independence and Liouville's Theorem

The class of ***elementary functions*** is the result of finitely many repeated or nested extensions of the field of rational functions F (possibly multivariate). The type of such extensions falls in the following two classes:

(1) ***Transcendental extensions***:

$\theta = \log(f)$, or

$\theta = \exp(f) = e^f$, where f arises from the field of extensions thus far;

(2) ***Algebriac extensions***:

an extension on the current field by an algebraic relation, i.e. by θ that satisfies an irreducible polynomial P such that $P(\theta)=0$, where the coefficients of P lies in the field of extensions thus far.

Such a function field **K** obtained by finitely many extensions of **F** is known as an ***elementary extension field of*** **F**, which can be written as

$$\mathbf{K} = \mathbf{F}(\theta_1,\theta_2,...,\theta_m).$$

An example is $\arccos z = \pm i \log (z + \sqrt{z^2-1}),$

which lies in the field $\mathbf{F}(\theta_1,\theta_2) = (\mathbf{F}(\theta_1))(\theta_2)$, where $\mathbf{F} = \mathbf{C}$, $\theta_1 = \sqrt{z^2-1}$ satisfying the polynomial relation $P(\theta_1) = \theta_1^2-z^2 + 1 = 0$, and and $\theta_2 = \log(z + \theta_1)$.

The class of elementary functions includes multivariate rational functions, trigonometric and hyperbolic functions (together with their inverses), since they can be expressed in terms of logarithmic, exponential, and square root functions.

The available space for this paper does not allow us to treat both types of extensions and their mixed case. In order to demonstrate that integration can be carried out algorithmically, we will restrict our attention for the remainder of this paper on only the transcendental elementary functions. Towards the desire of treating expressions as independent variables, we state the ***Risch Structure Theorem*** [RISC69], which demonstrates the essence of the technique for determining independence.

The theorem is used recursively to determine the independence of each newly extended transcendental function. That is, assuming the ground rational function field F has already been extended repeatedly by $\theta_1,\theta_2,...,\theta_m$, we denote the succesive extension fields by F_i. The set of θ's can be divided into two subsets $\{\theta_i = \log f_i,\ f_i \in F_{i-1}\}$ and $\{\theta_i = \exp f_i,\ f_i \in F_{i-1}\}$. Let L denote the set of indices for those θ that are log extensions and E for exp extensions. The Risch Structure Theorem states that (1) if the new extension $\theta_{m+1} = \log f_{m+1},\ f_{m+1} \in F_m$ then it is independent of the previous functional extensions if and only if the equation

$$c_{m+1}f_{m+1} = \prod_{j\in \mathrm{L}} f_j^{c_j} \prod_{j\in \mathrm{E}} \theta_j^{c_j}$$

has no solution c_j in $\mathbf{Q}$, for $j = 1,2,...,m+1$; (2) if the new extension $\theta_{m+1} = \exp f_{m+1},\ f_{m+1} \in F_m$ then it is independent of the previous functional extensions if and only if the equation

$$c_{m+1} + f_{m+1} = \sum_{j\in \mathrm{L}} c_j\theta_j + \sum_{j\in \mathrm{E}} c_j f_j$$

has no solution c_j in $\mathbf{Q}$, for $j = 1,2,...,m+1$.

The utility of the Risch Structure Theorem is not limited to determining extension independence while integrating elementary functions. It provides an algorithmic process to simplify elementary functions. Since simplification is an important aspect of computer algebra (as mentioned in the Introduction Section), we give an example here to illustrate this process. Consider the elementary function

$$f(x, y) = \log((e^x + \log x^2)e^{x+y}) + e^{x+y}\log x^2 - \log(e^x + \log x^2) - x - y.$$

The ground field in this case is $\mathbf{Q}(x, y)$. The first extension, as we work from inside out, is $\theta_1 = e^x$. This being the exponential extension, and it is clear that the equation $x + c = 0$ has no solution for c, thus θ_1 qualifies as an independent extension to be treated as a variable. Next, $\theta_2 = \log x^2$. Again, since the equation $cx^2 = (e^x)^{c_1}$ has no solution over $\mathbf{Q}$, θ_2 is independent over $\mathbf{Q}(x, y, \theta_1)$. Now let $\theta_3 = e^{x+y}$. And we find the equation $x + y + c = c_1 x + c_2 \log x^2$ has no solutions over $\mathbf{Q}$. Similarly, $\theta_4 = \log((e^x + \log x^2)e^{x+y}) = \log((\theta_1 + \theta_2)\theta_3)$ leads to the equation

$$c(\theta_1 + \theta_2)\theta_3) = \theta_1^{c_1}(x^2)^{c_2}\theta_3^{c_3}$$

with no solutions. However, the last extension $\theta_5 = \log(e^x + \log x^2) = \log(\theta_1 + \theta_2)$ leads to the equation

$$c(\theta_1 + \theta_2) = \theta_1^{c_1}(x^2)^{c_2}\theta_3^{c_3}[(\theta_1 + \theta_2)\theta_3]^{c_4},$$

which has solutions $c_1 = c_2 = 0$ and $c = c_4 = -c_3 = 1$. So the dependence of the new extension on the previous is actually determined as $\theta_5 = -(x + y) + \theta_4$. Now that all new extensions, and the independent variables, have been determined, the final simplification process is merely a multivariate rational function (in this case, polynomial) simplification operation: $f(x, y) = \theta_4 + \theta_3\theta_2 - (\theta_4 - (x + y)) - x - y = \theta_3\theta_2$, which is the simplified expression of the original elementary function $e^{x+y}\log x^2$.

The second desire, as mentioned before, is to provide a rigorous theoretical foundation for determining the possible form of the integral. The basic theorem was provided by Liouville during 1833-1841 and improved by Risch [RISC69]

Liouville's Theorem: Let f be a function from a differential field D of characteristic zero. Then there exists an elementary extension field of D with the common field of constants K and containing a function g such that $g = \int f$ if and only if

$$g = v_0 + \sum_{i=1}^{n} c_i \log v_i$$

where v_0, v_i in D and c_i in K.

The proof of this theorem can be found in [RISC69]. It provides precisely the necessary theoretical justification for concluding if a function is integrable it must have a prescribed form, or the integral does not exist. Furthermore, the form of the integral is much as expected from our experience with rational function integration. As we shall see from the following example, even the algorithm is very much reminiscent of rational function integration.

The algorithm we illustrate is recursive on the nested extensions, θ's. We will work out the last extension of the following example:

$$\int f dz = \int [2ze^{z^2}\log z + \frac{e^{z^2}}{z} + \frac{\log z - 2}{(\log^2 z + z)^2} + \frac{\frac{2}{z}\log z + \frac{1}{z} + 1}{\log^2 z + z}]dz$$

The extension concerned is $\theta = \log z$ over the previously extended field $\mathbf{D} = \mathbf{Q}(z, e^{z^2})$. The structure theorem should be applied to determine the independence of θ to $\gamma = e^{z^2}$. And indeed the independence is obvious. Now we are dealing with an integrand that is a function of three variables, z, γ, and θ:

$$f = 2z\gamma\theta + \frac{\gamma}{z} + \frac{\theta - 2}{(\theta^2 + z)^2} + \frac{\frac{2}{z}\theta + \frac{1}{z} + 1}{\theta^2 + z}$$

The first two terms form a polynomial in θ and the last two terms are rational functions in θ. The Liouville's theorem predicts the form of the integral, i.e.

$$f = \frac{d}{dz}[B_2\theta^2 + B_1\theta + B_0 + \frac{B_{11}}{\theta^2 + z} + c_1 \log(\theta^2 + z)].$$

Differentiating both sides, we get

$$f = B_2'\theta^2 + (\frac{2}{z}B_2 + B_1')\theta + (\frac{1}{z}B_1 + B_0')$$

$$+ \frac{-B_{11}(\frac{2}{z}\theta + 1)}{(\theta^2 + z)^2} + \frac{B_{11}'}{\theta^2 + z} + \frac{c_1(\frac{2}{z}\theta + 1)}{\theta^2 + z}$$

As we did with the examples in the last section before we encountered the theory, the remaining steps are to equate coefficients of powers of θ in the polynomial part and those with the same denominator in the rational part, and solve linear ordinary differential equations (a recursive integration problem) step by step.

Polynomial part: $B_2' = 0 \rightarrow B_2 = \text{constant} = b_2$

$$\frac{2}{z}b_2 + B_1' = 2z\gamma \rightarrow B_1 = -2b_2\theta + \gamma + b_1 \rightarrow b_1 = 0 \text{ and } B_1 = \gamma + b_1$$

$$\frac{\gamma + b_1}{z} + B_0' = \frac{\gamma}{z} \rightarrow B_0 = -b_1\theta \rightarrow b_1 = 0 \text{ and } B_0 = \text{constant} = b_0$$

Rational part: $-B_{11}(\frac{2}{z}\theta + 1) \equiv \theta - 2 \pmod{\theta^2 + z}$

Solving $P(\theta)(-(\frac{2}{z}\theta + 1)) + Q(\theta)(\theta^2 + z) = \theta - 2$ for $P(\theta), Q(\theta) \in \mathbf{Q}(z)$

$$\rightarrow \; P(\theta) = -\theta, \; Q(\theta) = \frac{-2}{z} \quad \rightarrow \; B_{11} = -\theta,$$

$$\frac{-\theta(\frac{2}{z}\theta + 1)}{(\theta^2 + z)^2} + \frac{\frac{1}{z}}{\theta^2 + z} + \frac{c_1(\frac{2}{z}\theta + 1)}{\theta^2 + z} = \frac{\theta - 2}{(\theta^2 + z)^2} + \frac{\frac{2}{z}\theta + \frac{1}{z} + 1}{\theta^2 + z}$$

$$\rightarrow \; c_1 = 1.$$

The final result after substituting all of the unknowns just found is

$$\int f \, dz = \gamma\theta + \frac{-\theta}{\theta^2 + z} + \log(\theta^2 + z)$$

$$= e^{z^2} \log z - \frac{\log z}{\log^2 z + z} + \log(\log^2 z + z).$$

V. Integration by Computer

Several computer algebra systems, namely MACSYMA, REDUCE, and SCRATCHPAD, have implemented the integration algorithm described and illustrated above. In the case of transcendental elementary functions, the resulting algorithm is known as a **decision procedure**. That is to say a problem posed in a particular domain can be decided to be solvable or not: when it is solvable, the result is given; when it is not solvable, a guarantee (or proof of unsolvability) is implicitly carried out. Such a program available on the computer provides an useful tool for working scientists and engineers. The following set of examples were carried out by SCRATCHPAD. Each example has one line of linearized input as the integrand; the system generated two-dimensional display of the problem; the answer in two-dimensional display; and the processing time it took the computer. The last example is the one carried out in the last section, where the difference in the solution is due to the computer leaving the answer in rational form.

```
"a most simple example, having one log and one exp extension"

(log(x)*(1-x)-1)/(exp(x)*log(x)**2)

   /  LOG(X)(1 - X) - 1
   |  -----------------
   /                2
   X     EXP(X)LOG (X)

INTEGRAL...
```

```
        X
   --------------
    EXP(X)LOG(X)

(END OF PROCESSING ON THIS EXPRESSION)
(631 MILLISECONDS)

"rational function integration is essentially a special case"
"even with symbolic parameters"

(a*x**3+b*x**2+c*x+d)/((x+1)*x*(x-3))

          3       2
   /  A*X   + B*X   + C*X + D
   | -----------------------
   /    ((X + 1)X)(X - 3)
   X

INTEGRAL...

                - D                27A + 9B + 3C + D
      A*X + (-----)LOG(X) + (-------------------)LOG(X - 3)
                 3                       12
    +
         - A + B - C + D
      (-----------------)LOG(X + 1)
                 4

(END OF PROCESSING ON THIS EXPRESSION)
(1251 MILLISECONDS)

"the classical non-integrable case -- the error function"

exp(x**2)

   /      2
   | EXP(X )
   /
   X

*** EXPRESSION IS NOT INTEGRABLE ***
(116 MILLISECONDS)

"a high degree integrand
 -- showing what can be done with an integrator available"
```

```
x**9*log(x)**11

   /  9   11
   | X LOG   (X)
   /
   X

INTEGRAL...

                            10                         10
     (-6237/156250000)*X     + (6237/15625000)*X   LOG(X)
   +
                          10    2                          10    3
     (-6237/3125000)*X   LOG  (X)  +  (2079/312500)*X   LOG  (X)
   +
                         10    4                          10    5
     (-2079/125000)*X   LOG  (X)  +  (2079/62500)*X   LOG  (X)
   +
                       10    6                      10    7
     (-693/12500)*X   LOG  (X)  +  (99/1250)*X   LOG  (X)
   +
                     10    8                     10    9
     (-99/1000)*X   LOG  (X)  +  (11/100)*X   LOG  (X)
   +
                    10    10                   10    11
     (-11/100)*X   LOG    (X)  +  (1/10)*X   LOG    (X)

(END OF PROCESSING ON THIS EXPRESSION)
(1796 MILLISECONDS)

"example with nested exponentials"

(2*x**3+x)*exp(x**2)**2*exp(1-x*exp(x**2))/(1-x*exp(x**2))**2

             3          2  2                    2
  /  ((2X   + X)EXP  (X  ))EXP(1  -  X*EXP(X  ))
  | ------------------------------------------
  /                                2   2
  X               (1  -  X*EXP(X  ))

INTEGRAL...
```

```
                       2
 (- 1)EXP(1 - X*EXP(X ))
-------------------------
               2
      X*EXP(X ) - 1

(END OF PROCESSING ON THIS EXPRESSION)
(1697 MILLISECONDS)

"a nested exponential that is not integrable"
exp(exp(exp(exp(x))))

   /
   | EXP(EXP(EXP(EXP(X))))
   /
   X

*** EXPRESSION IS NOT INTEGRABLE ***
(319 MILLISECONDS)

"the example illustrated in the last section"
2*x*exp(x**2)*log(x)+exp(x**2)/x+(log(x)-2)/(log(x)**2+x)**2+    _
  ((2/x)*log(x)+(1/x)+1)/(log(x)**2+x)

                                               2
                      2                 EXP(X )       LOG(X) - 2
         ((2X)EXP(X ))LOG(X) + --------- + ----------------
                                            X             2          2
      /                                                 (LOG (X) + X)
      |   +
      /          2             1
      X       (---)LOG(X) + --- + 1
                 X             X
             -----------------------
                        2
                   LOG (X) + X

INTEGRAL...

                                            3          2
 (- 1)LOG(X) + (X*LOG(X) + LOG (X))EXP(X )                   2
------------------------------------------ + LOG(LOG (X) + X)
                          2
                     LOG (X) + X
```

```
(END OF PROCESSING ON THIS EXPRESSION)
(2140 MILLISECONDS)
```

References

[HENR74] P. Henrici, *Applied and computational complex analysis*, John Wiley & Sons, New York, 1974.

[HERM12] C. Hermite, *Oeuvres de Charles Hermite*, E. Picard ed., Vol. III, Gauthier-Villars, Imprimeur-Libraire, Paris, 1912.

[HORO71] E. Horowitz, Algorithms for partial fraction decomposition and rational function integration, *Proceedings of the Second Symposium on Symbolic and Algebraic Manipulation*, S. R. Petrick ed., March 1971, 441-457.

[MOSE67] J. Moses, Symbolic integration, Project MAC TR-47, December 1967.

[MOSE71] J. Moses, Symbolic integration: the stormy decade, *Comm. ACM 14*, No. 8, Sept. 1971, pp. 548-560.

[RISC69] R. Risch, The problem of integration in finite terms, *Transactions of AMS*, Vol. 139, May 1969, 167-189.

[TRAG76] B. M. Trager, Algebraic factoring and rational function integration, *Proceedings of 1976 ACM Symposium on Symbolic and Algebraic Computation*, Aug. 1976, 219-116.

[YUND76] D. Y. Y. Yun, On square-free decomposition algorithms, *Proc. of the 1976 ACM Symposium on Symbolic and Algebraic Computation* R. Jenks (ed.), ACM, New York, Aug. 1976, pp. 26-35.

[YUND77] D. Y. Y. Yun, Fast Algorithm for Rational Function Integration. *Proc. IFIP Congress 1977*, Toronto, Canada, Aug. 1977, 493-498.

SIMULTANEOUS PADÉ APPROXIMATION TO CERTAIN HYPERGEOMETRIC SERIES: EXPLICIT FORMULAE

M.C. de Bruin

University of Amsterdam
The Netherlands

One of the generalizations of the ordinary Padé table to simultaneous approximation of (formal) power series can be defined by:

<u>given</u> formal power series $f_1,\dots,f_n$ and non-negative integers $\rho_0,\rho_1,\dots,\rho_n$ $(\sigma := \rho_0+\rho_1+\dots+\rho_n)$
<u>find</u> polynomials $P_0,P_1,\dots,P_n$ <u>such that</u>
$\deg P_i \le \sigma - \rho_i$ $(i = 0,1,\dots,n)$,
$\operatorname{ord}(P_0 f_i - P_i) \ge \sigma + 1$ $(i = 1,\dots,n)$

Using theorems for the evaluation of determinants one can derive explicit expressions for the coefficients of P_0 and $P_0 f_i - P_i$ for the n-tuples $\{{}_1F_1(1;c_j;x)\}^n_j$, $(c_j \notin \mathbb{Z} \setminus \mathbb{N}, c_i - c_j \notin \mathbb{Z})$ and $\{{}_2F_0(a_j,1;x)\}^n_j$, $(a_j \notin \mathbb{Z} \setminus \mathbb{N}, a_i - a_j \notin \mathbb{Z})$ under the condition $\rho_0 \ge \rho_i$ $(i = 1,\dots,n)$.

In this way one can prove some convergence results for certain sequence of Padé approximants P_i/P_0 using the connection with a Jacobi-Perron algorithm.

H. Werner et al. (eds.), Computational Aspects of Complex Analysis, 395.

EPSILON AND QD ALGORITHMS FOR THE MATRIX-PADÉ AND 2-D PADÉ PROBLEM

A. Bultheel

Universiteit Leuven
Belgium

Matrix continued fractions are used to derive QD algorithms corresponding to a staircase or a saw-tooth display in the matrix-Padé table. Also a proof of the cross-relation is extended to the matrix case. From this a matrix version of the epsilon algorithm is easily derived.

Finally, it is shown how certain rectangular types of two-variable Padé approximants can be computed as the convergents of coupled continued fractions much like the matrix-Padé case, so that the previous algorithms can also be used for this type of approximants.

H. Werner et al. (eds.), Computational Aspects of Complex Analysis, 396.

POLE DETECTION WITH PADÉ-TYPE APPROXIMANTS

L. Casasús

Universidad de la Laguna
Spain

We make use of sequences of Padé-type approximants (PTA's) to determine the poles of a meromorphic function $f(x)$. We prove a theorem which shows how to attain geometric convergence with an optimal sequence easily constructed from the results of previous paper. Finally, a numerical example from physics is exhibited, where the improvement of PTA's over ordinary Padé approximants becomes clear.

H. Werner et al. (eds.), Computational Aspects of Complex Analysis, 397.

MULTIVARIATE PADÉ-APPROXIMANTS: THEORY AND APPLICATION

A. Cuyt

University of Antwerp
Belgium

During the last few years many attempts have been made to generalize the notion of Padé-approximant to the multivariate case: one can make different choices for the polynomials and for the interpolation set. By means of the multivariate Padé-approximants introduced in [1], one can acceterate the convergence of a table with multiple entries. The computations are performed via the ε-algorithm, as in the univariate case. We have calculated some multidimensional integrals with this procedure [2].

[1] Cuyt, Annie A.M.: Multivariate Padé-Approximants. to appear in Journ. Math. Anal. Applcs.

[2] Cuyt, Annie A.M.: Acceterating the convergence of a table with multiple entry. submitted in Numerische Mathematik.

H. Werner et al. (eds.), Computational Aspects of Complex Analysis, 398.

SOLUTIONS OF ORDINARY DIFFERENTIAL EQUATIONS IN THE NEIGHBORHOOD OF SINGULAR POINTS (REGULAR AND IRREGULAR)

J. Della Dora

University of Grenoble
France

We present an algorithm which enables us to solve formally any ordinary differential equation in the neighborhood of singular points.

If the singular point is regular we use a Frobenius like algorithm. If the singularity is irregular then we can find a fundamental set of formal solutions of the form

$$Y_i = e^{Q_i(x)} x^{\nu_i} \varphi_i(x)$$

where Q_i is a polynomial in some fractional part of $\frac{1}{x}$ and φ_i is a Frobenius like solution. This part of the algorithm is described in [1].

[1] Della Dora, J. and Tournier , E.: Les bases d'un algorithme d'obtention des solutions formelles d'une E.D.O. en un point singulier irrégulier Rapport de recherche IRMA No. 266 (Laboratoire IMAG).

H. Werner et al. (eds.), Computational Aspects of Complex Analysis, 399.

SINGULARITIES OF REGULAR BRIOT-BOUQUET TYPE IN PLANE SYSTEMS

R. Farwig

University of Bonn
Federal Republic of Germany

No analogue of Painlevé's determinateness theorem exists for systems of differential equations. Rather the types of movable singularities of a two-dimensional system of first-order differential equations depend on the multiplicity of the zeros of a characteristic function of this system. Extending certain results of R.A. Smith to non-analytic, non-autonomous systems, the set of singular solutions is described and parameterized using the initial terms of a Taylor-series.

H. Werner et al. (eds.), Computational Aspects of Complex Analysis, 400.

ON THE NUMBER OF SOLUTIONS OF THE DISCRETE THEODORSEN-EQUATION

O. Hübner

University of Giessen
Federal Republic of Germany

The Theodorsen integral equation, which is important in conformal mapping, is known to have always one and only one solution. A certain way of discretization yields a system of 2N non-linear equations: the discrete Theodorsen-equation (T_d). It is also known that there is one and only one solution of (T_d), if a so-called ε-condition is fulfilled with $\varepsilon < 1$. Given $\varepsilon \geq 1$ and $N \in \mathbb{N} \setminus \{1\}$ we present examples for which (T_d) has infinitely many solutions. Other examples give a deeper understanding into what may happen if one applies iterative methods like JOR to (T_d), even in the case $\varepsilon < 1$.

H. Werner et al. (eds.), Computational Aspects of Complex Analysis, 401.

A CASE STUDY: THE LAPLACE EQUATION IN A PLANE DOMAIN BOUNDED BY A LINE AND N CIRCUMFERENCES

N. Germay and Mushamalirwa Daudi

Université Catholique de Louvain

The problem of calculating electric field or the thermal field between parallel circular cylinders anda parallel plane (usually the ground) arises in a variety of engineering problems. Several methods exist for solving it. Five of them will be compared: to obtain the stationary thermal field around parallel underground cables in a uniform or eventually piece-wise uniform soil.

The five methods investigated are:
1. The finite elements method
2. The Winders method
3. The Luoni-Morello-Holdup method
4. The successive images method
5. The Burnside automorphic functions method.

All of them except method 1 are based on suitable transformation of z-plane into w-plane in order to rectify one or all circular boundary curves into horizontal slits and solve the problem in w-plane.

Some improvements of the methods are also proposed.

H. Werner et al. (eds.), Computational Aspects of Complex Analysis, 402.

FAST ALGORITHMS FOR MICROELECTRONICS APPLICATIONS

H.T. Kung and E.W. Ng

Carnegie-Mellon University and
Jet Propulsion Laboratory

Advances in Very Large Scale Integration (VLSI) technology offer a great potential for the next generation of computing power as extensions to both computing software and hardware. The challenges to mathematicians lie in the prospects of implementing complex algorithms in VLSI chips and/or systems, which in turn can serve as peripheral devices to a conventional host computer, or used as a component in a larger special purpose system.

In this presentation we describe the interplay between mathematical algorithms and microelectronic architectures, via several examples of classical computational modules. It is primarily intended to highlight the issues in VLSI algorithmic design as distinct from that for conventional computers. These issues include functional architecture linking the microelectronic modules, communication geometries, data movements among modules, and utilization of parallelism. Both one-dimensional pipeline algorithms and two-dimensional systolic algorithms will be discussed, in the context of a few examples of the following kinds of mathematical applications:

Communication Geometry	Examples
1-dim linear arrays	Matrix-vector multiplication
	Real-time FIR filtering
	Convolution
	Discrete Fourier Transform (DFT)
	Pipeline arithmetic units
	Recurrence evaluation
	Solution of triangular linear systems
	Cartesian product

H. Werner et al. (eds.), Computational Aspects of Complex Analysis, 403–404.

2-dim square arrays	Pattern matching Relational database operations Dynamic programming for optimal parenthesization Numerical relaxation for PDE Merge sort Fast Fourier Transform (FFT) Graph algorithms using adjacency matrices Image processing
2-dim hexagonal arrays	Matrix multiplication Transitive closure LU-decomposition by Gaussian elimination without pivoting QR-decomposition

CONTOUR REPRESENTATIONS OF VARISOLVENT FAMILIES

H. Loeb

University of Oregon
USA

We discuss a re-parametrization of functions of the form,

$$f(x) = \sum_{i=1}^{p} \sum_{j=0}^{m_i-1} b_{ij} \frac{\partial^j}{\partial t_i^j} K(t_i,x) ,$$

where $\{b_{ij}\}$, $\{t_i\}$ are free parameters and for fixed x, K(t,x) is analytic in the variable t. A typical problem is: given a $g \in C[0,1]$. One wants to choose f to minimize $||g-f||_\infty$ or $||g-f||_2$. The new parametrization involves the use of a contour integral to represent f where these new parameters are the coefficients of the two polynomials which define a rational function appearing in the representation. This new form helps to alleviate the numerical problems which arise because of the multiple and complex $\{t_i\}$.

H. Werner et al. (eds.), Computational Aspects of Complex Analysis, 405.

THE FAST SOLUTIONS OF POISSONS AND THE BIHARMONIC EQUATIONS ON IRREGULAR REGIONS

A. Mayo

University of Stanford

We present fast methods for the numerical solution of Laplace's and the biharmonic equations on irregular regions. The methods used for solving both equations make use of fast Poisson solvers on a rectangular region in which the irregular region is embedded. They also both use an integral equation formulation of the problem. The main idea is to use the integral equation formulation to define a discontinuous extension of the solution to the rest of the rectangular region. Fast solvers are then used to compute the extended solution. Aside from solving the equations we have been able to compute derivatives of the solutions when the data was sufficiently smooth.

H. Werner et al. (eds.), Computational Aspects of Complex Analysis, 406.

ANALYTICAL REPRESENTATIONS FOR SOLUTIONS OF CERTAIN TRANSCENDENTAL EQUATIONS

G.P. Meyer

University of Regensburg
Federal Republic of Germany

We present a computational method for solving transcendental equations of the form

$$w(z) := \sum_{k=1}^{m} P_k(z)\, e^{a_k z} = 0$$

($0 \not\equiv P_k(z) \in \mathbb{C}[z]$, $a_k \in \mathbb{C}$, $a_i \neq a_j$ for $i \neq j$, $m \geq 2$). G. Pólya and E. Schwengeler have proved that the zeros of $w(z)$ asymptotically osculate a finite number of logarithmic lines.

We succeed in deriving convergent and asymptotic expansions for sequences of zeros of $w(z)$ "along" such a logarithmic line if on the corresponding side of the extended indicator diagram (in the sense of Pólya-Schwengeler) there are exactly two degree-points. The procedure uses a generalized Lagrange-Bürmann formula due to Herm. Schmidt.

H. Werner et al. (eds.), Computational Aspects of Complex Analysis, 407.

N-WIDTHS AND OPTIMAL INFORMATION FOR ANALYTIC FUNCTIONS

C. Micchelli

IBM-Research, Yorktown Heights

The material discussed in this talk appears in the two papers:

1. S. Fisher, C.A. Micchelli: The N-width of analytic functions, Duke Journal, 47 (1980), 789-801.
2. S. Fisher, C.A. Micchelli: Optimal sampling of holomorphic functions, to appear American Journal of Mathematics.

H. Werner et al. (eds.), Computational Aspects of Complex Analysis, 408.

THIELE-TYPE BRANCHED CONTINUED FRACTIONS

W. Siemaszko

University of Rzeszów
Poland

Continued fractions can be simply generalized if we use a new continued fractions instead of partial denominators. Such continued fractions are called branched continued fractions. Defining a two variables inverted differences we obtain a Thiele-type formula for rational interpolation of two variables function on a rectangular grid by using approximants of branched continued fractions. The error formula for such interpolation can be found. Considering the limit point case we get expansions of two variables functions into corresponding two variables branched continued fractions. Approximants of branched continued fraction obtained in this way can be used as Padé-type approximants for two varibles functions. Generalization to N-variables can be easily done.

H. Werner et al. (eds.), Computational Aspects of Complex Analysis, 409.

UNIFORM ASYMPTOTIC EXPANSIONS OF LAPLACE INTEGRALS

N.M. Temme

University of Amsterdam
The Netherlands

The Laplace integral

$$F_\lambda(z) = \frac{1}{\Gamma(\lambda)} \int_0^\infty t^{\lambda-1} e^{-zt} f(t)dt$$

is considered for $z \to \infty$. An asymptotic expansion is obtained which is uniformly valid with respect to $\mu = \lambda/z$ in an unbounded domain, containing the point $\mu = 0$. The asymptotic nature of the expansion is discussed and error bounds are given for the remainders in the expansions. A simple expansion with $f(t) = 1/(1 + t)$ is considered, giving expansions for the well-known exponential integral.

H. Werner et al. (eds.), Computational Aspects of Complex Analysis, 410.

CONTINUITY PROPERTIES OF PADÉ OPERATORS

H. Werner and L. Wuytack

University of Bonn
Federal Republic of Germany and
University of Antwerp
Belgium

Let T be the operator that associates a given power series f to its Padé approximant $r_{m,n}$. Let m', n' be the exact degree of the numerator p and denominator q of $r_{m,n}$, then $d_{m,n} = \min\{m-m', n-n'\}$.

It is shown that T satisfies a local Lipschitz continuity condition if and only if $d_{m,n} = 0$. If $d_{m,n} > 0$, then T is proved to be discontinuous. An example is given to illustrate this phenomenon.

H. Werner et al. (eds.), Computational Aspects of Complex Analysis, 411.

THE A-A-K ALGORITHM

F.-B. Yeh

Glasgow University

An alternative way to approach the A-A-K algorithm is described. We generalize the Sarason's result and Young's algorithm to the multichannel case.

H. Werner et al. (eds.), Computational Aspects of Complex Analysis, 412.

In Braunlage in August 82 there was a meeting
where the participants had to do a lot of eating.
Although the support of NATO was believed to be nice,
the meeting started with a surprise.
In fact because there was not enough money,
the organizers had to cancel the afternoon coffee.
At the meeting there were given a lot of talks
and in between people made all kinds of walks.
There were talks we could easily understand,
but others seemed to have no end.
The speakers had to use a projector and a small screen,
consequently not everything could sometimes be seen.
Several speakers didn't mind,
to have questions of all kind.
And, as you from mathematicians can expect,
the answers were, most of the time, correct.
There was also a session on problems that are unsolved,
in finding a solution a lot of work will be involved.
In the hotel it was possible to do many things, even swim
and sometimes it was difficult to decide where to begin.
At lunch and dinner we could have beer, wine or tea,
but only the ice-water was free.
During the day we could enjoy a lot of sun
and during the night there was a lot of fun.
So, even the meeting started with a surprise
we believe the result is quite nice.

Luc Wuytack
5. August 1982

H. Werner et al. (eds.), Computational Aspects of Complex Analysis, 413.

I was delighted to hear Dr. Wuytack deliver his fine ballad,
but it seems a few developments of a somewhat academic nature
should be added.
This week we've found that power series are sometimes best
viewed as formal,
and that the boundary of the DDR possesses only an inward-
pointing normal.
We've learned that integration can be flawlessly performed
symbolisch,
and that some of the best mathematicians in NATO are Polish.
We've seen that in conformal mapping, $\varepsilon > 1$ is a dangerous menace,
and we've examined the properties of many tables, including
Padé, CF, QD, and tennis.
We've studied Hankel matrices, convolutions, Blaschke products
and other mathematical fauna,
and found that Charles Micchelli has as many good ideas as
the rest of us combined, even after a couple of hours in the
sauna.
We've discussed the works of Frobenius, Lagrange, Kneser, Krein,
and Toeplitz,
and been informed that the problem of figuring out who really
invented the Fast Fourier Transform is simply hopelitz.
All in all, our meeting has indeed been a complex one,
and I look forward to meeting you all again soon at the next one!

Lloyd Trefethen
6 August 1982

H. Werner et al. (eds.), Computational Aspects of Complex Analysis, 414.

LIST OF PARTICIPANTS

Albrecht, J.	Institut für Mathematik, Techn. Universität Clausthal, Erzstr. 1, D-3392 Clausthal-Zellerfeld, W.-Germany.
Alpay, S.	Mathematics Dept. Middle East Techn. University, Ankara, Turkey.
Barkana, A.	State Academy of Engineering and Architecture Yunus Emre Cad. No. 96/4, Eskisehir, Turkey.
Barnard, R.W.	Math. Dept. Texas Tech University, Lubbock Tx 79409, USA.
Bozhüyük, M.E.	Atatürk Üniversitesi, Fen Fakultesi, Matematik Bölümü, Erzurum, Turkey.
Brezinski, C.	Université de Lille I.U.E.R.d'I.E.E.A.-Informatique B.P. 36, F-59650 Villeneuve d'Ascq, France.
de Bruin, M.G.	Dept. of Mathematics, University of Amsterdam, Box 20239, NL 1000 HE Amsterdam, Netherlands.
Bünger, H.J.	Institut für Angewandte Mathematik, Universität Bonn, Wegelerstr. 6, D-5300 Bonn 1, W.-Germany.
Bultheel, A.	Dept. Computerwetenschappen, Kath. Universiteit Leuven, Celestijnenlaan 200-A, B-3030 Heverlee-Leuven, Belgium.
Burbea, J.B.	Dept. of Mathematics, University of Pittsburgh, Pittsburgh, PA.15260, USA.
Byrne, C.	Springer-Verlag, 3000 Hannover, W.-Germany.
Casasus, L.	Facultad de Matematicas, Dept. Ecuaciones Funcionales, Universidad de la Laguna, Tenerife, Spain.
de Clerck, L.	Dept. Wiskunde, Katholieke Universiteit Leuven, Celestijnenlaan 200-B, B-3030 Heverlee-Leuven, Belgium.
Cuyt, A.A.M.	Dept. Mathematics, Universitaire Instelling Antwerpen, Universiteitsplein 1, B-2610 Wilrijk, Belgium.
Dahmen, W.	Fakultät f. Mathematik, Universität Bielefeld, Universitätsstr., D-4800 Bielefeld 1, W.-Germany.

Della Dora, J.	Université sci. et med. de Grenoble, IMAG, B.P. 53X, F-38041 Grenoble Cedex, France.
Egeli, S.A.	Temel Bil. Fak. Maslak, Istanbul Teknik Universitese, Istanbul, Turkey.
Eiermann, M.	Inst. f. Praktische Mathematik, Universität Karlsruhe, Englerstr. 2, D-7500 Karlsruhe, W.-Germany.
Elcrat, A.	Dept. of Mathematics, Wichita State University, 2825 W 17, Wichita, Kansas 67203, USA.
Farwig, R.	Inst. f. Angew. Mathematik, Sonderforschungsbereich 72, Universität Bonn, Wegelerstr. 6, D-5300 Bonn 1, W.-Germany.
Gaier, D.	Mathematisches Institut, Universität Gießen, Arndtstr. 2, D-6300 Gießen, W.-Germany.
Geatti, L.	Istituto Elaborazione dell'Informazione, Via S. Maria 46, I-56100 Pisa, Italy.
Gutknecht, M.	Seminar f. Angew. Mathematik, ETH Zürich, Rämistr. 101, CH-8092 Zürich,Switzerland.
Haverkamp, R.	Inst. f. Angew. Mathematik, Sonderforschungsbereich 72, Universität Bonn, Wegelerstr. 6, D-5300 Bonn 1, W.-Germany.
Henrici, P.	Seminar f. Angew. Mathematik,ETH Zürich, Rämistr. 101, CH-8092 Zürich,Switzerland.
Hübner, O.	Mathematisches Institut, Universität Gießen, Arndtstr. 2, D-6300 Gießen, W.-Germany.
Loeb, H.L.	Dept. of Mathematics, University of Oregon, Eugene Or 97405 USA.
Lübbe, W.	Inst. f. Praktische Mathematik, Universität Hannover, Welfengarten 1, D-3000 Hannover 1, W.-Germany.
Mayo, A.	Dept. of Computer Science, Stanford University, Stanford CA 94305, USA.
Meinguet, J.	Inst. de Mathematique, Université de Louvain, Chemin du Cyclotron 2, B-1348 Louvain-la-Neuve, Belgium.
Mercer, A.McD.	Dept. of Math. & Statistics, University of Guelph, Guelph, Ontario, Canada NIG 2W1.

Merz, G.	Fachbereich 17 Mathematik, Gesamthochschule Kassel, Wilhelmshöher Allee 73, D-3500 Kassel, W.-Germany.
Meyer, G.	Rechenzentrum, Universität Regensburg, Universitätsstr. 31, D-8400 Regensburg, W.-Germany.
Micchelli, Ch.	IBM Research, P.O. Box 218, Yorktown Heights, N.Y. USA.
Mühlbach, G.	Inst. f. Praktische Mathematik, Universität Hannover, Welfengarten 1, D-3000 Hannover 1, W.-Germany.
Mushamalirwa, D.	Groupe Electricité/Fort, Université Catholique de Louvain, Place du Levant 3, B-1348 Louvain-la-Neuve, Belgium.
Ng, E.	Jet Propulsion Lab.California Inst. of Technology, Pasadena, CA. 91109, USA.
Olver, F.W.J.	Institute f. Physical Science and Technology, University of Maryland, College Park, Maryland 20903, USA.
Pallaschke, D.	Inst. f. Statistik u. Math. Wirtschaftstheorie, Universität Karlsruhe, Kaiserstr. 12, D-7500 Karlsruhe, W.-Germany.
Papapanayotou, C.	University of Thessaloniki, 5, Mitropolitou Iossif str.,Thessaloniki,Greece.
Reimers, L.	Inst. f. Praktische Mathematik, Universität Hannover, Welfengarten 1, D-3000 Hannover 1, W.-Germany.
Ruppert, K.	Inst. f. Mathematik, Techn.Universität Clausthal, Erzstr. 1, D-3392 Clausthal-Zellerfeld, W.-Germany.
Ruscheweyh, S.	Mathematisches Institut, Universität Würzburg, Am Hubland, D-8700 Würzburg, W.-Germany.
Shaffer, D.B.	Dept. of Mathematics, Fairfield University, Fairfield CT. 06430, USA.
Shoaff, W.D.	Dept. of Mathematics, Murray State University, Murray, Kentucky 42071, USA.
Siemaszko, W.	Dept. of Math. and Physics, Technical University, Ul.W. Pola 2, 35959 Rzeszow, Poland.

da Silva, M.R.	Gr. d. Matematica Aplicada, Universidado do Porto, R. das Taipas 135, 4000 Porto, Portugal.
Solberg, R.	Norwegian Defence Research Establishment, P.O. Box 25, N 2007 Kjeller, Norway.
Symm, G.	Dept. of Industry, National Physical Laboratory, Teddington, Middlesex, TW 11 OLW, Great Britain.
Temme, N.M.	Mathematisch Centrum, Kruislaan 413, NL 1098 SJ Amsterdam, Netherlands.
Trefethen, L.N.	Courant Inst. of Math. Science, 251 Mercer St., New York, NY 10012, USA.
Verführt, R.	Mathematisches Institut, Ruhr-Universität Bochum, Universitätsstr. 150, D-4630 Bochum, W.-Germany.
Warby, M.K.	Dept. of Mathematics, Brunel University, Uxbridge, Middlesex, UB 83 PH,Great Brit.
Werner, H.	Inst. f. Angew. Mathematik, Universität Bonn, Wegelerstr. 6, D-5300 Bonn 1, W.-Germany.
Werner, I.	Universität Bonn, Studienkolleg, D-5300 Bonn 1, W.-Germany.
Wuytack, L.	Dept. of Mathematics, Universitaire Instelling Antwerpen, Universiteitsplein, B-2610 Wilrijk, Belgium.
Yeh, F.B.	Dept. of Mathematics, University of Glasgow, University Gardens, Glasgow G 12 8QW, Great Britain.
Yun, D.Y.Y.	IBM Research, P.O. Box 218, Yorktown Heights NY. 10598, USA.
Zwick, D.	Inst. f. Angew. Mathematik, Sonderforschungsbereich 72, Wegelerstr. 6, D-5300 Bonn 1, W.-Germany.

SUBJECT INDEX